Methods in Molecular Biology

For further volumes:
http://www.springer.com/series/7651

For over 35 years, biological scientists have come to rely on the research protocols and methodologies in the critically acclaimed *Methods in Molecular Biology* series. The series was the first to introduce the step-by-step protocols approach that has become the standard in all biomedical protocol publishing. Each protocol is provided in readily-reproducible step-by-step fashion, opening with an introductory overview, a list of the materials and reagents needed to complete the experiment, and followed by a detailed procedure that is supported with a helpful notes section offering tips and tricks of the trade as well as troubleshooting advice. These hallmark features were introduced by series editor Dr. John Walker and constitute the key ingredient in each and every volume of the *Methods in Molecular Biology* series. Tested and trusted, comprehensive and reliable, all protocols from the series are indexed in Pub Med.

The ESCRT Complexes

Methods and Protocols

Edited by

Emmanuel Culetto and Renaud Legouis

Diabetes and Islet Biology Group, Faculty of Medicine and Health, NHMRC Clinical Trials Centre, The University of Sydney, Camperdown, NSW, Australia

Humana Press

Editors
Emmanuel Culetto
Diabetes and Islet Biology Group
Faculty of Medicine and Health
NHMRC Clinical Trials Centre
The University of Sydney
Camperdown, NSW, Australia

Renaud Legouis
Diabetes and Islet Biology Group
Faculty of Medicine and Health
NHMRC Clinical Trials Centre
The University of Sydney
Camperdown, NSW, Australia

ISSN 1064-3745 ISSN 1940-6029 (electronic)
Methods in Molecular Biology
ISBN 978-1-4939-9491-5 ISBN 978-1-4939-9492-2 (eBook)
https://doi.org/10.1007/978-1-4939-9492-2

Cover Illustration Caption: D'Avino's lab (University of Cambridge, UK) used atomic force microscopy (AFM) to visualize the human ESCRT-III machinery and in particular the CHMP4C component to investigate the ability of CHMP4C to associate and remodel membranes. Purified human CHMP4C on mica surfaces, without lipid bilayers, could spontaneously assemble into spiral filaments indicating that CHMP4C can form polymers in vitro.

This Humana imprint is published by the registered company Springer Science+Business Media, LLC part of Springer Nature.
The registered company address is: 233 Spring Street, New York, NY 10013, U.S.A.

Preface

Imagine that you have the possibility of shrinking to enter any given cell, as in the 1966 movie *Fantastic Voyage*. There, you would see first-hand that the Endosomal Sorting Complexes Required for Transport (ESCRT) are key players in numerous membrane remodeling and scission events, as well as in performing many other cellular functions.

At the start of the twenty-first century, Scott Emr's research group discovered ESCRT proteins in yeast screens for genes required for vacuolar protein sorting [1, 2]. Further studies in many laboratories found that ESCRTs are required during endosome maturation to sort ubiquitinated membrane proteins into intraluminal vesicles (ILV) giving rise to multivesicular bodies (MVB). During this process, four ESCRT complexes (named ESCRT-0, ESCRT-I, ESCRT-II, ESCRT-III), each made by a set of distinct cytosolic proteins, and several ESCRT-associated proteins (ALIX/PDCD6IP, Vta1/LIP5, TOM1, VPS4/SKD1), are sequentially assembled onto endosomal membrane to incorporate membrane proteins into ILV for degradation [3]. During this initial characterization of ESCRT, it has been shown that many enveloped viruses bud out the infected cells by taking over the host ESCRT complexes [4–6]. Interestingly, even within the nucleus, ESCRTs promote budding of vesicles, containing virus, from the inner nuclear envelope. In this case, the local recruitment of ALIX, the ESCRT-III protein CHMP4, and VPS4 allows the virus to transit from the nucleus to the plasma membrane [7]. Similarly, various molecules that are required for cell-cell communication are either encapsulated in exovesicles under the control of ALIX, ESCRT-I, ESCRT-III, and the ESCRT-associated VPS4 [8], or are directly released as exosomes from MVB that fuse with the plasma membrane [9]. Generally, ALIX and/or ESCRT-I/ESCRT-II components were found to recruit ESCRT-III and VPS4 to membrane sites where lipid bilayer remodeling or scission occurs.

Beyond endosomal sorting and virus budding, recent discoveries have shed light on a broad range of additional ESCRT functions.

The ESCRT machinery is involved in cell division. In telophase, the thin cytoplasmic bridges still connecting the sister cells with the midbody are a place where the ESCRT machinery is in action. Indeed, filaments of ESCRT-III, along with ALIX, ESCRT-I, ESCRT-II, and VPS4, have been shown to constrict and sever the neck between the cells and the midbody to complete cytokinetic abscission [10, 11]. Interestingly, even some members of the archaeal domain of life divide in a process that is dependent upon orthologs of ESCRT-III and VPS4 [12]. Following chromosome segregation during mitosis, ESCRT-III proteins CHMP7 and CHMP4B, together with VPS4, are found to be recruited very transiently at the interface between the nuclear envelope and microfilaments to finalize nuclear envelope reformation at the end of anaphase [13, 14]. Several studies indicate that ESCRTs could make a system to control limited membrane damage. When deployed to the plasma membrane, ESCRT-III and VPS4 assemble to repair small fragments of damaged membrane by vesicle shedding [15]. Remarkably, endolysosomal compartments, often damaged by the cargoes that they incorporate, could also be mended by a ESCRTs-based repair system [16]. Also, repairing of injured nuclear pores involves the same set of ESCRTs [17].

Additional studies indicate emerging areas of ESCRT biology. In the cytosol, the formation of nascent peroxisomes from the endoplasmic reticulum depends on the

ESCRT-III proteins CHMP6 and CHMP4 [18]. Furthermore, in neuronal remodeling, ESCRT-I, ESCRT-III, and VPS4 act to remove cytoplasmic extensions [19]. Finally, different types of autophagy pathways also require ESCRT proteins. The ESCRT-III CHMP1/Vps46 and CHMP2A/Vps2 proteins are likely required for the closure of phagophores, the final step of autophagosome formation [20, 21]. The ESCRT machinery is also engaged in endosomal microautophagy and Nbr1-mediated vacuolar targeting [22–26].

Considering all these functions, if we could imagine visiting the interior of the cell, we would witness ESCRTs busy at bending cell and organelle membranes away from the cytosol.

Surprisingly, however, some ESCRTs appear to have cellular functions that are independent of the membrane remodeling activity. In fact, ESCRT-II proteins have been reported to participate in RNA trafficking and transcriptional regulation [27–30], and might be involved in homeostasis of the endoplasmic reticulum [31]. Furthermore, ESCRTs have been localized to the ciliary transition zone and could play a role in formation and/or function of cilia and centrosome [32, 33]. Finally, the ESCRT-III protein CHMP5 also has a noncanonical activity consisting in modulation of deubiquitinase recruitment to specific target proteins [34].

Most discoveries of ESCRT functions have been achieved using a combination of advanced technical approaches, with these often requiring development of new protocols. The purpose of this book, *The ESCRT Complex: Methods and Protocols*, is to gather both established and very recent technical procedures to study ESCRTs in a range of biological systems. Such comprehensive coverage of *in vitro* and *in vivo* techniques is presented in a way that is convenient to follow for a wide audience of experimentalists.

The collection of protocols that we have gathered in this book includes imaging, biochemistry, and genetics. Regarding imaging, protocols include imaging of the Archaeal ESCRT apparatus, visualization of ESCRT protein localization within *Drosophila melanogaster* cells, or of ESCRT-positive extracellular vesicles secreted from *D. melanogaster* wing imaginal disc cells. We also present correlative light-electron microscopy (CLEM) approaches to localize ESCRT proteins at the subcellular level in *Caenorhabditis elegans*. In addition, we list CLEM and proximity ligation assay procedures to analyze complex relationships and interactions between membrane structures, ESCRTs and viruses in virus-infected cells. Moreover, we show how fluorescence-based protocols can be used to measure ESCRT-dependent endolysosomal trafficking both in mammalian cells and in the budding yeast *S. cerevisiae*, as well as live imaging of ESCRT proteins in isolated axons using microfluidics. Finally, we describe high-resolution structured illuminated microscopy (SIM) and cryo-soft X-ray tomography imaging of ESCRTs in mammalian cells, as well as a guide to making 3D renderings of Z stacks of ESCRT proteins visualized by confocal microscopy.

Protocols based on biochemical approaches present strategies for production and characterization of recombinant ESCRT proteins, or of specific ESCRT protein domains from *C. elegans*, *D. melanogaster*, *Arabidopsis thaliana*, and mammalian cells. These protocols illustrate how to use recombinant proteins for crystallography and for protein-protein interaction studies by surface plasmon resonance and with fluorescence binding assays. In addition, this book describes methodologies to study ESCRT assembly and interaction with membranes, including gel filtration chromatography, glycerol density gradient analysis, multi-angle light scattering, liposome co-flotation, single liposome fluorescence microscopy, phosphorylation assays and atomic force microscopy, as well as protocols for assessing

ESCRT-dependent client protein ubiquitination and for ESCRT polyclonal antibody production.

This book also includes a final set of protocols describing genetic and proteomic experimental approaches. These protocols have been used either to identify ESCRT machinery interactors in various cell systems or to generate ESCRT mutant cells in a diverse set of model organisms, such as the fission yeast *Schizosaccharomyces pombe,* the model fungus *Ustilago maydis*, *A. thaliana*, and mammalian cells.

Altogether, this book provides a comprehensive overview of methodologies used by ESCRT researchers and constitutes a compact guide for people interested in advancing the study of ESCRT. We hope that this collection of protocols will be of use in a diverse set of laboratories, and that it will be of interest particularly for researchers interested in establishing an integrated approach to investigate cell biology.

We wish to thank all the contributors for sharing their hard-won experience and advice with the research community. Their work makes this book a state-of-the-art source of validated protocols to envisage experiments that test novel ideas and hypotheses in the constantly changing horizon of ESCRT research.

Gif-sur-Yvette, France

Emmanuel Culetto
Renaud Legouis

References

1. Babst M, Wendland B, Estepa EJ, Emr SD (1998) The Vps4p AAA ATPase regulates membrane association of a Vps protein complex required for normal endosome function. EMBO J 17:2982–2993
2. Katzmann DJ, Babst M, Emr SD (2001) Ubiquitin-dependent sorting into the multivesicular body pathway requires the function of a conserved endosomal protein sorting complex, ESCRT-I. Cell 106:145–155
3. Wollert T, Hurley JH (2010) Molecular mechanism of multivesicular body biogenesis by ESCRT complexes. Nature 464:864–869
4. Garrus JE et al (2001) Tsg101 and the vacuolar protein sorting pathway are essential for HIV-1 budding. Cell 107:55–65
5. Martin-Serrano J, Zang T, Bieniasz PD (2001) HIV-1 and Ebola virus encode small peptide motifs that recruit Tsg101 to sites of particle assembly to facilitate egress. Nat Med 7:1313–1319
6. VerPlank L et al (2001) Tsg101, a homologue of ubiquitin-conjugating (E2) enzymes, binds the L domain in HIV type 1 Pr55(Gag). Proc Natl Acad Sci USA 98:7724–7729
7. Arii J et al (2018) ESCRT-III mediates budding across the inner nuclear membrane and regulates its integrity. Nat Commun 9:3379
8. Matusek T et al (2014) The ESCRT machinery regulates the secretion and long-range activity of Hedgehog. Nature 516:99–103
9. Juan T, Fürthauer M (2018) Biogenesis and function of ESCRT-dependent extracellular vesicles. Semin Cell Dev Biol 74:66–77
10. Morita E et al (2007) Human ESCRT and ALIX proteins interact with proteins of the midbody and function in cytokinesis. EMBO J 26:4215–4227
11. Carlton JG, Agromayor M, Martin-Serrano J (2008) Differential requirements for Alix and ESCRT-III in cytokinesis and HIV-1 release. Proc Natl Acad Sci USA 105:10541–10546
12. Samson RY, Obita T, Freund SM, Williams RL, Bell SD (2008) A role for the ESCRT system in cell division in archaea. Science 322:1710–1713
13. Olmos Y, Hodgson L, Mantell J, Verkade P, Carlton JG (2015) ESCRT-III controls nuclear envelope reformation. Nature 522:236–239

14. Vietri M et al (2015) Spastin and ESCRT-III coordinate mitotic spindle disassembly and nuclear envelope sealing. Nature 522:231–235
15. Jimenez AJ et al (2014) ESCRT machinery is required for plasma membrane repair. Science 343:1247136
16. Skowyra ML, Schlesinger PH, Naismith TV, Hanson PI (2018) Triggered recruitment of ESCRT machinery promotes endolysosomal repair. Science 360(6384):eaar5078
17. Webster BM, Colombi P, Jäger J, Lusk CP (2014) Surveillance of nuclear pore complex assembly by ESCRT-III/Vps4. Cell 159:388–401
18. Mast FD et al (2018) ESCRT-III is required for scissioning new peroxisomes from the endoplasmic reticulum. J Cell Biol 217:2087–2102
19. Loncle N, Agromayor M, Martin-Serrano J, Williams DW (2015) An ESCRT module is required for neuron pruning. Sci Rep 5:8461
20. Spitzer C et al (2015) The endosomal protein charged multivesicular body protein1 regulates the autophagic turnover of plastids in Arabidopsis. Plant Cell 27:391–402
21. Takahashi Y et al (2018) An autophagy assay reveals the ESCRT-III component CHMP2A as a regulator of phagophore closure. Nat Commun 9:2855
22. Hammerling BC et al (2017) A Rab5 endosomal pathway mediates Parkin-dependent mitochondrial clearance. Nat Commun 8:14050
23. Liu X-M et al (2015) ESCRTs cooperate with a selective autophagy receptor to mediate vacuolar targeting of soluble cargos. Mol Cell 59:1035–1042
24. Mejlvang J et al (2018) Starvation induces rapid degradation of selective autophagy receptors by endosomal microautophagy. J Cell Biol 217(10):3640–3655. doi:10.1083/jcb.201711002
25. Mukherjee A, Patel B, Koga H, Cuervo AM, Jenny A (2016) Selective endosomal microautophagy is starvation-inducible in Drosophila. Autophagy 12:1984–1999
26. Sahu R et al (2011) Microautophagy of cytosolic proteins by late endosomes. Dev Cell 20:131–139
27. Ghoujal B, Milev MP, Ajamian L, Abel K, Mouland AJ (2012) ESCRT-II's involvement in HIV-1 genomic RNA trafficking and assembly. Biol Cell 104:706–721
28. Irion U, St Johnston D (2007) bicoid RNA localization requires specific binding of an endosomal sorting complex. Nature 445:554–558
29. Jin Y, Mancuso JJ, Uzawa S, Cronembold D, Cande WZ (2005) The fission yeast homolog of the human transcription factor EAP30 blocks meiotic spindle pole body amplification. Dev Cell 9:63–73
30. Kamura T et al (2001) Cloning and characterization of ELL-associated proteins EAP45 and EAP20. a role for yeast EAP-like proteins in regulation of gene expression by glucose. J Biol Chem 276:16528–16533
31. Lefebvre C et al (2016) The ESCRT-II proteins are involved in shaping the sarcoplasmic reticulum in C. elegans. J Cell Sci 129:1490–1499
32. Diener DR, Lupetti P, Rosenbaum JL (2015) Proteomic analysis of isolated ciliary transition zones reveals the presence of ESCRT proteins. Curr Biol 25:379–384
33. Ott C et al (2018) VPS4 is a dynamic component of the centrosome that regulates centrosome localization of γ-tubulin, centriolar satellite stability and ciliogenesis. Sci Rep 8:3353
34. Adoro S et al (2017) Post-translational control of T cell development by the ESCRT protein CHMP5. Nat Immunol 18:780–790

Contents

Contributors

MARIA ALBA ABAD • *Wellcome Trust Centre for Cell Biology, Institute of Cell Biology, University of Edinburgh, Edinburgh, UK*

SHAI ADAR-LEVOR • *Department of Life Sciences and NIBN, Ben Gurion University of the Negev, Beer Sheva, Israel*

STANLEY ADORO • *Department of Pathology, School of Medicine, Case Western Reserve University, Cleveland, OH, USA*

PIRJO M. APAJA • *Department of Physiology, McGill University, Montréal, QC, Canada; Nutrition and Metabolism Theme, HCN and EMBL Australia, South Australian Health and Medical Research Institute, Adelaide, SA, Australia*

ANJON AUDHYA • *Department of Biomolecular Chemistry, School of Medicine and Public Health, University of Wisconsin-Madison, Madison, WI, USA*

SUDEEP BANJADE • *Department of Molecular Biology and Genetics, Weill Institute for Cell and Molecular Biology, Cornell University, Ithaca, NY, USA*

RINA BAROUCH-BENTOV • *Division of Infectious Diseases and Geographic Medicine, Department of Medicine, Stanford University School of Medicine, Stanford, CA, USA*

STEPHEN D. BELL • *Department of Molecular and Cellular Biochemistry, Indiana University, Bloomington, IN, USA; Department of Biology, Indiana University, Bloomington, IN, USA*

MUSAB S. BHUTTA • *Henry Wellcome Laboratory of Cell Biology, Institute of Molecular, Cell and Systems Biology, College of Medical, Veterinary and Life Sciences, University of Glasgow, Glasgow, UK*

MICHELE W. BIANCHI • *Institute for Integrative Biology of the Cell (I2BC), CEA, CNRS, Université Paris-Sud, Université Paris-Saclay, Gif-sur-Yvette, France; Faculté des Sciences et Technologie, Université Paris-Est Créteil, Créteil, France*

VERONICA BIRDSALL • *Neurobiology and Behavior PhD Program, Columbia University, New York, NY, USA*

SAMUEL BLOCK • *Department of Biomolecular Chemistry, School of Medicine and Public Health, University of Wisconsin-Madison, Madison, WI, USA*

KATHERINE BOWERS • *Division of Biosciences, Institute of Structural and Molecular Biology, University College London, London, UK*

SHAUN K. BREMNER • *Henry Wellcome Laboratory of Cell Biology, Institute of Cardiovascular and Medical Studies, College of Medical, Veterinary and Life Sciences, University of Glasgow, Glasgow, UK*

WENHAN CAO • *State Key Laboratory of Agrobiotechnology, Centre for Cell and Developmental Biology, School of Life Sciences, The Chinese University of Hong Kong, Hong Kong, China*

LUISA CAPALBO • *Department of Pathology, University of Cambridge, Cambridge, UK*

BALA CHANDRAN • *H. M. Bligh Cancer Research Laboratories, Department of Microbiology and Immunology, Chicago Medical School, Rosalind Franklin University of Medicine and Science, North Chicago, IL, USA; Department of Molecular Medicine, Morsani College of Medicine, University of South Florida, Tampa, FL, USA*

YONG CUI • *State Key Laboratory of Agrobiotechnology, Centre for Cell and Developmental Biology, School of Life Sciences, The Chinese University of Hong Kong, Hong Kong, China*

EMMANUEL CULETTO • *Diabetes and Islet Biology Group, Faculty of Medicine and Health, NHMRC Clinical Trials Centre, The University of Sydney, Camperdown, NSW, Australia*
PIER PAOLO D'AVINO • *Department of Pathology, University of Cambridge, Cambridge, UK*
IAIN G. DUGGIN • *The iThree Institute, University of Technology Sydney, Sydney, NSW, Australia*
J. MICHAEL EDWARDSON • *Department of Pharmacology, University of Cambridge, Cambridge, UK*
SHIRIT EINAV • *Division of Infectious Diseases and Geographic Medicine, Department of Medicine, Stanford University School of Medicine, Stanford, CA, USA; Department of Microbiology and Immunology, Stanford University School of Medicine, Stanford, CA, USA*
MICHAEL ELBAUM • *Department of Materials and Interfaces, Weizmann Institute of Science, Rehovot, Israel*
NATALIE ELIA • *Department of Life Sciences and NIBN, Ben Gurion University of the Negev, Beer Sheva, Israel*
SCOTT D. EMR • *Department of Molecular Biology and Genetics, Weill Institute for Cell and Molecular Biology, Cornell University, Ithaca, NY, USA*
MICHAEL FELDBRÜGGE • *Cluster of Excellence on Plant Sciences (CEPLAS), Institute for Microbiology, Heinrich-Heine University Düsseldorf, Düsseldorf, Germany*
MAXIMILIAN FÜRTHAUER • *Université Côte d'Azur, CNRS, INSERM, iBV, France*
INNA GOLIAND • *Department of Life Sciences and NIBN, Ben Gurion University of the Negev, Beer Sheva, Israel*
GWYN W. GOULD • *Henry Wellcome Laboratory of Cell Biology, Institute of Molecular, Cell and Systems Biology, College of Medical, Veterinary and Life Sciences, University of Glasgow, Glasgow, UK*
MARCO GUALTIERI • *Dipartimento di Bioscienze, Università degli Studi di Milano, Milan, Italy*
CARL HAAG • *Cluster of Excellence on Plant Sciences (CEPLAS), Institute for Microbiology, Heinrich-Heine University Düsseldorf, Düsseldorf, Germany*
ULRICH HENGST • *The Taub Institute for Research on Alzheimer's Disease and the Aging Brain, Columbia University, New York, NY, USA; Department of Pathology and Cell Biology, Columbia University, New York, NY, USA*
SHUXIAN HUANG • *State Key Laboratory of Agrobiotechnology, Centre for Cell and Developmental Biology, School of Life Sciences, The Chinese University of Hong Kong, Hong Kong, China*
ERIKA ISONO • *Plant Physiology and Biochemistry, Department of Biology, University of Konstanz, Konstanz, Germany*
YVES JACOB • *Département de Virologie, Unité de Génétique Moléculaire des Virus ARN (GMVR), Institut Pasteur, Centre national de la recherche scientifique, Université Paris Diderot, Paris, France*
A. AROCKIA JEYAPRAKASH • *Wellcome Trust Centre for Cell Biology, Institute of Cell Biology, University of Edinburgh, Edinburgh, UK*
LIWEN JIANG • *State Key Laboratory of Agrobiotechnology, Centre for Cell and Developmental Biology, School of Life Sciences, The Chinese University of Hong Kong, Hong Kong, China; CUHK Shenzhen Research Institute, The Chinese University of Hong Kong, Shenzhen, China*
JALAL M. KAZAN • *Department of Biochemistry, McGill University, Montréal, QC, Canada; Goodman Cancer Research Centre, McGill University, Montréal, QC, Canada*

THOMAS KLEIN • *Institute of Genetics, Heinrich-Heine University Düsseldorf, Düsseldorf, Germany*

RIEKO KOJIMA • *Faculty of Pharmaceutical Sciences, University of Toyama, Toyama, Japan; Department of Material and Biological Chemistry, Faculty of Science, Yamagata University, Yamagata, Japan*

BINOD KUMAR • *H. M. Bligh Cancer Research Laboratories, Department of Microbiology and Immunology, Chicago Medical School, Rosalind Franklin University of Medicine and Science, North Chicago, IL, USA*

CÉLINE LARGEAU • *Institute for Integrative Biology of the Cell (I2BC), CEA, CNRS, Univ. Paris-Sud, Université Paris-Saclay, Gif-sur-Yvette, France*

ROMAIN LE BARS • *Institute for Integrative Biology of the Cell (I2BC), CEA, CNRS, Université Paris-Sud, Université Paris-Saclay, Gif-sur-Yvette, France*

CHRISTOPHE LEFEBVRE • *Institute for Integrative Biology of the Cell (I2BC), CEA, CNRS, Université Paris-Sud, Université Paris-Saclay, Gif-sur-Yvette, France*

RENAUD LEGOUIS • *Diabetes and Islet Biology Group, Faculty of Medicine and Health, NHMRC Clinical Trials Centre, The University of Sydney, Camperdown, NSW, Australia*

GERGELY L. LUKACS • *Department of Physiology, McGill University, Montréal, QC, Canada*

JOSE C. MARTINEZ • *Medical Scientist Training Program, Columbia University, New York, NY, USA*

TAMÁS MATUSEK • *Université Côte d'Azur, CNRS, INSERM, iBV, France*

CHRISTOPHER J. MCINERNY • *Henry Wellcome Laboratory of Cell Biology, Institute of Molecular, Cell and Systems Biology, College of Medical, Veterinary and Life Sciences, University of Glasgow, Glasgow, UK; Henry Wellcome Laboratory of Cell Biology, School of Life Sciences, College of Medical, Veterinary and Life Sciences, University of Glasgow, Glasgow, UK*

IOANNA MELA • *Department of Pharmacology, University of Cambridge, Cambridge, UK*

MINEYUKI MIZUGUCHI • *Faculty of Pharmaceutical Sciences, University of Toyama, Toyama, Japan*

EIJI MORITA • *Department of Biochemistry and Molecular Biology, Faculty of Agriculture and Life Science, Hirosaki University, Hirosaki, Japan*

MARIE-KRISTIN NAGEL • *Plant Physiology and Biochemistry, Department of Biology, University of Konstanz, Konstanz, Germany*

ATSUKI NARA • *Nagahama Institute of Bio-Science and Technology, Nagahama, Japan*

TAKAYUKI OBITA • *Faculty of Pharmaceutical Sciences, University of Toyama, Toyama, Japan*

HIROKO OMORI • *Core Instrumentation Facility, Research Institute for Microbial Diseases, Osaka University, Osaka, Japan*

MARISA S. OTEGUI • *Department of Botany, University of Wisconsin-Madison, Madison, WI, USA; Laboratory of Molecular and Cellular Biology, University of Wisconsin-Madison, Madison, WI, USA; Department of Genetics, University of Wisconsin-Madison, Madison, WI, USA*

JULIO PAEZ-VALENCIA • *Department of Botany, University of Wisconsin-Madison, Madison, WI, USA; Laboratory of Molecular and Cellular Biology, University of Wisconsin-Madison, Madison, WI, USA*

ARNIM PAUSE • *Department of Biochemistry, McGill University, Montréal, QC, Canada; Goodman Cancer Research Centre, McGill University, Montréal, QC, Canada*

LISA RANDOLPH • *Neurobiology and Behavior PhD Program, Columbia University, New York, NY, USA*

IMANE M. REZIG • *Henry Wellcome Laboratory of Cell Biology, Institute of Molecular, Cell and Systems Biology, College of Medical, Veterinary and Life Sciences, University of Glasgow, Glasgow, UK*
ARUNAVA ROY • *H. M. Bligh Cancer Research Laboratories, Department of Microbiology and Immunology, Chicago Medical School, Rosalind Franklin University of Medicine and Science, North Chicago, IL, USA; Department of Molecular Medicine, Morsani College of Medicine, University of South Florida, Tampa, FL, USA*
IAN P. SALT • *Henry Wellcome Laboratory of Cell Biology, Institute of Cardiovascular and Medical Studies, College of Medical, Veterinary and Life Sciences, University of Glasgow, Glasgow, UK*
RACHEL Y. SAMSON • *Department of Molecular and Cellular Biochemistry, Indiana University, Bloomington, IN, USA*
AMBER L. SCHUH • *Department of Biomolecular Chemistry, School of Medicine and Public Health, University of Wisconsin-Madison, Madison, WI, USA*
JINBO SHEN • *State Key Laboratory of Agrobiotechnology, Centre for Cell and Developmental Biology, School of Life Sciences, The Chinese University of Hong Kong, Hong Kong, China*
JAE-HYUCK SHIM • *Division of Rheumatology, Department of Medicine, University of Massachusetts Medical School, Worcester, MA, USA*
FRANCHESKA SON • *Department of Pathology, School of Medicine, Case Western Reserve University, Cleveland, OH, USA*
KEISUKE TABATA • *Department of Infectious Diseases, Molecular Virology, Heidelberg University, Heidelberg, Germany*
SHAOGENG TANG • *Department of Molecular Biology and Genetics, Weill Institute for Cell and Molecular Biology, Cornell University, Ithaca, NY, USA; Department of Biochemistry, Stanford University, Stanford, CA, USA*
PASCAL THÉROND • *Université Côte d'Azur, CNRS, INSERM, iBV, France*
KATHARINE UMPHRED-WILSON • *Department of Pathology, School of Medicine, Case Western Reserve University, Cleveland, OH, USA*
THOMAS VACCARI • *Dipartimento di Bioscienze, Università degli Studi di Milano, Milan, Italy*
MOHANAN VALIYA VEETTIL • *H. M. Bligh Cancer Research Laboratories, Department of Microbiology and Immunology, Chicago Medical School, Rosalind Franklin University of Medicine and Science, North Chicago, IL, USA; Department of Biotechnology, Cochin University of Science and Technology, Cochin, Kerala, India*
KARIN VOGEL • *Plant Physiology and Biochemistry, Department of Biology, University of Konstanz, Konstanz, Germany*
CLARISSA L. WAITES • *Department of Pathology and Cell Biology, Columbia University, New York, NY, USA; Department of Neuroscience, Columbia University, New York, NY, USA*
QIONG ZHAO • *State Key Laboratory of Agrobiotechnology, Centre for Cell and Developmental Biology, School of Life Sciences, The Chinese University of Hong Kong, Hong Kong, China*
YING ZHU • *State Key Laboratory of Agrobiotechnology, Centre for Cell and Developmental Biology, School of Life Sciences, The Chinese University of Hong Kong, Hong Kong, China*

Chapter 1

Analysis of the Archaeal ESCRT Apparatus

Rachel Y. Samson, Iain G. Duggin, and Stephen D. Bell

Abstract

Members of the archaeal domain of life that lack homologs of actin and tubulin divide by binary fission in a process that is dependent upon orthologs of eukaryotic ESCRT components. Many of these archaeal organisms are hyperthermophilic acidophiles with unique cell wall structures, which create technical challenges for performing traditional cell biological techniques. Here, we describe the "baby machine" method for synchronizing microorganisms at high temperatures in order to study cell cycle-related processes. We also provide details for fixing, permeabilizing, and staining archaeal cells and ESCRT assemblies for observation by light microscopy.

Key words Archaea, *Sulfolobus*, Cell cycle, Baby machine, Synchronization, Immunofluorescence, ESCRT, Cell division

1 Introduction

Genome sequencing in the early 2000s revealed a surprising absence of actin and tubulin homologs in the genomes of many hyperthermophilic crenarchaea. In organisms lacking these nearly ubiquitous cytoskeletal proteins, cell division is driven by archaeal homologues of ESCRT proteins (for a review, see [1]). A series of expression profiling, genetic, biochemical, structural, and imaging studies in a well-characterized model organism, *Sulfolobus acidocaldarius*, have revealed that an archaeal-specific, membrane-binding protein, called cell division protein A, or CdvA, forms a ring-like structure that localizes to the mid-cell position [2, 3]. Archaeal orthologs of ESCRT-III then assemble on this CdvA platform [3]. A particular ESCRT-III paralog encoded by *saci1373* serves as a substrate for the ATPase Vps4 during the process of membrane constriction [4]. Details of the assembly and disassembly of the ESCRT apparatus in *Sulfolobus* have been aided by microscopy studies of cells undergoing cell division. The process of cell division is believed to occur rapidly at the end of the cell cycle, which reduces the probability of observing dividing cells in asynchronous

Emmanuel Culetto and Renaud Legouis (eds.), *The ESCRT Complexes: Methods and Protocols*, Methods in Molecular Biology, vol. 1998, https://doi.org/10.1007/978-1-4939-9492-2_1, © Springer Science+Business Media, LLC, part of Springer Nature 2019

populations. Synchronizing *Sulfolobus* cultures and fixing cells at the time of cell division produce samples enriched with cells at various stages of the division process [2, 4]. The first procedure in the Methods section describes general growth conditions and tips for handling *S. acidocaldarius.* The second subsection provides detailed instructions for assembling, running, and calibrating a membrane elution apparatus, called the "baby machine", for synchronizing cultures at high temperature [5–8]. The baby machine technique provides a convenient way of enriching a population of freshly divided cells from an immobilized asynchronous population. The population of cells is attached to a membrane and, under constant flow conditions, progeny cells from cell division are eluted from the membrane and collected. While the approach we describe has been optimized for *Sulfolobus,* it should be applicable to other species in which daughter cells separate cleanly upon completion of cell division. The third subsection outlines the procedure for fixing, permeabilizing, and staining *Sulfolobus* cells for immunofluorescence microscopy.

2 Materials

Sulfolobus cells are highly vulnerable to trace amounts of detergent, so all surfaces should be thoroughly rinsed with ultrapure water to ensure that they are free of detergent. Prepare all media and solutions with ultrapure water.

2.1 Cell Culturing

1. 100× solution: Dissolve 130 g $(NH_4)_2SO_4$, 25 g $MgSO_4{\cdot}7\,H_2O$, 2 g $FeCl_3{\cdot}6\,H_2O$, and 3 ml of 18 N H_2SO_4 in 800 ml of water. Adjust the volume to 1 l with water. Filter through a 0.22 μm filter into a sterile bottle. Store at 4 °C.
2. 200× solution: Dissolve 56 g KH_2PO_4, 36 ml $MnCl_2$ (10 mg/ml), 4.4 ml $ZnSO_4$ (10 mg/ml), 1 ml $CuCl_2$ (10 mg/ml), 0.6 ml $VOSO_4$ (10 mg/ml), 0.2 ml $CoSO_4$ (10 mg/ml), 9 ml $Na_2B_4O_7$ (10 mg/ml), 0.6 ml Na_2MoO_4 (10 mg/ml), and 5 ml of 50% H_2SO_4 in 800 ml of water. Adjust the volume to 1 l with water. Filter through a 0.22 μm filter into a sterile bottle. Store at 4 °C.
3. 1000× solution: Dissolve 14 g $CaCl_2{\cdot}2\,H_2O$ in water. Adjust the volume to 200 ml with water. Autoclave to sterilize. Store at room temperature.
4. 1× Brock's Medium: Add 10 ml of the 100× solution, 5 ml of the 200× solution, 1 ml of the 1000× solution, and 1 g of tryptone to 900 ml of water. Adjust the pH to 3.2 with 50% H_2SO_4. Bring up the volume to 1 l with ultrapure water. Filter through a 0.22 μm filter into a sterile bottle. Store at 4 °C.

5. Shaking water bath.
6. Sterile 25-ml and 250-ml Erlenmeyer flasks.
7. Spectrophotometer.

2.2 Baby Machine Synchronization

1. Shaking water bath (78 °C).
2. Large fan-forced incubator, pre-warmed to 78 °C.
3. Baby machine apparatus.
4. Static water bath (78 °C).
5. 1× Brock's Medium (*see* **step 4** in Subheading 2.1): 1 l in a 1-l bottle and 250 ml in a 250-ml bottle.
6. 100 ml of ultrapure water in a bottle.
7. Waste beaker.
8. Sterile graduated cylinder.
9. Peristaltic pump and rubber tubing connected to a rubber stopper.
10. Nitrocellulose membrane (0.22 μm pore size, 142 mm diameter).
11. Side-arm flask (1 l capacity, which fits the spout of the baby machine funnel).
12. 1× PBS: 10 mM Na_2HPO_4, 2 mM KH_2PO_4, 137 mM NaCl, 2.7 mM KCl, pH 7.4. Autoclave or filter-sterilize.
13. Poly-D-lysine (molecular weight 30,000–70,000) solution: Dissolve 10 mg/ml in 1× PBS. Store aliquots at −20 °C.
14. 80% ethanol: Prepare 900-μl aliquots in 1.5-ml tubes and store at 4 °C.
15. Vacuum pump.
16. 1.5-ml microfuge tubes.
17. 50-ml conical tubes, prechilled in ice.
18. Styrofoam beverage cups.
19. Sterile Erlenmyer flask (150–250 ml, depending on the volume of cells to be collected).

2.3 Immunofluorescence Microscopy

1. 1.5-ml and 2.0-ml microfuge tubes.
2. Centrifuge.
3. 1 M sodium phosphate buffer pH 7.6. Mix 8.45 ml of 1 M Na_2HPO_4 with 1.55 ml of 1 M NaH_2PO_4.
4. FM4-64X: Dissolve in water at 1 mg/ml. Store protected from light at −20 °C.
5. 16% (w/v) paraformaldehyde: Dissolve 16 g of paraformaldehyde in 80 ml of ultrapure water + 1 ml of 1 N NaOH, slowly stirring in a fume hood on a hotplate set to approximately

65 °C. After the paraformaldehyde dissolves, add 10 ml of 10× PBS and cool to room temperature. Adjust the pH to 7.4 using 1 N HCl and bring up the volume to 100 ml using ultrapure water. Pass through a 0.45 μm filter to remove particulate matter and store aliquots at −20 °C. Discard thawed aliquots.

6. Microfuge tube mixer.
7. 1× PBS: recipe (*see* **step 12** of Subheading 2.2).
8. GTET: 50 mM glucose, 20 mM Tris, pH 7.5, 10 mM EDTA, 0.2% Tween-20.
9. Glass coverslips: 25 mm × 25 mm, no. 1 thickness.
10. Poly-D-lysine [10 mg/ml]: (*see* **step 13** of Subheading 2.2).
11. Humid box (*see* **Note 8**).
12. Blocking buffer: 1× PBS, 2% (w/v) BSA.
13. Primary antibody.
14. Fluorescent secondary antibody.
15. Antibody binding buffer: 1× PBS, 0.05% (v/v) Tween-20, 2% (w/v) BSA.
16. Glass microscope slides.
17. Vectashield Antifade Mounting Medium, Vector Laboratories.
18. DAPI (4′,6-Diamidino-2-phenylindole dihydrochloride): Dissolve 1 mg/ml in ultrapure water. Store small aliquots at −20°. Throw away thawed aliquots.
19. Clear nail polish.
20. FM4-64X solution: FM4-64X [11.9 ng/μl], 35 mM sodium phosphate, pH 7.6.

3 Methods

3.1 Growth of *Sulfolobus acidocaldarius*

1. Revive *S. acidocaldarius* DSM 639 from a frozen glycerol stock in 10 ml of 1× Brock's medium. Incubate in a shaking water bath at 78 °C (100–200 rpm).
2. Maintain the culture in a steady state of growth for at least 2 days, by dilution into preheated 1× Brock's medium, as necessary to keep the OD_{600} below 0.5. Monitor growth (OD_{600}) at regular intervals to measure the growth rate.
3. Inoculate 100 ml of preheated 1× Brock's medium with the log-phase *S. acidocaldarius* starter culture, so that the OD_{600} of the culture will be 0.1–0.2 the next morning at a time when the baby machine will be started (*see* **Note 1**.)

3.2 Baby Machine

1. The day before running the baby machine, preheat the static water bath and the fan-forced incubator to 78 °C. Place the tripod and lower funnel of the baby machine apparatus, the

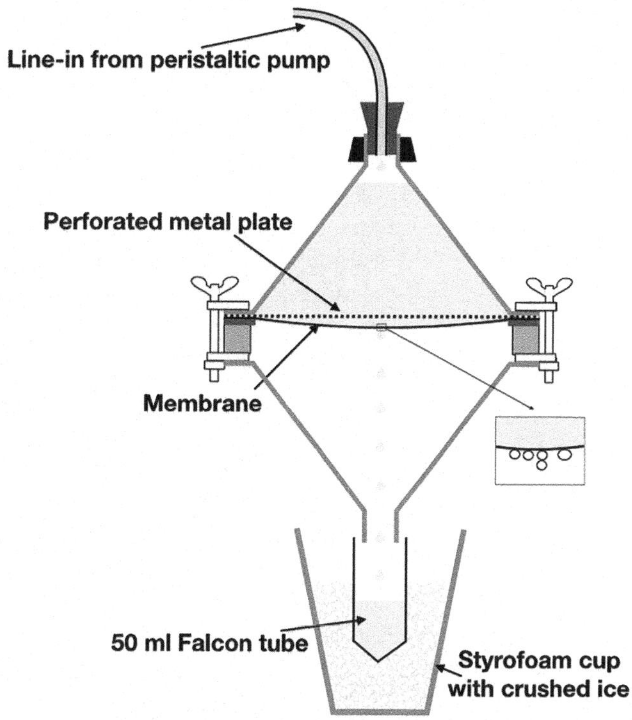

Fig. 1 Schematic of the design of a Baby Machine. The structure of a baby machine during the elution of newborn cells is illustrated. Medium is slowly pumped into the upper chamber and flows through the membrane. The liquid flow captures newborn cells and allows their collection in the tube placed below the lower chamber

waste collection beaker, and the sterile graduated cylinder into the fan-forced incubator to pre-warm overnight (or for at least 1 h on the day of the run).

2. At least 1 h before commencing the baby machine experiment, place the 250-ml bottle and the 1-l bottle of Brock's medium and the 100-ml bottle of water into the static water bath. Load the silicone tubing into the peristaltic pump and place the pump in the fan-forced incubator to warm to 78 °C.

3. To assemble the baby machine apparatus (Fig. 1), place the funnel section of the apparatus (dry) on a clean flat surface (spout-side down). Place the perforated disk onto the funnel (Teflon-coated side facing up), followed by the nitrocellulose membrane, silicone rubber gasket, and then the acrylic spacer ring, ensuring that the gasket and spacer ring are well-centered. Clamp the upper reservoir together, avoiding overtightening of the nuts. Wet and rinse the membrane and apparatus thoroughly with ultrapure water.

4. Install the baby machine funnel on top of the side-arm flask. Dilute 50 μl of the 10 mg/ml poly-D-lysine solution into 10 ml of 1× PBS. Pipet the solution onto the membrane and allow to drain through at room temperature. Swirl regularly to ensure uniform coating of the membrane.
5. Move the preheated 1-l bottle of Brock's medium into the fan-forced incubator. Place the end of the peristaltic pump tubing into the bottle, ideally using a ported screw-cap and air-filter, and then prime the pump lines with the Brock's medium and switch off the pump. Place the baby machine and side-arm flask into the fan-forced incubator. Allow the incubator to stabilize at 78 °C (~45 min). The membrane may dry during this time.
6. Measure the OD_{600} of the *S. acidocaldarius* overnight culture (between 0.1 and 0.2). Calculate the volume of culture that contains 7×10^9 cells (*see* **Note 2**). Withdraw a sample of 100 μl of the starter culture and add to 900 μl of ice-cold 80% ethanol to fix the cells for later flow cytometry analysis. Store the sample at 4 °C (*see* **Note 3**.)
7. Attach the vacuum pump to the side-arm flask. Pour the pre-warmed 100 ml of water onto the membrane and apply a vacuum to draw the water through the membrane (~10 ml/s).
8. Using the pre-warmed graduated cylinder, measure the volume of culture that contains 7×10^9 cells. Pour the cells onto the membrane and apply the vacuum to give a flow rate of ~1 ml/s. Leave the last few milliliters of culture to prevent the cells from drying out.
9. Immediately detach the baby machine from the side-arm flask and then invert and center it over the lower funnel of the apparatus (with waste collection beaker in place). Immediately fill the upper reservoir with the 250 ml of pre-warmed Brock's medium (flow will begin by gravity), and then insert the perforated stopper on the end of the peristaltic pump tubing into the spout of the inverted funnel.
10. Immediately run the peristaltic pump at 10 ml/min for 1 min to wash off loosely attached cells. Reduce the speed of the pump to 1 ml/min. Check that the liquid drains to the center of the membrane and drops cleanly through the hole in the lower funnel; adjust the position of the funnels to achieve this, if necessary.
11. Collect small samples of the liquid effluent every 30 min to monitor the synchronization by Coulter and flow cytometry. Fix 100 μl of cells in 900 μl of ice-cold 80% ethanol. Store samples at 4 °C (*see* **Note 4**.)

12. To collect G1-enriched cells for growing synchronized cultures, place a 50-ml conical tube in an ice bucket made from two Styrofoam drinking cups. Place the conical tube under the baby machine to catch the cells. Replenish the ice as it melts, every 20–30 min, by swapping in a new conical tube and ice bucket (*see* **Note 5**.)
13. When the desired volume of G1 cells has been collected, pool the contents of the conical tubes into an ice-cold Erlenmyer flask. Transfer the flask from the ice bucket into the 78 °C shaking water bath and immediately start a timer. Ensure that the culture is below the waterline of the water bath. Shake the culture at 120 rpm.
14. Collect 100 μl of cells every 30 min to fix for flow cytometry analysis in order to monitor the synchronous progression of the culture (*see* **Note 6**.)

3.3 Immunofluorescence Microscopy

1. When the synchronized *S. acidocaldarius* culture is most enriched with dividing cells, usually around 150–180 min of grow-out, centrifuge 2 ml of culture at 6000 × *g* for 3 min. Remove the supernatant.
2. Stain the cell membrane by gently resuspending the pellet in 421.8 μl of FM4-64X buffer. Incubate on ice for 5 min in the dark.
3. Add 78.2 μl of fresh 16% paraformaldehyde to fix the cells. Incubate in a microfuge tube mixer in the dark at 25 °C. Shake at 600 rpm for 45 min.
4. Pellet the cells and wash one time with 500 μl of 1× PBS. All spins are done at 6000 × *g* for 3 min.
5. Gently resuspend the cells in 20 μl of GTET buffer and shake in a microfuge tube mixer at 600 rpm for 15 min at 25 °C in the dark.
6. Pellet the cells and wash 2–3 times with 500 μl of 1× PBS.
7. Gently resuspend the cells in 20–30 μl of 1× PBS.
8. Use a pipet tip to spread 10 μl of the cell suspension onto a freshly prepared poly-D-lysine-coated coverslip (*see* **Note** 7). Allow to air-dry slightly. Coverslips may be stored in a dark, humid box overnight, but it is recommended to proceed through **step 19** (*see* **Note 8**).
9. Wash the coverslip with 1× PBS. Do not spray the cells directly (*see* **Note 9**).
10. Cover the cells by pipetting 200–300 μl of blocking buffer onto the coverslip. Incubate for 15–30 min at room temperature in the dark (*see* **Note 10**).

11. Do not allow the cells to dry out. Wash once with 1× PBS.
12. Dilute the desired primary antibody 1:10–1:50 in antibody binding buffer (*see* **Note 11**). Apply 200–300 μl to the cells on the surface of the coverslip and incubate at room temperature for 1 h in the dark.
13. Wash the coverslip three times with 1× PBS. Allow the PBS to sit for 5 min each time. Keep the coverslip in the dark.
14. Dilute the appropriate secondary antibody 1:1000 in antibody binding buffer. Apply 200–300 μl to the cells on the surface of the coverslip and incubate in the dark at room temperature for 1 h.
15. Wash the coverslip three times with 1× PBS. Allow the PBS to sit for 5 min in the dark each time.
16. Allow the coverslip to air-dry approximately 5 min before mounting or else the cells will not adhere well to the glass.
17. Spot 25 μl of mounting medium containing DAPI [1.5 ng/μl] onto a glass slide (*see* **Note 12**).
18. Invert the coverslip onto the mounting medium and seal edges with a light coating of nail polish. Allow the nail polish to dry completely before viewing. Keep the slides in the dark.
19. The slides can be stored at 4 °C for up to a year. Examples of immunolocalization results are shown in Fig. 2.

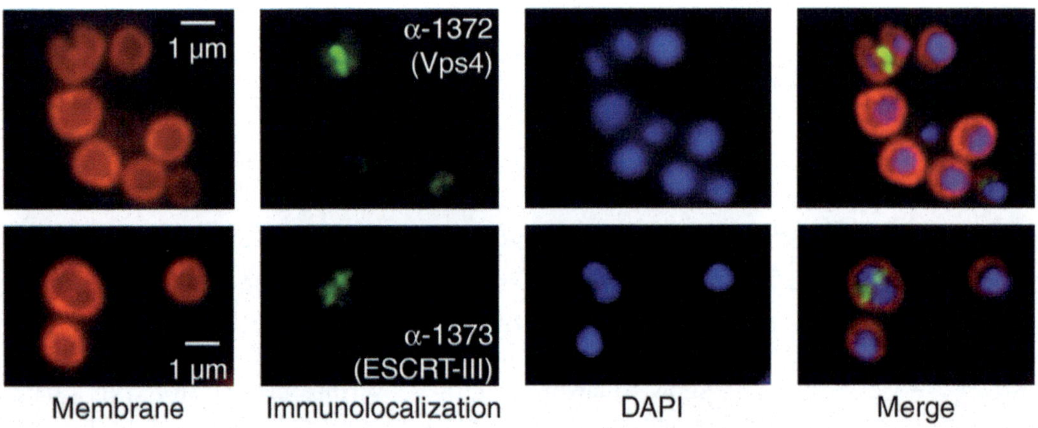

Fig. 2 Examples of immunolocalization studies with antisera raised against *Sulfolobus acidocaldarius* Vps4 (upper panels) or the ESCRT-III paralog, Saci1373 (lower panels). From left to right, panels show membrane staining with FM4-64X, immunolocalization with ESCRT machinery antibodies, DNA staining by DAPI, and a merged image of the three preceding channels. Figure was first published in ref. 4 and is reprinted with permission

4 Notes

1. The growth rate of *S. acidocaldarius* can vary depending on laboratory conditions and should be determined in the hands of the experimentalist. *S. acidocaldarius* has a doubling-time of approximately 3 h in our laboratory. The number of generations between the dilution time and baby machine starting time, the culture's doubling-time (T_d), and the starter culture's OD_{600} immediately before its dilution should be used to determine the appropriate dilution for inoculating the 100 ml culture for application to the baby machine.
2. For *S. acidocaldarius*, an OD_{600} = 0.1 corresponds to 9.6×10^7 cells/ml in our laboratory. However, preliminary experiments should be performed to calibrate the OD_{600} with cell number per ml for the particular spectrophotometer to be used. This is best done by determining cell numbers for samples, using a sensitive Coulter counter/cytometer that can accommodate a 15–20μm aperture tube, or may be done by dilution and spread-plate colony counting (if the plating efficiency is close to 100%).
3. Various protocols can be found in the literature for preparing cells for flow cytometry. Most commonly, we fix *Sulfolobus* cells in cold ethanol at a final concentration of 72% and incubate the cells for a minimum of 1 h at 4 °C. Cells can be stored at this stage for several months at 4 °C. When ready to process the samples, centrifuge them for 5 min at 16,000 × *g*. Remove the supernatant and gently resuspend the cell pellet in 1 ml of filtered 10 mM Tris–HCl, pH 7.4, 10 mM $MgCl_2$. Centrifuge again for 5 min at 16,000 × *g*. Remove the supernatant and gently resuspend the cell pellet in 500 μl of filtered 10 mM Tris–HCl, pH 7.4, 10 mM $MgCl_2$ + 100 ng/μl RNase A + 10 μM SYTOX Green. Protect the samples from light and incubate for 20 min at room temperature before analyzing with a flow cytometer.
4. Perform a pilot experiment to determine the window of time which yields the most G1 cells (*see* **Note 1**). The maximum yield, and optimum collection time, of synchronized cells (i.e., those that have just divided and were shed from the membrane-bound culture) is achieved at one generation time after loading the cells onto the membrane [8]. Under our laboratory conditions, collection of cells during the 100–250 min period after loading provides good purity and allows larger volumes to be collected and analyzed.

5. Ensure that the conical tubes used to collect the G1 cells and the Erlenmeyer flask used to grow the synchronized culture are well-chilled in ice before **step 12**. Monitor the conical tubes during the collection step to ensure that the cells are encased in ice and that the position of the tubes has not shifted as the ice melts.
6. Perform a pilot experiment to determine the window of time in which most cells are dividing.
7. To prepare the coverslips, wash them first with 100% ethanol and rinse with ultrapure water. Mark the corner of the coverslip to indicate which side will be coated. Dilute the poly-D-lysine stock in ultrapure water to 0.1%. Pipet the poly-D-lysine onto the surface of the coverslips until it reaches the edges (approximately 300–500 μl) and incubate at room temperature for 10 min. Pipet the liquid off the surface of the glass and allow the coverslips to air-dry. Rinse with ultrapure water and air-dry again. Poly-D-lysine-coated coverslips must be prepared fresh the day of use.
8. To prepare a humid box, line the bottom of a plastic freezer box with a pad of wet paper towels, approximately 1 cm thick. Place a foam tube holder on top of the paper towels and rest the coverslips on top of the foam insert. Cover the plastic lid with aluminium foil.
9. Wash steps are performed by angling the coverslip and gently squirting the top edge of the glass with 1× PBS using a squirt bottle. Allow the buffer to run down the surface of the coverslip.
10. Coverslips can be incubated on a lab bench with a foil-covered cardboard box inverted over them.
11. We always affinity-purify our primary antibodies to reduce background fluorescence in our images.
12. Do not use varieties of mounting media that hard-set. These products cause *Sulfolobus* cells to lyse. Mounting media containing DAPI can be purchased, or alternatively, DAPI can be added to the mounting medium immediately before spotting onto the glass slide.

Acknowledgments

S.D.B. and R.Y.S. are funded by the College of Arts and Sciences, Indiana University, S.D.B.'s lab is funded by grant R01 GM125579-01 from the National Institutes of Health, I.G.D. was supported by the Australian Research Council (FT160100010).

References

1. Samson RY, Dobro MJ, Jensen GJ, Bell SD (2017) The structure, function and roles of the archaeal ESCRT apparatus. Subcell Biochem 84:357–377. https://doi.org/10.1007/978-3-319-53047-5_12
2. Samson RY, Obita T, Hodgson B, Shaw MK, Chong PL, Williams RL, Bell SD (2011) Molecular and structural basis of ESCRT-III recruitment to membranes during archaeal cell division. Mol Cell 41(2):186–196. https://doi.org/10.1016/j.molcel.2010.12.018
3. Lindas AC, Karlsson EA, Lindgren MT, Ettema TJG, Bernander R (2008) A unique cell division machinery in the Archaea. Proc Natl Acad Sci U S A 105(48):18942–18946. https://doi.org/10.1073/pnas.0809467105
4. Samson RY, Obita T, Freund SM, Williams RL, Bell SD (2008) A role for the ESCRT system in cell division in archaea. Science 322 (5908):1710–1713. https://doi.org/10.1126/science.1165322
5. Helmstetter CE, Eenhuis C, Theisen P, Grimwade J, Leonard AC (1992) Improved bacterial baby machine: application to *Escherichia coli* K-12. J Bacteriol 174(11):3445–3449
6. Helmstetter CE, Thornton M, Romero A, Eward KL (2003) Synchrony in human, mouse and bacterial cell cultures–a comparison. Cell Cycle 2(1):42–45. https://doi.org/10.4161/cc.2.1.185
7. Thornton M, Eward KL, Helmstetter CE (2002) Production of minimally disturbed synchronous cultures of hematopoietic cells. BioTechniques 32(5):1098–1100, 1102, 1105
8. Duggin IG, McCallum SA, Bell SD (2008) Chromosome replication dynamics in the archaeon Sulfolobus acidocaldarius. Proc Natl Acad Sci U S A 105(43):16737–16742. https://doi.org/10.1073/pnas.0806414105

Chapter 2

Genetic and Cell Biology Methods to Study ESCRTs in *Drosophila melanogaster*

Marco Gualtieri and Thomas Vaccari

Abstract

Mosaic analysis in *Drosophila* represents a convenient entry point for studying the role of ESCRT (Endosomal Sorting Complex Required for Transport) genes in multiple cell processes crucial for organ development and homeostasis. Here, we describe the procedure to generate populations of ESCRT-mutant cells within *Drosophila* larval epithelial organs and to study them in whole-mount preparations using confocal microscopy. The use of antibodies directed to endocytic cargoes, vesicular trafficking, cell proliferation, death, and polarity markers allows one to investigate the consequences of loss of ESCRT activity at the subcellular and tissue level. The protocols described here can be used in fixed tissue as well as in unfixed tissue using endocytic uptake assays.

Key words ESCRT function, Endocytosis, *Drosophila melanogaster*, Imaginal discs, Whole-mount immunohistochemistry, Confocal microscopy

1 Introduction

A conserved set of ESCRT complexes, named ESCRT-0, -I, -II, -III, controls a wide range of cellular processes occurring at membranes that are crucial for tissue development and homeostasis [1]. These include, among others, endosomal sorting of ubiquinated proteins destined to lysosomal degradation during endocytosis, generation of exovesicles during secretion, regulation of cytokinesis, and nuclear envelope reformation during cell division. Owing to such a pleiotropic function, ESCRT genes are essential and their alteration causes a number of congenital syndromes, neurodegenerative diseases, and is associated to a wide range of cancers for review [2, 3].

Because of its genetic tractability and accessibility for microscopic analysis, *Drosophila melanogaster* is a premier model to study gene function in health and disease [4]. In particular, the possibility of generating animals containing ESCRT-mutant cells, or misexpressing ESCRT cDNAs, or depleted ESCRT tissues, specifically, allows the study of the consequences of altered ESCRT activity to

Emmanuel Culetto and Renaud Legouis (eds.), *The ESCRT Complexes: Methods and Protocols*, Methods in Molecular Biology, vol. 1998, https://doi.org/10.1007/978-1-4939-9492-2_2,

tissue or organ development. Such studies in *Drosophila* revealed that, by controlling endosomal sorting, ESCRT is crucial to silence a large set of signaling receptors, ultimately preventing tumorigenesis [5]. To alter experimentally ESCRT activity in *Drosophila*, a number of lines are available from public stock centers, most of which are listed in Table 1. These include null mutants generated on FRT (Flippase Recognition Target) chromosomes for clonal analyses [26], UAS lines for misexpression, or for in vivo RNAi to be used in conjunction with tissue-specific GAL4 drivers [27]. A multiplicity of genetic backgrounds allowing spatiotemporal control of gene expression in multiple fly tissues are available [28–30]. Given the space constraints, they are not all covered in this chapter. Here, we will only provide an example of genetic crosses to generate clones of ESCRT-mutant cells within otherwise wild-type imaginal disc epithelia or entirely mutant imaginal disc epithelia as shown in Fig. 1. Using such genetic backgrounds, we restrict our focus on immunolocalization of Notch, a signalling receptor subjected to endosomal sorting in eye imaginal discs cells, as well as on visualization of epithelial cell shape, both of which are impaired in cells lacking ESCRT-I, -II, -III components [10, 16, 22, 32, 33] (Fig. 2). We also provide as part of Table 2 an extensive list of antibodies that can be used to immuno-detect other endocytic cargoes and proteins that help to determine the morphology of the membrane trafficking and polarity apparatus of mutant and wild-type cells. In addition, several fluorescent compounds or compounds directly conjugated to fluorophores can be used to identify a number of cell structures, such as the cortical actin cytoskeleton and the nucleus (Fig. 2).

The purpose of this practical chapter is to explain in detail immunohistochemical procedures that can be applied to whole-mount preparations of *Drosophila* imaginal discs for confocal microscopy analysis. The procedure is routinely used in our lab and in most other labs that study ESCRTs in the context of their tumor suppressor activity, or in the control of degradation of signaling molecules, such as the Notch receptor [10, 22]. Due to the impossibility of providing detailed preparation procedures for the many *Drosophila* tissues and organs amenable to immunolabeling, below we describe only preparations of eye-antennal imaginal discs, including dissection, culturing for endocytic uptake assays, fixation, and immunostaining, and we refer the reader to other sources for detailed descriptions of preparation of other organs. Similarly, given the wide diversity of microscopy setups employed in cell biology labs, we do not provide guidelines for image acquisition.

Table 1
The table provides stock number, type (UAS, misexpression; RNAi, UAS-based knock down; Mutant, loss of function mutation), genotype, and recent examples of use of a selected list of *Drosophila* lines used to study genes encoding ESCRT components. Tabulated stocks are available from the Bloomington Drosophila Stock Center (bdsc.indiana.edu/) and the Vienna Drosophila Research Center (stockcenter.vdrc.at/)

Complex	Gene	Type	Stock N°	Flybase Genotype	Reference
ESCRT-0	Hrs	UAS	42692	y^1,w*; P{UAS-Hrs.L}12	[6]
		RNAi	28026	y^1,v^1;P{TRiP.JF02860}attP2	[7]
			34086	y1,sc*,v1; P{TRiP.HMS00840}attP2	[8]
			33900	y^1, sc*,v^1; P{TRiP.HMS00841}attP2	[9]
		Mutant	42691	Hrs^{D28}; P{UAS-arr.flu}3/T(2;3)B3, CyO: TM6B, Tb^+	[10]
			54574	P{hsFLP}12, y^1 w*; Hrs^{D28} P{neoFRT}40A/In(2LR)Gla, wg^{Gla-1} $PPO1^{Bc}$	[11]
	Stam	RNAi	v22497	w^{1118}; P{GD11948}v22497	[12]
			35016	y^1 sc* v^1; P{TRiP.HMS01429}attP2	[11]
		Mutant	41804	$Stam^{2L2896}$ P{neoFRT}40A/CyO, P{ftz/lacB}E3	[13]
ESCRT-I	TSG101	RNAi	v23944	w^{1118}; P{GD14295}v23944/CyO	[14]
			35710	y^1 sc* v^1; P{TRiP.GLV21075}attP2	[9]
	Vps28	Mutant	11182	y^1 w^{67c23}; P{lacW}$Vps28^{k16503}$/CyO	[15]
			39624	w*; P{neoFRT}42D $Vps28^{D2}$/CyO, P{GAL4-twi.G}2.2, P{UAS-2xEGFP}AH2.2	[16]
			39634	w*; $Vps28^{B9}$ P{neoFRT}42D/CyO, P{GAL4-twi.G}2.2, P{UAS-2xEGFP}AH2.2	[9]
ESCRT-II	Vps36	RNAi	38286	y^1 sc* v^1; P{TRiP.HMS01739}attP40	[9]
			v16847	w^{1118}; P{GD5901}v16847	[7]
		Mutant	10176	y^1 w^{1118}; P{lacW}$Vps36^{L5212}$/TM3, Ser^1	[7],
	Vps22	RNAi	v21658	w^{1118}; P{GD10787}v21658	[9]
		Mutant	39632	w*; lsn^{NN31} P{neoFRT}82B/TM6B, Tb^1	[16]
			39631	w*; wg^{Sp-1}/CyO; P{neoFRT}82B lsn^{SS6}/TM6C, Sb^1 Tb^1	[16]
	Vps25	RNAi	26286	y^1 v^1; P{TRiP.JF02055}attP2	[9]
		Mutant	39622	w*; P{hs-Vps25.V}11	[10]
			39633	w*; $Vps25^{A3}$ P{neoFRT}42D/CyO, P{GAL4-twi.G}2.2, P{UAS-2xEGFP}AH2.2	[10]
			10839	y^1 w^{67c23}; P{lacW}$Vps25^{k08904}$/CyO	[10]

(continued)

Table 1
(continued)

Complex	Gene	Type	Stock N°	Flybase Genotype	Reference
ESCRT-III	Vps2	RNAi	v24869	w^{1118}; P{GD8363}v24869/TM3	[17]
			38995	y^1 sc* v^1; P{TRiP.HMS01911}attP40	[9]
		Mutant	39630	w*; $Vps2^{PP6}$ P{neoFRT}82B/TM6B, Tb^1	[16]
			203012 (KSC)	y^1 w^{67c23}; P{GSV6}$Vps2^{GS11024}$/TM3, Sb^1 Ser^1	[17]
	Vps24	RNAi	38281	y^1 sc* v^1; P{TRiP.HMS01733}attP40	[9]
	Vps32	RNAi	v106823	P{KK108557}VIE-260B	[18]
			38305	y^1 v^1; P{TRiP.HMS01767}attP40	[19]
		UAS	32559	w*; P{UAS-shrb-GFP}2; P{NIG.4699R}3	[20],
		Mutant	11016	y^1 w^{67c23}; P{lacW}$shrb^{k11201}$/CyO	[20]
			19705	y^1 w^{67c23}; P{EPgy2}$shrb^{EY05194}$/CyO	[20]
			14372	y^1; P{SUPor-P}$shrb^{KG01481}$/SM6a; ry^{506}	[21]
			39623	w*; P{neoFRT}42D $shrb^{O3}$/CyO, P{GAL4-twi.G}2.2, P{UAS-2xEGFP}AH2.2	[16]
			39635	w*; $shrb^{G5}$ P{neoFRT}42D/CyO, P{GAL4-twi.G}2.2, P{UAS-2xEGFP}AH2.2	[16]
	Vps20	RNAi	v47653	w^{1118}; P{GD16875}v47653/TM3	[12]
			40894	y^1 v^1; P{TRiP.HMS02142}attP40	[9]
		Mutant	67706	w*; $Vps20^{I3}$/CyO, P{GAL4-twi.G}2.2, P{UAS-2xEGFP}AH2.2	[22]
	Chmp1	RNAi	v21788	w^{1118}; P{GD11219}v21788/CyO	[23]
			28906	y^1 v^1; P{TRiP.HM05117}attP2	[23]
	Chmp5	RNAi	v101422	P{KK109120}VIE-260B	[9]
Vps4	Vps4	RNAi	v35126	w^{1118}; P{GD12054}v35126	[12]
			31751	y^1 v^1; P{TRiP.HM04061}attP2	[24]
		Mutant	63798	y^1 w* $Vps4^{3B1}$ P{neoFRT}19A/FM7i, P{ActGFP}JMR3	[25]
			26594	w* P{EP}$Vps4^{G524}$	[24]

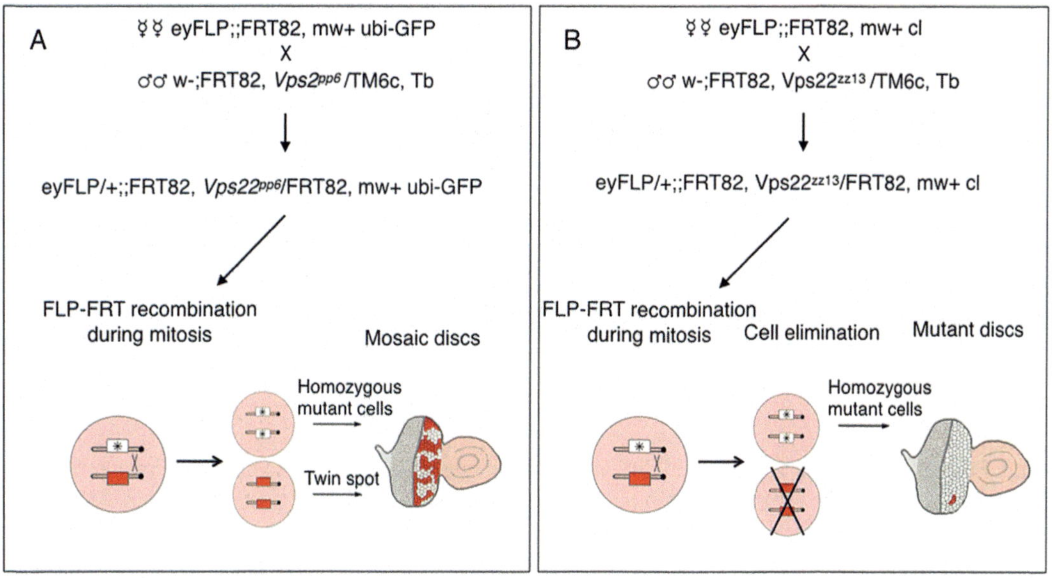

Fig. 1 Schematics of sample generation using the eye FLP-FRT system [26]. Crossing scheme of virgins (☿☿) of the tester lines and males (♂♂) heterozygous for the *Vps2 pp6* mutation (ESCRT-III) (**a**) or *Vps22 zz13* (ESCRT-II) mutation (**b**) to generate third instar larvae carrying eye-antennal discs containing clones of ESCRT-mutant cells surrounded by wild-type tissue (mosaic discs, **a**), or predominantly mutant eye-antennal discs (mutant discs, **b**). Larvae with mosaic or mutant discs have to be selected by scoring against a dominant marker visible (in this case TM6c, Tb). An illustrative representation of FLP/FRT-mediated recombination to generate chimeric animals is shown below each genetic cross. Reporter genes encoding the White pigment (mw+) or GFP (ubi-GFP) allow recognition of nonmutant cells in larval or adult tissue, respectively. Cl denotes the presence of a recessive cell lethal mutation that allows elimination of nonmutant cells after mitotic recombination. EyFLP indicates the presence of a transgene expressing the FLP enzyme downstream of the *eyeless* promoter, specifically in eye-antennal discs. 82 represents the cytologic position of the inserted FRT element used to induce recombination, corresponding to the base of right arm of *Drosophila* chromosome 3. For a simple explanation of fly genetics and husbandry, please refer to [31]

2 Materials

2.1 Dissection

1. Dissecting stereomicroscope.
2. Two pair of Dumont forceps #5 and #55.
3. Glass watch for dissection (Corning or similar).
4. Glass Pasteur pipettes, 150 mm.
5. Gentle specimen mixer for microcentrifuge tubes.
6. Directed light source (ideally, with dual gooseneck light guides).
7. Phosphate Buffer Saline 1× (PBS 1×): 137 mM NaCl; 2.7 mM KCl; 4.3 mM Na_2HPO_4; 1.47 mM KH_2PO_4. Adjust to a final pH of 7.4.
8. Notch uptake solution: Schneider's Drosophila medium containing 1% fetal calf serum.

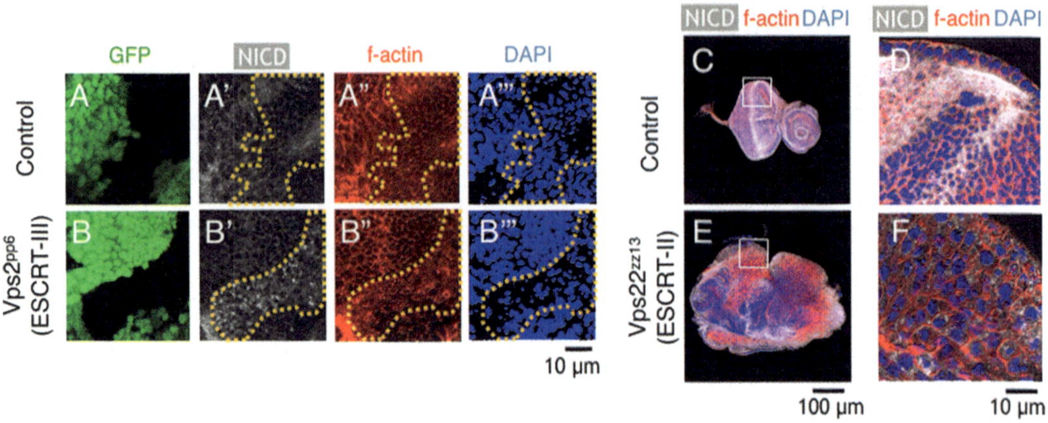

Fig. 2 Examples of immunolabeling of ESCRT mosaic (**a**–**b**) and mutant (**c**–**f**) imaginal discs. Immunolabeling of a prominent endocytic cargo, the Notch receptor, using anti-NICD (Notch intracellular domain, Table 2) in eye-antennal imaginal discs of third instar larvae in confocal sections. Panels **d** and **f** are high magnifications of areas approximately corresponding to the boxed regions in **c** and **e**. Discs have been colabeled with Phalloidin-TRITC to detect filamentous Actin (f-actin) and with DAPI to detect cell nuclei. Absence of GFP expression in **a**, **b** marks the mutant tissue, whose outline is highlighted by dotted lines in non-GFP channels. Note that Notch accumulates at high levels intracellularly in puncta (**b**′, **f**) that correspond to enlarged ESCRT-mutant endosomes [16]. Also, compared to controls, the cortical f-actin cytoskeleton is altered in ESCRT mutant cells (**b**″, **f**), denoting a loss of epithelial polarity. Finally, ESCRT mutant cells, when surrounded by wild-type cells, tend to be eliminated, as indicated by the picnotic aspect of their nuclei (**b**‴). Such effect is extremely reduced in the absence of nonmutant cells (f)

2.2 Notch Uptake Assay

1. Notch uptake solution: (*see* step 8, Subheading 2.1).
2. Anti-Notch antibody (*see* Table 2).
3. Fixation solution: 4% (v/v) paraformaldehyde (PFA) in phosphate-buffered saline pH 7.4 (PBS). Dilute 10% PFA in 1× PBS. To make a 100 mL solution, add 40 mL 10% PFA with 60 mL 2× PBS pH 7.4 (*see* **Note 1**).

2.3 Immunohistochemistry

1. Fixation solution: 4% (v/v) paraformaldehyde (PFA) in phosphate-buffered saline pH 7.4 (*see* **step 3**, Subheading 2.2).
2. Washing solution (PBT): 0.1% (v/v) Triton X-100 in 1× PBS. To prepare the PBT solution, dilute 1 mL of TritonX-100 in 1 L of 1× PBS. Store at room temperature (*see* **Note 2**).
3. Blocking solution: 3% (w/v) bovine serum albumin (BSA) in PBT. To prepare the blocking solution, dissolve 0.5 g BSA in 10 mL PBT (*see* **Note 3**).
4. Primary antibody: Primary antibodies at the appropriate dilution in freshly prepared blocking solution (*see* **Note 4**, Table 2; Fig. 2 for examples).

Table 2
The table provides sources, experimental conditions, and recent examples of use of a selected list of the most used antibodies that recognize ESCRT-associated proteins. DSHB: Developmental Studies Hybridoma Bank, University of Iowa

Name of antibody	Source	Dilution for immunolabeling and species	Epitope	Example of use
(a) Antibodies directed to endocytic cargoes				
Anti-NICD	DSHB C17.9C6	1:50 Mouse	Intracellular	[16]
Anti-NECD	DSHB C458.2H	1:50 Mouse	Extracelullar	[16]
Anti-delta	DSHB C594-9B	1:100 Mouse monoclonal	Extracellular	[34]
Anti-Unpaired	[35]	1:50 Mouse	Extracellular	[35]
Anti-Thick veins	[36]	1:150 Rabbit	Intracellular	[32]
(b) Antibodies directed to endocytic proteins				
Anti-avalanche/Syntaxin 7	[37]	1:100 Rabbit	Early endosome	[10]
Anti-Rab5	ab31261 Abcam	1:500 Rabbit	Early endosome	[17]
Anti-Rab5	[38]	1:200 Rabbit	Early endosome	[38]
Anti-smad anchor for receptor activation (Sara)	[39]	n.d. Rabbit	SARA endosome	[40]
Anti-hepatocyte growth factor-regulated tyrosine substrate (Hrs)	[6]	1:600 Guinea pig	Sorting endosome	[34]
Anti-vacuolar protein sorting 2 (Vps2)	[17]	1:500 Guinea pig	Late endosome	[17]
Anti-Rab7	[41]	1:5000 Rabbit	Late endosome	[17]
Anti-Rab11	[42]	1:1000 Rat	Recycling endosome	[43]
(c) Antibodies directed to polarity, proliferation, and cell death proteins				
Anti-atypical protein kinase C (aPKC)	Santa Cruz sc-216	1:100 Rabbit	Apical Par complex	[10]
Anti-crumbs (Crb)	DSHB CQ4 Supernatant	1:20 Mouse monoclonal	Subapical Crumbs complex	[44]
Anti-armadillo (Arm)	DSHB N2-7A1 Supernatant	1:100 Mouse monoclonal	Adherens junctions	[43]
Anti-DE-cadherin (DE-Cad)	DSHB DCAD2 Supernatant	1:25Rat Monoclonal	Adherens junctions	[10]
Anti-discs large (Dlg)	DSHB4F3 Supernatant	1:100 Mouse monoclonal	Lateral membrane	[10]
Anti-cleaved caspase 3	Cell signaling	1:200 Rabbit	Intracellular	[45]
Anti-phosphohistone H3	Millipore	1:10,000 Mouse	Nuclear	[46]

(continued)

Table 2
(continued)

Name of antibody	Source	Dilution for immunolabeling and species	Epitope	Example of use
(d) Miscellaneous antibody to membrane compartment components				
Anti-ubiquitinylated protein (FK2)	Enzo life sciences BML-PW8819	1:1000 Mouse	Mono- and polyubiquitinylated conjugates	[10]
Anti-lava lamp (Lva)	[47]	1:1000 Guinea pig	Cis-Golgi apparatus	[43]
Anti-GP120	Calbiochem[a]	1:100 Mouse	Medial-Golgi apparatus	[48]
Anti-Sec15	[49]	1:2000 Guinea pig	Exocyst complex	[43]

5. Secondary antibody: Fluorescently conjugated secondary antibodies at the appropriate dilution in PBT (e.g., Alexa Fluor-conjugated polyclonal antibodies). Discard it after use. Be careful to match the secondary antiserum to the species of the primary antisera (*see* **Note 5**).
6. Other Markers (Optional): DNA stain: DAPI (4′,6-diamidino-2-phenylindole). The stock solution is made at 1 mg/mL in PBS 1× and used 1:10,000 in PBS 1×.
7. F-actin stain: Phalloidin-TRITC (tetramethylrhodamine B isothiocyanate). The stock solution is made at 0.5 mg/mL in DMSO and used 1:100 in PBT.

2.4 Mounting

1. Mounting medium: 1.5% (w/v) DABCO (1,4-diazabicyclo [2.2.2]octane) and 70% glycerol in 1×PBS. For a 50 mL solution, add 0.75 g DABCO, 15 mL 1× PBS, and 35 mL glycerol and mix on rocking platform until the solution is homogeneous. It's possible to prepare aliquots in 1.5 mL tubes and store at −20 °C (*see* **Note 6**).
2. Microscope slides, 26 × 76 mm.
3. Cover slips, Nr. 1/mm 24 × 24.
4. Nail polish fast dry.

3 Methods

3.1 Dissection of Larval Eye-Antennal Imaginal Discs

1. To begin, gently transfer third instar larvae to be dissected into a glass watch filled with PBS using #5 forceps, under a stereoscope with side lighting. Dissect the larvae in ice-cold Notch uptake solution, if you plan to perform a Notch uptake assay (Subheading 3.2) or in PBS 1×, if you proceed directly to immunohistochemistry (Subheading 3.3) (*see* **Note 7**).
2. The imaginal discs are located on the anterior portion of the larva, near the mouth hooks. Tear the larva in half and discard the posterior. Invert the anterior like a sleeve by pushing in delicately on the mouth hooks with the #55 forceps. Hold the body steady near the severed part with the #5 forceps. The eye-antennal discs are located between the surface of the optical lobe of the larval brain and the mouth parts. (For dissection of other larval organs *see* **Note 9**). Imaginal discs are kept attached to carcasses to facilitate handling. To prepare carcasses for fixation, clean them of the gut, fat tissue, and salivary glands using the #55 forceps while holding them with the #5 forceps.

3. Transfer carcasses to a 1.5 mL tube using a Pasteur pipette. Up to 20 carcasses can be put in one tube. It is important to restrict dissection time to 20 min or less to avoid decay of the sample. For a detailed video description of imaginal discs, *see* [50]. Carefully remove most of the dissection solution using a Pasteur pipette. Proceed to Subheadings 3.2.1 or 3.3.1 (*see* **Note 8**).

3.2 Notch Uptake Assay

The endocytic uptake of the Notch receptor is monitored by tissue fixation at different time points after the initial incubation with the NECD antibody that recognizes an extracellular Notch epitope.

3.2.1 Notch Antibody Incubation

1. Rinse carcasses with ice-cold Notch uptake solution. After adding the solution, let the carcasses settle by gravity and remove most of the medium with a Pasteur pipette (*see* **Note 9**).
2. Add ice-cold Notch uptake solution containing 1:5 anti-Notch extracellular domain (NECD) antibody (Table 2) (*see* **Note 10**).
3. Incubate the samples at 4 °C for 2 h on a nutator, then wash five times for 5 min at 4 °C on a nutator with ice-cold Notch uptake solution. Remove most of the solution.

3.2.2 Fixation 0 min Time Point

1. Remove 1/4 of the carcasses and transfer to a new tube. Add approximately 0.5 mL ice-cold PFA 4% and leave the sample on the nutator for 20 min at 4 °C to allow uniform fixation (*see* **Note 11**).
2. Remove the fixation solution and wash three times with PBS 1× 5 min each at 4 °C. Store at 4 °C until further processing.

3.2.3 Notch Uptake

Add warm Notch uptake solution at RT to the remaining carcasses. Incubate for 15 min at RT on a nutator.

3.2.4 Fixation 15 min Time Point

Remove 1/4 of the carcasses and transfer to a new tube. Process the new tube as in Subheading 3.2.2 (*see* **Note 12**).

3.2.5 Fixation 45 min Time Point

Incubate for further 30 min at RT on a nutator. Remove 1/4 of the carcasses and transfer to a new tube. Process the new tube as in Subheading 3.2.2.

3.2.6 Fixation 4 h Time Point

1. Incubate for further 195 min at RT on a nutator. Remove 1/4 of the carcasses and transfer to a new tube. Process the new tube as in Subheading 3.2.2.
2. Collect tubes with the different time points and process them in parallel for immunohistochemistry starting at Subheading 3.3.2.

3.3 Immunohistochemistry

3.3.1 Fixation

1. If you are not performing an uptake assay, add approximately 0.5 mL PFA 4% and leave the tubes with the dissected carcasses (Subheading 3.1) on the nutator for 20 min to allow uniform fixation.
2. Remove the tube from the nutator and allow the carcasses to settle to the bottom of the tubes before removing the fixative.
3. After removal of the fixative, add PBT and place on the nutator to rinse for 5–10 min. Repeat this **step 3** times to remove all traces of fixative.

3.3.2 Blocking

1. Remove PBT (or PBS if samples are those of the Notch uptake assay) and add blocking solution. Reposition samples on the nutator to allow homogenous blocking for 30 min at RT (*see* **Note 13**).

3.3.3 Antibody Incubation

1. Remove the blocking solution and add the desired primary antibodies, diluted in the blocking solution. Do not add a mouse antibody if the samples are derived from the Notch uptake experiment because these are already labeled with the mouse anti-Notch antibody. Incubate primary antibodies overnight on the nutator at 4 °C or 2 h at room temperature (*see* **Note 14**).
2. Remove the primary antibody solution and store it at 4 °C (*see* **Note 15**).
3. Wash three times for 5 min with PBT for 5–10 min each.
4. Remove the wash and add the fluorophore-conjugated secondary antibodies diluted in PBT. Incubate for 1–2 h at room temperature on the nutator in the dark (*see* **Note 16**).
5. Wash tissues three times for 5 min with PBT and, if you wish to stain the cell nuclei, add PBT containing 1× DAPI and incubate at room temperature on the nutator for 20 min in the dark.
6. Wash the samples once with PBS for 5 min.

3.4 Mounting

3.4.1 Mounting Eye-Antennal Discs

1. Transfer carcasses from the tube to a glass watch filled with PBS under the dissection scope with side lighting.
2. To remove eye-antennal imaginal discs, gently rip the nerves connecting the ventral ganglion with the carcass wall by sliding the forceps tips between them (*see* **Note 17**).
3. Grab the basal part of the mouth hooks with one forcep and rip it away from the rest of the body with the other forceps. Eye-antennal imaginal discs and brain will be removed from the carcass as a single mass, together with the mouth hooks. To sever the brain from the eye-antennal discs, use one forcep to carefully pinch the nerve connecting each optic lobe to its disc.

4. Transfer discs to the slide by holding them by the attached mouth hooks.
5. Blot excess liquid with paper tissue or absorbing paper. This procedure must be done carefully to avoid damage.
6. Quickly add a couple of drops of mounting medium using the Pasteur pipette. The amount changes depending on the number of discs on the slide; however, 20 μl of mounting medium should be enough for most preps.
7. Detach the eye-antennal discs from the mouth hooks by pinching the narrow connection between the antennal disc and the mouth hooks. Discard the mouth hooks.

3.4.2 Sealing of the Slide

1. Make sure that the tissue is flat and unwanted tissue parts are removed from the slide, before covering with a coverslip. In particular, parts thicker than the tissue to be analyzed should be removed to avoid excess spacing between slide and coverslip, which creates movement/vibration of the sample during imaging (*see* **Note 18**).
2. Seal the edges with nail polish. Store the slide at 4 °C in the dark.
3. When using Mowiol-containing medium (*see* **Note 6**), allow 12–24 h before imaging to ensure hardening of the resin.
4. Analyze the sample at a fluorescent microscope. Whole-mount preparations are perfectly suited for confocal microscopy.

4 Notes

1. The PFA solution is essential to fix tissues by preserving the physical structures and preventing digestion by enzymes and bacteria. The PFA solution is toxic and it should be carefully manipulated under the hood. PFA is a suspected carcinogen and should be handled as such with appropriate PPE (Personal Protective Equipment). Aliquot in 2 mL tubes and store at −20 °C where it is stable for few months.
2. Pure Triton X-100 is very viscous. 10 mL of a 10% solution can be used to reduce pipetting errors due to viscosity. A low detergent concentration preserves membrane structures for imaging such as small endocytic vesicles. Moreover, a harsh treatment with detergents might affect tissue integrity.
3. Other blocking reagents such as 5% normal goat serum in PBT might be used to reduce background staining. Store blocking solutions at 4 °C for a short period of time.
4. In order to reduce aspecific staining, it is possible to pre-adsorb primary antisera. To this end, sera can be pre-adsorbed to fixed

tissues which do not express the desired antigen. If this is not possible, antigen-expressing tissue can be used.

5. Fluorescent labels for the secondary antibodies can be chosen depending on the availability of the microscope lasers. Note that the staining protocol preserves the properties of the fluorescently tagged proteins, a very important feature when one wants to perform immunolocalizations in tissues expressing fluorescent proteins such as GFP (Fig. 2a–b).
6. DABCO is an antifade reagent. It acts as a reactive oxygen species scavenger and it prevents the bleaching process. The mounting medium can be supplemented with a clariant resin. In our lab, we use also Mowiol® 4-88. Mowiol-containing mounting medium hardens and has the same refractive index as the immersion oil. The inclusion of Mowiol in the mounting medium depends on the experiment. For colocalization experiments, it is better to mount the sample in glycerol since it maintains the 3D conformation of the tissue, while the hardening properties of Mowiol help long-term storage of samples. Commercial preparations are also available which achieve similar effects as the DABCO/Mowiol mixture.
7. By positioning incident light perpendicular to the path of view (side lighting), the contrast between the tissues and the environment increases and identification of the different tissues becomes easier.
8. The dissection protocol for imaginal discs refers to third instar (L3) larvae. For general fly husbandry and for protocols to obtain L3 larvae, please refer to [51]. Other most studied organs that can be harvested for fixation and staining in larvae are the optic lobes and ventral ganglion, the salivary glands, the gut, the lymph gland, and the hemocytes. Detailed dissection protocols and examples of immunolocalizations are described in [52–58].
9. It is possible to inhibit the lysosomal degradation to accumulate endocytic cargoes such as Notch in lysosomes. For this, substitute the Schneider's medium with one of the following solutions: Schneider's medium supplemented with 200 μM chloroquine (Sigma C-6628; 100 mM stock in ddH_2O); or with 50 mM NH_4Cl (1.5 M stock in ddH_2O); or with 200 μM leupeptin (20 mM stock in ddH_2O). All the three compounds block lysosomal degradation and cause Notch accumulation after 4 h in discs, to the same degree. During the 4 h uptake, replace the medium at least once to ensure maximum compound activity at RT.
10. General endocytic uptake can also be followed by adding 0.25 mM Texas-Red–dextran MW3000, lysine-fixable (Molecular Probes, Eugene, OR) [37].

11. Fixed tissue can be stored at 4 °C for a week to 10 days at this stage. In this case, make sure to wash and store with 1xPBS. The detergent in PBT will prevent preservation overtime.
12. The uptake protocol allows following Notch internalization at different time points, which can be chosen at the discretion of the researcher. We have found that 15 min is enough to find anti-NECD in early endosomes and 45 min is enough to find anti-NECD in late endosomes, suggesting that ESCRT activity in endosomal sorting occurs in such timeframe. 4–6 h allows complete clearance of anti-NECD [10, 22]. Longer incubation is not recommended because the medium is not suitable for long-term culturing.
13. Pretreatment with 1×PBS, 1% Triton X-100 for 30 min or 1 h may be needed to ameliorate the permeabilization of the membrane and increase the penetration of some antibodies. However, due to extraction of soluble proteins, this treatment may reduce detection of some antigens.
14. The number of antibodies used at this step is dictated ultimately by nature of the lasers on your confocal system. In a typical setup, up to three different primary antibodies can be used.
15. Overnight incubations return a sharper signal and allow the staining experiment to cover 2 days, with an image acquisition session in the afternoon of day 2. Primary antibody solutions can be reused a couple of times or more.
16. At this step, together with the secondary antibody, it is possible to add other fluorescent compounds that work in fixed tissue, such as phalloidin-TRITC, to mark F-actin and visualize the overall morphology of cells (Fig. 2).
17. Repetitive incubations and washes could have moved or detached the discs from their original position. Be ready to search for them close to the mouth hooks but not attached to the optical lobe, or attached to the lobes but not connected to the mouth hooks anymore.
18. Conversely, when *z* dimensions need to be preserved (i.e., for *z*-confocal sectioning), a Dakopen or another hydrophobic barrier marker can be used to ensure appropriate spacing between slide and coverslip.

Acknowledgments

M.G. and T.V. are supported by AIRC (Associazione Italiana Ricerca contro il Cancro) and WCR (Worldwide Cancer Research).

References

1. Hurley JH (2015) ESCRTs are everywhere. EMBO J 34(19):2398–2407. Available from: http://www.ncbi.nlm.nih.gov/pubmed/26311197
2. Horner DS, Pasini ME, Beltrame M, Mastrodonato V, Morelli E, Vaccari T (2018) ESCRT genes and regulation of developmental signaling. Semin Cell Dev Biol 74:29–39. Available from: http://www.ncbi.nlm.nih.gov/pubmed/28847745
3. Alfred V, Vaccari T (2016) When membranes need an ESCRT: endosomal sorting and membrane remodelling in health and disease. Swiss Med Wkly 146:w14347. Available from: http://doi.emh.ch/smw.2016.14347
4. Jennings BH (2011) Drosophila-a versatile model in biology & medicine. In: Materials today, vol 14. Elsevier, Amsterdam, pp 190–195. Available from: http://www.sciencedirect.com/science/article/pii/S1369702111701134
5. Vaccari T, Bilder D (2009) At the crossroads of polarity, proliferation and apoptosis: the use of Drosophila to unravel the multifaceted role of endocytosis in tumor suppression. Mol Oncol 3:354–365
6. Lloyd TE, Atkinson R, Wu MN, Zhou Y, Pennetta G, Bellen HJ (2002) Hrs regulates endosome membrane invagination and tyrosine kinase receptor signaling in drosophila. Cell 108(2):261–269. Available from: https://www.sciencedirect.com/science/article/pii/S0092867402006116
7. Yang X, Mao F, Lv X, Zhang Z, Fu L, Lu Y et al (2013) Drosophila Vps36 regulates Smo trafficking in Hedgehog signaling. J Cell Sci 126 (18):4230–4238. Available from: http://www.ncbi.nlm.nih.gov/pubmed/23843610
8. Corrigan L, Redhai S, Leiblich A, Fan S-J, Perera SMW, Patel R et al (2014) BMP-regulated exosomes from Drosophila male reproductive glands reprogram female behavior. J Cell Biol 206(5):671–688. Available from: http://jcb.rupress.org/content/206/5/671
9. Loncle N, Agromayor M, Martin-Serrano J, Williams DW (2015) An ESCRT module is required for neuron pruning. Sci Rep 5:8461. Available from: http://www.ncbi.nlm.nih.gov/pubmed/25676218
10. Vaccari T, Bilder D (2005) The Drosophila tumor suppressor vps25 prevents nonautonomous overproliferation by tegulating Notch trafficking. Dev Cell 9(5):687–698. Available from: http://www.sciencedirect.com/science/article/pii/S1534580705003813?via%3Dihub
11. Sheng Z, Yu L, Zhang T, Pei X, Li X, Zhang Z et al (2016) ESCRT-0 complex modulates *Rbf* -mutant cell survival by regulating Rhomboid endosomal trafficking and EGFR signaling. J Cell Sci 129(10):2075–2084. Available from: http://www.ncbi.nlm.nih.gov/pubmed/27056762
12. Troost T, Jaeckel S, Ohlenhard N, Klein T (2012) The tumour suppressor Lethal (2) giant discs is required for the function of the ESCRT-III component Shrub/CHMP4. J Cell Sci 125 (3):763–776. Available from: http://jcs.biologists.org/content/125/3/763
13. Tognon E, Wollscheid N, Cortese K, Tacchetti C, Vaccari T (2014) ESCRT-0 is not required for ectopic notch activation and tumor suppression in drosophila. PLoS One 9 (4):e93987. Available from: http://dx.plos.org/10.1371/journal.pone.0093987
14. Fan J, Jiang K, Liu Y, Jia J, Payre F (2013) Hrs promotes ubiquitination and mediates endosomal trafficking of smoothened in drosophila hedgehog signaling. Xie J, editor. PLoS One 8(11):e79021. Available from: http://dx.plos.org/10.1371/journal.pone.0079021
15. Sevrioukov EA, Moghrabi N, Kuhn M, Krämer H (2005) A mutation in dVps28 reveals a link between a subunit of the endosomal sorting complex required for transport-I complex and the actin cytoskeleton in Drosophila. Mol Biol Cell 16(5):2301–2312. Available from: http://www.ncbi.nlm.nih.gov/pubmed/15728719
16. Vaccari T, Rusten TE, Menut L, Nezis IP, Brech A, Stenmark H et al (2009) Comparative analysis of ESCRT-I, ESCRT-II and ESCRT-III function in Drosophila by efficient isolation of ESCRT mutants. J Cell Sci 122 (14):2413–2423. Available from: http://www.ncbi.nlm.nih.gov/pubmed/19571114
17. Aoyama N, Yamakawa T, Sasamura T, Yoshida Y, Ohori M, Okubo H et al (2013) Loss- and gain-of-function analyses of vacuolar protein sorting 2 in Notch signaling of *Drosophila melanogaster*. Genes Genet Syst, Humana Press 88(1):45–57. Available from: http://www.ncbi.nlm.nih.gov/pubmed/23676709
18. Aradhya R, Zmojdzian M, Da Ponte JP, Jagla K (2015) Muscle niche-driven Insulin-Notch-Myc cascade reactivates dormant adult muscle precursors in drosophila. elife 4:e08497. Available from: http://www.ncbi.nlm.nih.gov/pubmed/26650355

19. Matias NR, Mathieu J, Huynh J-R, Bellen H, Glover D (2015) Abscission is regulated by the ESCRT-III protein shrub in drosophila germline stem cells. PLoS Genet 11(2):e1004653. Available from: http://dx.plos.org/10.1371/journal.pgen.1004653
20. Sweeney NT, Brenman JE, Jan YN, Gao F-B (2006) The coiled-coil protein shrub controls neuronal morphogenesis in drosophila. Curr Biol 16(10):1006–1011. Available from: http://www.ncbi.nlm.nih.gov/pubmed/16713958
21. Irion U, St Johnston D (2007) bicoid RNA localization requires specific binding of an endosomal sorting complex. Nature 445 (7127):554–558. Available from: http://www.ncbi.nlm.nih.gov/pubmed/17268469
22. Vaccari T, Lu H, Kanwar R, Fortini ME, Bilder D (2008) Endosomal entry regulates Notch receptor activation in *Drosophila melanogaster*. J Cell Biol 180(4):755–762. Available from: http://www.ncbi.nlm.nih.gov/pubmed/18299346
23. Valentine M, Hogan J, Collier S (2014) The drosophila Chmp1 protein determines wing cell fate through regulation of epidermal growth factor receptor signaling. Dev Dyn 243(8):977–987. Available from: http://www.ncbi.nlm.nih.gov/pubmed/24753138
24. Jia T, Macare C, Desrivières S, Gonzalez DA, Tao C, Ji X et al (2016) Neural basis of reward anticipation and its genetic determinants. Proc Natl Acad Sci 113(14):3879–3884. Available from: http://www.ncbi.nlm.nih.gov/pubmed/27001827
25. Legent K, Liu HH, Treisman JE (2015) Drosophila Vps4 promotes Epidermal growth factor receptor signaling independently of its role in receptor degradation. Development 142 (8):1480–1491. Available from: http://www.ncbi.nlm.nih.gov/pubmed/25790850
26. Theodosiou NA, Xu T (1998) Use of FLP/FRT system to studydrosophiladevelopment. Methods 14(4):355–365. Available from: http://www.ncbi.nlm.nih.gov/pubmed/9608507
27. Schulz JG, David G, Hassan BA (2009) A novel method for tissue-specific RNAi rescue in Drosophila. Nucleic Acids Res 37(13):e93. Available from: http://www.ncbi.nlm.nih.gov/pubmed/19483100
28. Michaud S, Marin R, Tanguay RM (1997) Regulation of heat shock gene induction and expression during Drosophila development. Cell Mol Life Sci 53(1):104–113. Available from: http://www.ncbi.nlm.nih.gov/pubmed/9117990
29. Lee T, Luo L (2001) Mosaic analysis with a repressible cell marker (MARCM) for Drosophila neural development. Trends Neurosci 24(5):251–254. Available from: http://www.ncbi.nlm.nih.gov/pubmed/11311363
30. Suster ML, Seugnet L, Bate M, Sokolowski MB (2004) Refining GAL4-driven transgene expression inDrosophila with a GAL80 enhancer-trap. Genesis 39(4):240–245. Available from: http://www.ncbi.nlm.nih.gov/pubmed/15286996
31. Roote J, Prokop A (2013) How to design a genetic mating scheme: a basic training package for Drosophila genetics. G3 (Bethesda) 3 (2):353–358
32. Thompson BJ, Mathieu J, Sung HH, Loeser E, Rørth P, Cohen SM (2005) Tumor suppressor properties of the ESCRT-II complex component Vps25 in Drosophila. Dev Cell 9 (5):711–720
33. Moberg KH, Schelble S, Burdick SK, Hariharan IK (2005) Mutations in erupted, the drosophila ortholog of mammalian tumor susceptibility gene 101, elicit non-cell-autonomous overgrowth. Dev Cell 9(5):699–710
34. Le Borgne R, Bardin A, Schweisguth F (2005) The roles of receptor and ligand endocytosis in regulating Notch signaling. Development 132 (8):1751–1762. Available from: http://eutils.ncbi.nlm.nih.gov/entrez/eutils/elink.fcgi?dbfrom=pubmed&id=15790962&retmode=ref&cmd=prlinks&holding=FindIt@Stanford%5Cnpapers3://publication/doi/10.1242/dev.01789
35. Harrison DA, McCoon PE, Binari R, Gilman M, Perrimon N (1998) Drosophila unpaired encodes a secreted protein that activates the JAK signaling pathway. Genes Dev 12 (20):3252. Available from: https://www.ncbi.nlm.nih.gov/pmc/articles/PMC317220/
36. Teleman AA, Cohen SM (2000) Dpp gradient formation in the drosophila wing imaginal disc. Cell 103(6):971–980. Available from: https://www.sciencedirect.com/science/article/pii/S0092867400001999
37. Lu H, Bilder D (2005) Endocytic control of epithelial polarity and proliferation in Drosophila. Nat Cell Biol 7(12):1232–1239. Available from: http://www.nature.com/doifinder/10.1038/ncb1324
38. Wucherpfennig T, Wilsch-Bräuninger M, González-Gaitán M (2003) Role of Drosophila Rab5 during endosomal trafficking at the synapse and evoked neurotransmitter release. J Cell Biol 161(3):609–624. Available from: http://jcb.rupress.org/content/161/3/609

39. Bökel C, Schwabedissen A, Entchev E, Renaud O, González-Gaitán M (2006) Sara endosomes and the maintenance of dpp signaling levels across mitosis. Science 314 (5802):1135–1139. Available from: http://science.sciencemag.org/content/314/5802/1135.full
40. Daskalaki A, Shalaby NA, Kux K, Tsoumpekos G, Tsibidis GD, Muskavitch MAT et al (2011) Distinct intracellular motifs of Delta mediate its ubiquitylation and activation by Mindbomb1 and Neuralized. J Cell Biol 195(6):1017–1031. Available from: http://jcb.rupress.org/content/195/6/1017
41. Tanaka T, Nakamura A (2008) The endocytic pathway acts downstream of Oskar in Drosophila germ plasm assembly. Development 135(6):1107–1117. Available from: http://dev.biologists.org/cgi/doi/10.1242/dev.017293
42. Dollar G, Struckhoff E, Michaud J, Cohen RS (2002) Rab11 polarization of the Drosophila oocyte: a novel link between membrane trafficking, microtubule organization, and oskar mRNA localization and translation. Development 129(2):517–526. Available from: http://www.ncbi.nlm.nih.gov/pubmed/11807042
43. Jafar-Nejad H, Andrews HK, Acar M, Bayat V, Wirtz-Peitz F, Mehta SQ et al (2005) Sec15, a component of the exocyst, promotes notch signaling during the asymmetric division of drosophila sensory organ precursors. Dev Cell 9(3):351–363. Available from: https://www.sciencedirect.com/science/article/pii/S153458070500256X
44. Wang W, Li Y, Zhou L, Yue H, Luo H (2011) Role of JAK/STAT signaling in neuroepithelial stem cell maintenance and proliferation in the Drosophila optic lobe. Biochem Biophys Res Commun 410(4):714–720. Available from: https://www.sciencedirect.com/science/article/pii/S0006291X11008941
45. Petzoldt AG, Gleixner EM, Fumagalli A, Vaccari T, Simons M (2013) Elevated expression of the V-ATPase C subunit triggers JNK-dependent cell invasion and overgrowth in a Drosophila epithelium. Dis Model Mech 6 (3):689–700. Available from: http://dmm.biologists.org/content/6/3/689
46. Markstein M, Dettorre S, Cho J, Neumüller RA, Craig-Müller S, Perrimon N (2014) Systematic screen of chemotherapeutics in Drosophila stem cell tumors. Proc Natl Acad Sci U S A 111(12):4530–4535. Available from: http://www.ncbi.nlm.nih.gov/pubmed/24616500
47. Sisson JC, Field C, Ventura R, Royou A, Sullivan W (2000) Lava lamp, a novel peripheral golgi protein, is required for *Drosophila melanogaster* cellularization. J Cell Biol 151 (4):905–918. Available from: http://www.ncbi.nlm.nih.gov/pubmed/11076973
48. Emery G, Hutterer A, Berdnik D, Mayer B, Wirtz-Peitz F, Gaitan MG et al (2005) Asymmetric Rab11 endosomes regulate delta recycling and specify cell fate in the drosophila nervous system. Cell 122(5):763–773. Available from: http://www.cell.com/abstract/S0092-8674(05)00821-4
49. Mehta SQ, Hiesinger PR, Beronja S, Zhai RG, Schulze KL, Verstreken P et al (2005) Mutations in Drosophila sec15 reveal a function in neuronal targeting for a subset of exocyst components. Neuron 46(2):219–232. Available from: http://www.cell.com/neuron/abstract/S0896-6273(05)00237-0
50. Blair SS (2007) Dissection of imaginal discs in drosophila. Cold Spring Harb Protoc 2007 (24):pdb.prot4794–pdb.prot4794
51. Stocker H, Gallant P (2008) Getting started. In: Methods in molecular biology, Clifton, NJ, pp 27–44. Available from: http://www.ncbi.nlm.nih.gov/pubmed/18641939
52. Spratford CM, Kumar JP (2014) Dissection and immunostaining of imaginal discs from *Drosophila melanogaster*. J Vis Exp 91:51792. Available from: http://www.ncbi.nlm.nih.gov/pubmed/25285379
53. Kyriacou CP (2000) Drosophila protocols edited by William Sullivan, Michael Ashburner and R. Scott Hawley. Trends Genet 16 (11):524. Available from: http://linkinghub.elsevier.com/retrieve/pii/S0168952500020941
54. Chassefeyre R, Martínez-Hernández J, Bertaso F, Bouquier N, Blot B, Laporte M et al (2015 Feb) Regulation of postsynaptic function by the dementia-related ESCRT-III subunit CHMP2B. J Neurosci 35 (7):3155–3173
55. Daul AL, Komori H, Lee CY (2010) Immunofluorescent staining of Drosophila larval brain tissue. Cold Spring Harb Protoc 5:7
56. Egger B, Gold KS, Brand AH (2010) Notch regulates the switch from symmetric to asymmetric neural stem cell division in the Drosophila optic lobe. Development 137(18):2981–2987
57. Kennison JA (2008) Dissection of larval salivary glands and polytene chromosomes preparation. Cold Spring Harb Protoc 3(5):pdb.prot4708
58. Lebestky T, Jung SH, Banerjee U (2003) A Serrate-expressing signaling center controls Drosophila hematopoiesis. Genes Dev 17 (3):348–353

Chapter 3

Functional Analysis of ESCRT-Positive Extracellular Vesicles in the *Drosophila* Wing Imaginal Disc

Tamás Matusek, Pascal Thérond, and Maximilian Fürthauer

Abstract

A large number of studies have shown that proteins of the Endosomal Sorting Complex Required for Transport (ESCRT) can trigger the biogenesis of different types of Extracellular Vesicles (EV). The functions that these vesicular carriers exert in vivo remain, however, poorly understood. In this chapter, we describe a series of experimental approaches that we established in the *Drosophila* wing imaginal disc to study the importance of ESCRT-positive EVs for the extracellular transport of signaling molecules, as exemplified by a functional analysis of the mechanism of secretion and propagation of the major developmental morphogen Hedgehog (Hh).

Through the combined use of genetic, cell biological, and imaging approaches, we investigate four important aspects of exovesicle biology: (1) The genetic identification of ESCRT proteins that are specifically required for Hh secretion. (2) The imaging of ESCRT and Hh-positive EVs in the lumenal space of both living and fixed wing imaginal discs. (3) The receptor-mediated capture of Hh-containing EVs on the surface of Hh-receiving cells. (4) The effect of manipulations of ESCRT function on the extracellular pool of Hh ligands.

Key words ESCRT, Hedgehog, *Drosophila*, Extracellular vesicles, Exosomes, Ectosomes

1 Introduction

Virtually all aspects of organismal development and physiology require a tight control of cellular membrane shape. The proteins of the Endosomal Sorting Complex Required for Transport (ESCRT) have attracted particular interest due to their ability to promote the outward bending of cellular membranes away from the cytoplasm [1, 2]. Pioneering genetic studies in yeast first identified ESCRT proteins as key players in endosomal biogenesis. In this context, ESCRTs direct the budding of Intra-Lumenal Vesicles (ILV) into the endosome lumen, promoting, therefore, the formation of so-called Multi-Vesicular Bodies (MVB). Throughout the last years, an ever-increasing number of studies have revealed that ESCRT proteins function not only in the endo-lysosomal system

Emmanuel Culetto and Renaud Legouis (eds.), *The ESCRT Complexes: Methods and Protocols*, Methods in Molecular Biology, vol. 1998, https://doi.org/10.1007/978-1-4939-9492-2_3,

but are also important for numerous other biological processes that require the generation of negative membrane curvature, such as cytokinetic abscission, viral budding, or the elimination and resealing of damaged membrane patches [3].

One such process that requires the outward bending of cellular membranes is the biogenesis of Extracellular Vesicles (EVs). The existence of membrane-bound vesicles in the extracellular space of different organs has long been recognized. It is, however, only relatively recently that EVs have started to attract considerable interest as potential extracellular carriers of biological information in both normal physiology and pathological conditions [4, 5]. EVs are generally classified according to their mode of biogenesis: small EVs named exosomes originate from the fusion of MVBs with the plasma membrane and liberation of their internal ILVs into the extracellular space. A second class of EVs, most often designated as ectosomes or microvesicles, arises through direct budding from the plasma membrane. ESCRT proteins have been implicated in both exo- and ectosomal biogenesis, highlighting the particular interest of these factors for EV biology [6].

In the present paper, we report a series of methods that were designed to study the function of ESCRT-dependent EVs in the secretion and transport of Hedgehog (Hh) signaling molecules. Hh ligands are morphogenetic growth factors of major importance for development and disease [7]. While the downstream signaling events triggered by Hh are increasingly well understood, the mechanism through which these signaling molecules are secreted remains highly debated. The highly hydrophobic nature of these proteins, which are dually lipid-modified, requires them indeed to be solubilized to be able to travel through the hydrophilic extracellular environment. Cells have been reported to use various strategies to secrete proteins of the Hh morphogen family, including the formation of soluble multimers [8] and the association with lipoprotein particles [9]. Several recent studies have provided evidence that in *Drosophila*, chicken and humans, Hh ligands can be loaded onto EVs that act as extracellular carriers of these lipophilic morphogens [10–13].

A major methodological challenge is to devise strategies that allow impairing the biogenesis and function of ESCRT-dependent EVs in vivo without generating pleiotropic effects that arise from simultaneous interference with other cellular ESCRT functions. In the following, we describe a number of experimental protocols that were designed to dissect in vivo the importance of ESCRT-dependent EVs for the secretion and extracellular transport of Hh ligands in the *Drosophila* wing imaginal disc [10]. Given the potentially pleiotropic nature of ESCRT loss-of-function conditions, assessing ESCRT function in developmental signaling requires a spatially and temporally controlled manipulation of their protein level. To achieve this, we use tissue-specific Gal4 driver lines together with the temperature-sensitive Gal4-antagonist Gal80^{ts} to control the place, time, and level of transgene, which express double-stranded RNA for RNA interference (RNAi) against ESCRT.

We start by describing a genetic setup to identify ESCRT proteins involved in Hh secretion (Subheading 3.1). To this aim, we use a genetically engineered tester strain that allows raising the levels of Hh expression in posteriorly localized Hh-secreting cells, inducing thereby an overproliferation of Hh-receiving cells that is easily scoreable as a visible wing disc outgrowth. In case an ESCRT candidate is necessary for Hh secretion, its specific knockdown in Hh-producing cells is expected to reduce the outgrowth of the disc that is normally observed in the tester strain. As ESCRT proteins are well known for their role in the ubiquitin-dependent trafficking of endocytosed signaling molecules [2], we also provide a second protocol to determine whether the knockdown of a given candidate protein leads to an excessive accumulation of ubiquitinated endocytic cargo (Subheading 3.2).

In order to allow the visualization of ESCRT-dependent EVs in the tissue of interest, we then describe methods to image Hh- and ESCRT-positive particles in the extracellular space of living (Subheading 3.3) and fixed (Subheading 3.4) wing imaginal discs. To provide functional evidence that Hh ligands that are transported by ESCRT-positive EVs are able to interact with the Hh-receptor Patched (Ptc), we describe a protocol to specifically capture Hh/ESCRT-EVs on the surface of Hh-receiving cells (Subheading 3.5).

Finally, we report a method aimed at addressing the cellular mechanism of EV-dependent Hh secretion. If the generation of Hh EVs proceeds through an exosomal, MVB-dependent route, the inactivation of ESCRT proteins would be expected to result in an accumulation of Hh ligands in intracellular, endosomal compartments. If, on the other hand, Hh-positive EVs are generated by plasma membrane budding, ESCRT depletions should cause Hh to accumulate at the outer leaflet of the plasma membrane. We provide a protocol that allows testing this second hypothesis by measuring the consequences of ESCRT knockdown on the extracellular pool of Hh (Subheading 3.6).

Taken together, the described methods provide an experimental toolset that allows to study the contribution of the ESCRT machinery to exovesicular Hh secretion [10] and may be further adapted for the analysis of other signaling pathways.

2 Materials

2.1 Genetic Identification of ESCRT Genes Required for Hh Secretion

Fly lines

1. tub-Gal80^{ts}; hh-Gal4<UAS-Hh/TM6b, Tb tester line [10].
2. UAS-ESCRT RNAi lines: for strength of different lines, *see* Table 1.

Reagents and equipment for the dissection of wing imaginal discs

3. Phosphate-Buffered Saline (PBS): 137 mM NaCl, 2.7 mM KCl, 10 mM Na_2HPO_4, 1.8 mM KH_2PO_4.
4. Dissecting scope.
5. Glass well dissection dish.
6. 1 pair of Dumont #55 Forceps

2.2 Assessing ESCRT Loss-of-Function by Ubiquitin Epitope Accumulation

Fly lines

1. dpp-lacZ^{P10638}<tub-Gal80^{ts}; hh-Gal4/TM6, Tb tester line [10].

Table 1
Strength of ESCRT RNAi and Vps4^{DN} transgenic fly lines

ESCRT subcomplex	Gene name	CG number	Transgenic lines	Strength
I.	TSG101 / Vps23	CG9712	Blo#35710	Strong (very strong Ubi accumulation after 24 h of induction)
			VDRC#23944	Weak (weak Ubi accumulation after 3 days induction)
	Vps28	CG12770	VDRC#105124	Strong (mild Ubi accumulation at 30 h, strong at 48 h without apoptotic defects)
II.	Vps22 / larsen	CG6637	VDRC#21658	Strong (causing apoptosis in wing disc outgrowth assay)
III.	Chmp3 / Vps24	CG9779	VDRC#29275	Weak (only mild Ubi accumulation is detected at 25 °C even without tub-Gal80^{ts})
			VDRC#100295	Weak (no Ubi accumulation after 24 h induction)
	Chmp1 / Vps46	CG4108	Blo#28906	Moderately strong (mild Ubi accumulation is detected with 30 h knockdown at 29 °C)
	Shrub / Vps32 / Chmp4B	CG8055	RNAi, Gift from Thomas Vaccari	Moderately strong (Ubi accumulation is ranging from barely detectable to strong after 30 h of induction)
			VDRC#106823	Very strong, 24 h of depletion in our hands leads to apoptotic defects
	Vps4	CG6842	UAS-Vps4^{DN}	Very strong, maximum 8 h induction, very strong Ubi accumulation even after 6 h of induction

2. UAS-ESCRT RNAi lines: for strength of different lines, *see* Table 1.

Fixative, buffers, and antibodies

3. Phosphate-Buffered Saline (PBS): 137 mM NaCl, 2.7 mM KCl, 10 mM Na2HPO4, 1.8 mM KH2PO4.
4. Fixative: PBS, 4% formaldehyde (freshly prepared).
5. Staining buffer: PBS, 0.3% Triton-X100.
6. Antibody blocking solution: PBS, 10 mg/ml Bovine Serum Albumin (BSA).
7. Primary antibody solution: PBS, 0.3% Triton-X100, 1:1000 Mouse@Ubiquitin antibody (FK2 clone, ENZO Life Sciences BML-PW8810).
8. Fluorescent Goat@Mouse secondary antibody.

Material for dissection, staining, and microscopic mounting

9. Fluorescent dissecting scope.
10. Glass well dissection dishes: Rinse under the fume hood with Plus One Repel-Silane ES (GE Healthcare), wash off excess Silane, and let dry. This treatment will avoid that dissected wing discs stick to the glass.
11. 1 pair of Dumont #55 Forceps.
12. Eppendorf tubes.
13. Standard set of lab pipets (e.g., 0–10, 10–100 & 100–1000 μl).
14. Ice bucket.

2.3 Live Imaging of ESCRT-Positive EVs in Wing Imaginal Discs

Fly lines

1. hh-Gal4<tub-Gal80^{ts}/TM6B,Tb driver line [10].
2. UAS-Vps32-GFP effector line [14].

Tissue culture medium and fluorescent dye for live imaging

3. Dissection medium: *Drosophila* Shield and Sang M3 medium (Sigma).
4. Imaging medium: *Drosophila* Shield and Sang M3 medium (Sigma), 5 μM FM4-64 membrane dye (*N*-(3-Triethylammoniumpropyl)-4-(6-(4-(Diethylamino) Phenyl) Hexatrienyl) Pyridinium Dibromide, Invitrogen).

Material for dissection and microscopic mounting

5. Fluorescent dissecting scope.
6. Glass well dissection dishes: Rinse under the fume hood with Plus One Repel-Silane ES (GE Healthcare), wash off excess Silane, and let dry. This treatment will avoid that dissected wing discs stick to the glass.

7. 1 pair of Dumont #55 Forceps.
8. Eppendorf tubes.
9. Standard set of lab pipets (e.g., 0–10, 10–100 & 100–1000 μl).
10. Ice bucket.
11. Standard microscope slides and #1.5 coverslips (22 × 22 and 24 × 50 mm).
12. Nail polish.
13. Glass bottom dishes: 7 mm uncoated glass bottom, #1.5 coverslips, MatTek corporation.
14. 0.75% Low Melting Point Agarose (Sigma) in PEM: dissolve by heating in microwave.
15. Bench top heating block set to around 55 °C.

2.4 Imaging of Hh/ESCRT-Positive EVs in Fixed Wing Imaginal Discs

Fly lines

1. hh-Gal4<tub-Gal80^{ts}/TM6B,Tb driver line [10].
2. UAS-Hh-GFP effector line [15].

Buffers, fixatives, and antibodies for immuno-histochemistry

3. Dissection medium: *Drosophila* Shield and Sang M3 medium (Sigma).
4. Pipes EGTA $MgCl_2$ (PEM) buffer: 80 mM Na-Pipes, 5 mM EGTA, 1 mM $MgCl_2$, pH 7.4.
5. Fixative: PEM, 4% PFA. Dissolve PFA powder on a hotplate magnetic stirrer under the fume hood, pass through 0.22 μm filter, aliquot, and store at −20 °C.
6. Staining buffer: PEM, 0.2% TritonX-100.
7. PEM-NH_4Cl: PEM, 50 mM NH_4Cl.
8. Primary antibody blocking solution: PEM, 0.2% Triton-X100, 20 mg/ml Bovine Serum Albumin (BSA).
9. Primary antibody solution: PEM, 0.2% Triton-X100, 20 mg/ml BSA, 1:300 Rabbit@zCHMP1B antibody [10].
10. Secondary antibody blocking solution: PEM, 0.2% Triton-X100, 20 mg/ml BSA, 2% Normal Goat Serum (NGS).
11. Secondary antibody solution: PEM, 0.2% Triton-X100, 20 mg/ml BSA, 2% NGS, fluorescent Goat@Rabbit secondary antibody, fluorescent phalloidin.

Materials for dissection and microscopic mounting

12. Fluorescent dissecting scope.
13. Glass well dissection dishes: Rinse under the fume hood with Plus One Repel-Silane ES (GE Healthcare), wash off excess Silane, and let dry. This treatment will avoid that dissected wing discs stick to the glass.
14. 1 pair of Dumont #55 Forceps.

15. Eppendorf tubes.
16. Standard set of lab pipets (e.g., 0–10, 10–100 & 100–1000 μl).
17. Ice bucket.

2.5 Capturing Hh/ESCRT-Positive EVs on the Surface of Hh-Receiving Cells

Fly lines

1. ptc-Gal4, tub-Gal80^{ts}/CyO, act-GFP driver line [10].
2. UAS-Ptc^{1130X} effector line (Bloomington Drosophila Stock Center #44612).

Reagents for immuno-histochemistry

3. Dissection medium: *Drosophila* Shield and Sang M3 medium (Sigma).
4. Pipes EGTA $MgCl_2$ (PEM) buffer: 80 mM Na-Pipes, 5 mM EGTA, 1 mM $MgCl_2$, pH 7.4.
5. Fixative: PEM, 4% PFA. Dissolve PFA powder on a hotplate magnetic stirrer under the fume hood, pass through 0.22 μm filter, aliquot, and store at −20 °C.
6. Staining buffer: PEM, 0.2% TritonX-100.
7. PEM-NH_4Cl: PEM, 50 mM NH_4Cl.
8. Primary antibody blocking solution: PEM, 0.2% Triton-X100, 20 mg/ml Bovine Serum Albumin (BSA).
9. Primary antibody solution: PEM, 0.2% Triton-X100, 20 mg/ml BSA, 1:300 Rabbit@zCHMP1B antibody [10], 1:250 Rat@dHh antibody [10].
10. Secondary antibody blocking solution: PEM, 0.2% Triton-X100, 20 mg/ml BSA, 2% Normal Goat Serum (NGS).
11. Secondary antibody solution: PEM, 0.2% Triton-X100, 20 mg/ml BSA, 2% NGS, fluorescent Goat@Rabbit and Goat@Rat secondary antibodies, fluorescent phalloidin.

Equipment for dissection and microscopic mounting

12. Fluorescent dissecting scope.
13. Glass well dissection dishes: Rinse under the fume hood with Plus One Repel-Silane ES (GE Healthcare), wash off excess Silane, and let dry. This treatment will avoid that dissected wing discs stick to the glass.
14. 1 pair of Dumont #55 Forceps.
15. Eppendorf tubes.
16. Standard set of lab pipets (e.g., 0–10, 10–100 & 100–1000 μl).
17. Ice bucket.
18. Standard microscope slides and 22 × 22 mm #1.5 coverslips.

19. Mowiol mounting medium: Add 60 g Mowiol4–88 (Sigma) to 150 g Glycerol, stir. Add 150 ml H_2O, leave at room temperature for 2 h. Add 300 ml Tris pH 8.5 and incubate at around 55 °C, stirring occasionally until dissolution. Transfer to 50 ml Falcon tubes and centrifuge for 20 min at 3200× *g*. Aliquot the obtained supernatant and freeze at −20 °C.

2.6 Assessing the Effect of ESCRT Depletion on Extracellular Hh

Fly lines

1. ap-Gal4<tub-Gal80^{ts} [16].
2. UAS-ESCRT RNAi lines: for strength of different lines, *see* Table 1.
3. UAS-Vps4^{DN} [17].

Media and reagents for live antibody incubation and subsequent staining

4. Phosphate-Buffered Saline (PBS): 137 mM NaCl, 2.7 mM KCl, 10 mM Na2HPO4, 1.8 mM KH2PO4.
5. Fixative: PBS, 2% formaldehyde (freshly prepared).
6. Staining buffer: PBS, 0.3% Triton-X100.
7. Antibody blocking solution: PBS, 20 mg/ml Bovine Serum Albumin (BSA).
8. Primary antibody solution: PBS, 0.3% Triton-X100, 14.2 ng/μ l affinity purified Rabbit@dHh antibody [18].
9. Goat@Rabbit secondary antibody.

Equipment for dissection and microscopic mounting

10. Fluorescent dissecting scope.
11. Glass well dissection dishes: Rinse under the fume hood with Plus One Repel-Silane ES (GE Healthcare), wash off excess Silane, and let dry. This treatment will avoid that dissected wing discs stick to the glass.
12. 1 pair of Dumont #55 Forceps.
13. Eppendorf tubes.
14. Standard set of lab pipets (e.g., 0–10, 10–100 & 100–1000 μl).
15. Ice bucket.

3 Methods

3.1 Genetic Identification of ESCRT Genes Required for Hh Secretion

In this assay, a genetic tester line is used to overexpress Hh together with different ESCRT RNAi lines, specifically in posterior, Hh-producing cells. While Hh overexpression induces tissue overgrowth, disc outgrowth is suppressed by the inactivation of ESCRT genes that are involved in exovesicular Hh secretion.

1. Cross the *tub-Gal80*^{ts}*;hh-Gal4< UAS-Hh/TM6b,Tb* "tester line" with either control or UAS-ESCRT RNAi lines (*see* **Notes 1–4**).
2. Flip starter crosses every 12 h at 25 °C, to synchronize the progeny.
3. Transfer vials with early L2 instar larvae to 29 °C.
4. After 72 h incubation at 29 °C, collect non-Tb L3 larvae into ice-cold PBS.
5. Carefully clean them with a brush within the PBS solution.
6. Rinse the larvae with ice-cold PBS two times.
7. Dissect individual animals in ice-cold PBS using the inside-out technique [19] (*see* **Notes 5** and **6**).
8. Score the imaginal discs according to the size of the anterior outgrowth: categories are based on the comparison of the wing pouch area to the size of the outgrowth.
   (a) No outgrowth detected: classified as complete suppression.
   (b) Discs are classified in the moderate outgrowth category when the size of the anterior outgrowth does not reach the size of the wing pouch region.
   (c) In the strong outgrowth category, the size of the anterior outgrowth exceeds the size of the wing pouch.

3.2 Assessing ESCRT Loss-of-Function by Ubiquitin Epitope Accumulation

1. Set up crosses between the *dpp-lacZ*^{P10638}*<tub-Gal80*^{ts}*;hh-Gal4/TM6,Tb* tester line, and UAS-ESCRT RNAi lines (*see* **Note 7**).
2. Flip crosses every 12 h.
3. Transfer vials to 29 °C at early L3 stage (*see* **Note 8**).
4. Collect non-Tb L3 larvae into ice-cold PBS.
5. Carefully clean them with a brush within the PBS solution.
6. Rinse the larvae with ice-cold PBS two times.
7. Dissect individual animals in ice-cold PBS using the inside-out technique *see* **Notes 9–11**.
8. Transfer the individual carcasses to Eppendorf tubes containing PBS on ice.
9. Fix for 20 min in PBS + 4% formaldehyde at room temperature with gentle rotation.
10. Fixed tissue is then washed with PBS + 0.3% Triton-X100 at room temperature, three times for 10 min.
11. Block by incubation for 1 h in PBS + 1% BSA at room temperature *see* **Note12**.

12. Incubate with Mouse@Ubiquitin primary antibody overnight at 4 °C.
13. Apply secondary antibody the next day for 1 h at room temperature.

3.3 Live Imaging of ESCRT-Positive EVs in Wing Imaginal Discs

1. Cross the *hh-Gal4, tub-Gal80^{ts} / TM6b, Tb* driver line with the UAS-Vps32-GFP effector line. Overexpression of high levels of Vps32-GFP is toxic to the organism. For this reason, the cross is kept at 23 °C until the animals reach the late third larval instar. The evening before dissection, the vial is shifted to 29 °C to induce Vps32-GFP expression for 10–14 h.
2. Dissect wing discs in Shields and Sang M3 medium at room temperature.
3. Transfer dissected wing discs to a new dissection dish with 500 μl Shields and Sang M3 medium supplemented with FM4-64 membrane label. Leave in there 5 min for labeling.

Depending on the experimental aim, wing imaginal discs can be mounted in two different ways (*see* also **Notes 13** and **14**):

Flat mount in a microscope slide chamber with coverslip (Fig. 1, steps 4–8):

In this configuration, the discs will be imaged from the top (i.e., the microscope objective will be facing the peripodial membrane). The mounting of the disc between microscope slide and cover-slip flattens the disc in a way that makes it ideal for analyzing EV spreading away from the source of secretion.

4. The evening before the experiment, use nail polish to glue two 22 × 22 mm coverslips on a microscope slide, leaving a space of approx. 15 mm between them.
5. Following the dissection of the wing disc, fill the space between the two coverslips with Shields and Sang M3 medium.
6. Transfer the wing disc to the microscope slide with the peripodial membrane facing upwards.

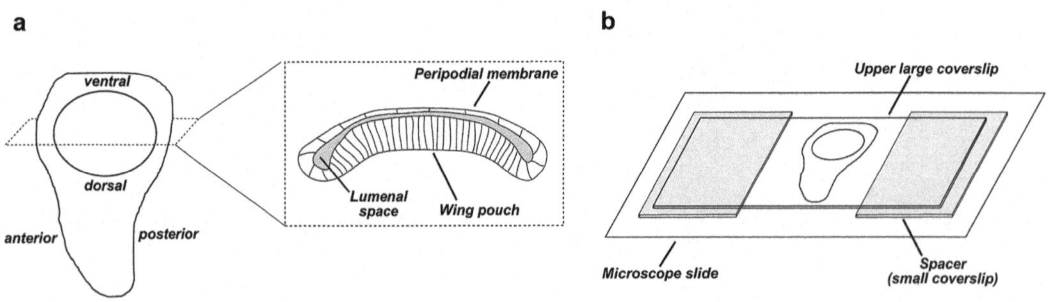

Fig. 1 (**a**) Schematic representation of a *Drosophila* wing imaginal disc. (**b**) Flat mount of a wing disc for live imaging in a microscope slide chamber with coverslip

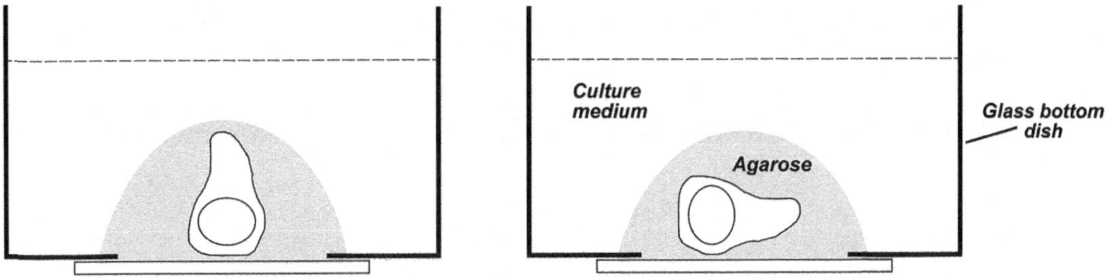

Fig. 2 Transversal mounting of *Drosophila* wing imaginal discs in low melting point agarose

7. Put a 24 × 50 mm cover-slip on the two square coverslips to close the imaging chamber and seal with nail polish.
8. Proceed to the microscope for imaging (*see* **Notes 15–17**).

*Transversal mounting of wing discs in low melting point agarose (Fig. 2, **steps 9–11**)*

In this procedure, the wing disc is mounted in an open configuration "standing", on its lateral or distal edge. The advantage is that there is no compression by a cover-slip, facilitating, therefore, the visualization of EVs in the lumenal space.

9. Put 65 μl of 0.75% LMP-Agarose/PEM in the imaging well of the glass bottom dish. The agarose is sufficiently cool if there is no more vapor condensing on the dish.
10. Transfer the dissected wing disc into the drop of agarose and use the forceps (or a fine needle) to hold the disc in the desired orientation until the agarose solidifies.
11. Cover the solidified agarose with Shields and Sang M3 medium.
12. Proceed to the microscope for imaging (*see* **Notes 15–17**).

3.4 Imaging of Hh/ESCRT-Positive EVs in Fixed Wing Imaginal Discs

1. Cross the *hh-Gal4, tub-Gal80^{ts}/TM6b, Tb* driver line with the UAS-Hh-GFP effector line. The cross is kept at 23 °C until the animals reach the late third larval instar. The evening before dissection, the animals are shifted to 29 °C to induce Vps32-GFP expression.
2. Dissect wing discs from non-Tb larvae in Shields and Sang M3 medium at room temperature.
3. Fix wing discs for 20 min at room temperature in 0.5 ml PEM – 4% PFA. Do not put samples on the rocker at this stage to avoid discs sticking to the glass in the absence of detergent.
4. Postfix wing discs for 20 min at room temperature in 0.5 ml PEM – 4% PFA – 0.2% TritonX100.
5. Rinse once quickly and wash 2 × 5 min in 1 ml PEMT.

6. Incubate 10 min in 1 ml PEM-NH_4Cl. Do not put samples on the rocker at this stage to avoid discs sticking to the glass in the absence of detergent.
7. Rinse once quickly and wash 2 × 5 min in 1 ml PEMT.
8. Before exposure to the primary antibodies, preincubate with 0.5 ml PEMT – 2% BSA, 2 h at room temperature or overnight at 4 °C.
9. Incubate in 0.5 ml PEMT – 2% BSA with Rabbit@zCHMP1B primary antibody, 2 h at room temperature or overnight at 4 °C.
10. Rinse with 1 ml PEMT, then wash in PEMT for 5, 10, 15, and 20 min.
11. Before exposure to the secondary antibodies, preincubate with 0.5 ml PEMT – 2% BSA – 2% NGS, 2 h at room temperature or overnight at 4 °C.
12. Incubate with fluorescent secondary antibodies and phalloidin in 0.5 ml PEMT – 2% BSA – 2% NGS, 2 h at room temperature or overnight at 4 °C.
13. Rinse with 1 ml PEMT, then wash in PEMT for 5, 10, 15, and 20 min.
14. Following immunostaining, discs can be mounted in Low Melting Point Agarose for transversal viewing of the lumenal space, as described above in Subheading 3.3, or flat-mounted between microscope slides and coverslips as described below in Subheading 3.5, **step 14**.

3.5 Capturing Hh/ESCRT-Positive EVs on the Surface of Hh-Receiving Cells

Misexpression of a mutant form of the Hh-receptor Patched (Ptc) that cannot be endocytosed sequesters Hh ligands at the surface of receptor-expressing cells [20]. In the following assay, we use this approach to provide evidence that the Hh-containing ESCRT-positive EVs that are present in the lumen of the wing imaginal disc can interact with Hh receptors.

Unless mentioned otherwise, all incubations are performed on a rocker under gentle agitation. Protect from light all stages in which fluorescent proteins or labels are involved.

1. Cross the *ptc-Gal4, tub-Gal80^{ts}/CyO, act-GFP* driver with the UAS-Ptc^{1130X} effector line. The cross is kept at 23 °C for 24 h, then shifted to 29 °C until late third larval instar.
2. Dissect wing discs from GFP-negative larvae in Shields and Sang M3 medium at room temperature.
3. Fix wing discs for 20 min at room temperature in 0.5 ml PEM – 4% PFA. Do not put samples on the rocker at this stage to avoid discs sticking to the glass in the absence of detergent.

4. Postfix wing discs for 20 min at room temperature in 0.5 ml PEM – 4% PFA – 0.2% TritonX100.
5. Rinse once quickly and wash 2 × 5 min in 1 ml PEMT.
6. Incubate 10 min in 1 ml PEM-NH_4Cl. Do not put samples on the rocker at this stage to avoid discs sticking to the glass in the absence of detergent.
7. Rinse once quickly and wash 2 × 5 min in 1 ml PEMT.
8. Before exposure to the primary antibodies, preincubate with 0.5 ml PEMT – 2% BSA, 2 h at room temperature or overnight at 4 °C.
9. Incubate in 0.5 ml PEMT – 2% BSA with Rabbit@zCHMP1B and Rat@dHh primary antibodies, 2 h at room temperature or overnight at 4 °C.
10. Rinse with 1 ml PEMT, then wash in PEMT for 5, 10, 15, and 20 min.
11. Before exposure to the secondary antibodies, preincubate with 0.5 ml PEMT – 2% BSA – 2% NGS, 2 h at room temperature or overnight at 4 °C.
12. Incubate with fluorescent secondary antibodies and phalloidin in 0.5 ml PEMT – 2% BSA – 2% NGS, 2 h at room temperature or overnight at 4 °C.
13. Rinse with 1 ml PEMT, then wash in PEMT for 5, 10, 15, and 20 min.
14. Exchange PEMT for Mowiol mounting medium and keep at 4 °C for at least 12 h.
15. Put a drop of Mowiol on a microscope slide, transfer wing discs to the drop of Mowiol, orient discs with the peripodial membrane facing upwards, and cover with coverslip.
16. Let solidify for at least 12 h (Mowiol solidifies upon exposure to air) at 4 °C, then proceed to imaging.

3.6 Assessing the Effect of ESCRT Depletion on Extracellular Hh

Through the use of Hh antibodies under non-permeabilizing conditions, the following protocol allows to assess the impact of ESCRT depletion on extracellular Hh levels.

1. Set up crosses between the *ap-Gal4<tub-Gal80^{ts};+/SM6^TM6b,Tb* tester line, and UAS-ESCRT RNAi or UAS-Vps4^{DN} lines (*see* **Notes 18** and **19**).
2. Transfer vials containing early third instar larvae to 29 °C.
3. Collect L3 larvae into ice-cold PBS.
4. Carefully clean them with a brush within the PBS solution.
5. Rinse the larvae with ice-cold PBS two times.

6. Replace the PBS solution with ice-cold Schneider's medium and let larvae there for 1 min.
7. Dissect individual animals in ice-cold Schneider's medium (Sigma) using the inside-out technique [19] (*see* **Notes 20** and **21**).
8. Transfer the individual carcasses to Eppendorf tubes containing Schneider's medium on ice; maximum dissection time to collect samples should not exceed 15 min.
9. Rinse the collected samples with ice-cold Schneider's medium twice.
10. Incubate with Rabbit@dHh affinity purified primary antibody in Schneider's medium at 4 °C for 1 h on a rocker placed at 4 °C (*see* **Note 22**).
11. Following this step, rinse samples three times with ice-cold PBS.
12. Fix for 15 min in PBS + 2% formaldehyde at 4 °C with gentle rotation.
13. Fixed tissue can then be washed with PBS at room temperature, three times for 10 min (*see* **Note 23**).
14. Block by incubation for 30 min in PBS + 2% BSA at room temperature.
15. Add secondary antibodies in PBS + 2% BSA and incubate the samples for 1 h at room temperature.
16. Wash in PBS two times for 10 min.
17. Finally, before mounting, wash the samples with PBS + 0.3% Triton-X100 (*see* **Note 24**).

4 Notes

4.1 Genetic Identification of ESCRT Genes Required for Hh Secretion

1. The induction of UAS-Hh expression in this setup leads to strong pharate adult lethality, the amount of eclosed escaper flies in the control *w;tub-Gal80*^{ts}*/+; hh-Gal4< UAS-Hh /+* genotype is below 1%.
2. Hh overexpression is causing delays in development [21], therefore L3 wandering larvae can be easily recovered following the 72 h incubation period.
3. In this sensitized system, only weak UAS-ESCRT RNAi lines can be used as strong knockdown may result in cell polarity defects and apoptosis.
4. In this experimental setup, 90% of the dissected discs show a strong outgrowth phenotype, and 10% present a moderate outgrowth [10].

5. During dissection, we typically remove all the intestines and fat body, leaving only the imaginal discs attached to the larval carcass.
6. Replace the dissection medium after every 5 larvae.

4.2 Assessing ESCRT Loss-of-Function by Ubiquitin Epitope Accumulation

7. The tub-Gal80^{ts} transgene is not fully efficient at 25 °C; we detected a considerable amount of leakage while using strong RNAi constructs, therefore we advise to keep starter crosses at 22 °C or 18 °C and constantly monitor larval development as well as letting one copy develop at such restrictive temperature until adulthood to look at Mendelian ratios and possible signs of UAS-ESCRT RNAi leakage.
8. Typically, strong ESCRT knockdown by RNAi in *Drosophila* using the hh-Gal4 driver should not exceed 30 h at 29 °C.
9. During dissection we typically remove all the intestines and fat body, leaving only the imaginal discs attached to the larval carcass.
10. Replace the dissection medium after every 5 larvae.
11. Maximum dissection time to collect samples should not exceed 15 min.
12. After **step 11**, samples can be stored for further processing for at least a week at 4 °C.

4.3 Live Imaging of ESCRT-Positive EVs in Wing Imaginal Discs

13. The mounting of wing discs in a closed chamber between microscope slide and coverslip is appropriate for upright and inverted microscopes. Mounting in the open configuration (glass bottom dishes) needs to be performed on inverted microscopes where the solidified drop of agarose can be covered with culture medium.
14. Both mounting procedures described in Subheading 3.3 allow keeping wing imaginal discs alive for over an hour. For extended periods of imaging, we strongly favor the mounting in glass bottom dishes which warrants better oxygenation of the dissected tissue. Long-term imaging experiments should benefit if Shields and Sang M3 medium is replaced by more elaborate formulations. Both overcall tissue morphology and the occurrence of cell divisions should be used to monitor tissue health.
15. Due to the thickness of the tissue, imaging is best performed on regular scanning confocal systems rather than on spinning disc microscopes, which give only limited access to deeper tissue layers.
16. The imaging of wing discs that are viewed transversally in the open (glass bottom dish) configuration is best achieved with water immersion objectives of long working distance (e.g.,

40× NA1.1 0.6 mm WD). For objectives with a shorter working distance (around 0.2 mm), it can be desirable to use glass bottom dishes with a thinner (#0) coverslip to be able to optically access structures that are located farther from the glass.

17. The imaging of EV in the extracellular space requires capturing the rapid diffusive movement of a small particle in a comparatively very large three-dimensional space. To facilitate EV detection, we suggest acquiring relatively thick (e.g., 2 μm) optical sections. To maximize the speed of acquisition, 488 nm laser illumination is used to simultaneously excite and record the fluorescence of the FM4-64 membrane label and the GFP-tagged EV component. The rate of image acquisition can be further increased through bidirectional scanning or the use of resonant scanning systems.

4.4 Assessing the Effect of ESCRT Depletion on Extracellular Hh

18. ESCRT knockdown by RNAi in *Drosophila* using the ap-Gal4 driver should not exceed 30 h at 29 °C; the tub-Gal80^{ts} transgene is not fully efficient at 25 °C; we detected considerable amount of leakage while using strong RNAi constructs, therefore we advise to keep starter crosses at 22 °C or 18 °C and constantly monitor larvae development and let one copy develop at restrictive temperature till adulthood to look at Mendelian ratios and possible signs of UAS-ESCRT RNAi leakage.
19. The UAS-Vps4^{DN} transgene is very strong, starter crosses should be kept at 18 °C, and induction time should not exceed 8 h; in this case, transfer wandering third instar larvae to 29 °C for induction of Vps4^{DN} expression.
20. During dissection, we typically remove all the intestines and fat body, leaving only the imaginal discs attached to the larval carcass.
21. Replace the dissection medium after every 3 larvae.
22. For **step 10**: preferably already use a cold room to add the primary antibodies.
23. After **step 13** samples can be stored for further processing at last for a day at 4 °C.
24. For **step 17**: this will help to prevent unwanted folding during the mounting process.

Acknowledgments

This work was supported by the LABEX SIGNALIFE (ANR-11-LABX-0028-01), by the Fondation pour la Recherche Médicale, and by ANR ANR-15-CE13-0002. T. M was supported by the Fondation ARC pour la Recherche Contre le Cancer and Ligue

Nationale Contre le Cancer. PT is supported by Ligue Nationale Contre le Cancer "Équipe labellisée". M.F. was supported by the CNRS/INSERM ATIP/Avenir program, ARC (SFI20111203750), and HFSP (CDA 00036/2010).

References

1. Henne WM, Stenmark H, Emr SD (2013) Molecular mechanisms of the membrane sculpting ESCRT pathway. Cold Spring Harb Perspect Biol 5(9):a016766
2. Henne WM, Buchkovich NJ, Emr SD (2011) The ESCRT pathway. Dev Cell 21(1):77–91
3. Campsteijn C, Vietri M, Stenmark H (2016) Novel ESCRT functions in cell biology: spiraling out of control? Curr Opin Cell Biol 41:1–8
4. Colombo M, Raposo G, Théry C (2014) Biogenesis, secretion, and intercellular interactions of exosomes and other extracellular vesicles. Annu Rev Cell Dev Biol 30:255–289
5. Tkach M, Théry C (2016) Communication by extracellular vesicles: where we are and where we need to go. Cell 164(6):1226–1232
6. Juan T, Fürthauer M (2018) Biogenesis and function of ESCRT-dependent extracellular vesicles. Semin Cell Dev Biol 74:66–77
7. Briscoe J, Thérond PP (2013) The mechanisms of Hedgehog signalling and its roles in development and disease. Nat Rev Mol Cell Biol 14 (7):416–429
8. Zeng X, Goetz JA, Suber LM, Scott WJ, Schreiner CM, Robbins DJ (2001) A freely diffusible form of Sonic hedgehog mediates long-range signalling. Nature 411 (6838):716–720
9. Panáková D, Sprong H, Marois E, Thiele C, Eaton S (2005) Lipoprotein particles are required for Hedgehog and Wingless signalling. Nature 435(7038):58–65
10. Matusek T, Wendler F, Polès S, Pizette S, D'Angelo G, Fürthauer M, Thérond PP (2014) The ESCRT machinery regulates the secretion and long-range activity of Hedgehog. Nature 516(7529):99–103
11. Gradilla AC, González E, Seijo I, Andrés G, Bischoff M, González-Mendez L, Sánchez V, Callejo A, Ibáñez C, Guerra M, Ortigão-Farias JR, Sutherland JD, González M, Barrio R, Falcón-Pérez JM, Guerrero I (2014) Exosomes as Hedgehog carriers in cytoneme-mediated transport and secretion. Nat Commun 5:5649
12. Parchure A, Vyas N, Ferguson C, Parton RG, Mayor S (2015) Oligomerization and endocytosis of Hedgehog is necessary for its efficient exovesicular secretion. Mol Biol Cell 26(25):4700–4717
13. Vyas N, Walvekar A, Tate D, Lakshmanan V, Bansal D, Lo Cicero A, Raposo G, Palakodeti D, Dhawan J (2014) Vertebrate Hedgehog is secreted on two types of extracellular vesicles with different signaling properties. Sci Rep 4:7357
14. Sweeney NT, Brenman JE, Jan YN, Gao FB (2006) The coiled-coil protein shrub controls neuronal morphogenesis in Drosophila. Curr Biol 16(10):1006–1011
15. Torroja C, Gorfinkiel N, Guerrero I (2004) Patched controls the Hedgehog gradient by endocytosis in a dynamin-dependent manner, but this internalization does not play a major role in signal transduction. Development 131 (10):2395–2408
16. Marois E, Mahmoud A, Eaton S (2006) The endocytic pathway and formation of the Wingless morphogen gradient. Development 133 (2):307–317
17. Rusten TE, Vaccari T, Lindmo K, Rodahl LM, Nezis IP, Sem-Jacobsen C, Wendler F, Vincent JP, Brech A, Bilder D, Stenmark H (2007) ESCRTs and Fab1 regulate distinct steps of autophagy. Curr Biol 17(20):1817–1825
18. Gallet A, Rodriguez R, Ruel L, Therond PP (2003) Cholesterol modification of hedgehog is required for trafficking and movement, revealing an asymmetric cellular response to hedgehog. Dev Cell 4(2):191–204
19. Hafer N, Schedl P (2006) Dissection of larval CNS in *Drosophila melanogaster*. J Vis Exp 1:85
20. Ayers KL, Gallet A, Staccini-Lavenant L, Thérond PP (2010) The long-range activity of Hedgehog is regulated in the apical extracellular space by the glypican Dally and the hydrolase Notum. Dev Cell 18(4):605–620
21. Rodenfels J, Lavrynenko O, Ayciriex S, Sampaio JL, Carvalho M, Shevchenko A, Eaton S (2014) Production of systemically circulating Hedgehog by the intestine couples nutrition to growth and development. Genes Dev 28 (23):2636–2651

Chapter 4

Subcellular Localization of ESCRT-II in the Nematode *C. elegans* by Correlative Light Electron Microscopy

Céline Largeau, Emmanuel Culetto, and Renaud Legouis

Abstract

In this chapter, we report a protocol to perform correlative light electron microscopy (CLEM) on adult *Caenorhabditis elegans*. We use a specific fixation protocol, which preserves both the GFP fluorescence and the structural integrity of the samples. Thin sections are first analyzed by light microscopy to detect GFP-tagged proteins and, subsequently, with transmission electron microscopy (TEM) to characterize the ultrastructural anatomy of cells. The superimposition of light and electron images allows determining the subcellular localization of the fluorescent protein.

We used CLEM to characterize the subcellular localization of the *C. elegans* ESCRT-II component VPS-36. VPS-36 protein localization in *C. elegans* muscle cell is strongly correlated with the sarcoplasmic reticulum network. Together with genetic evidences, the CLEM data support a role for ESCRT-II proteins in sarcoplasmic reticulum membrane shaping.

Key words ESCRT-II, Freeze substitution, GMA resin, Green fluorescent protein, High pressure freezing, Electron microscopy, Muscle, Sarcoplasmic reticulum, VPS-36

1 Introduction

The *C. elegans* body-wall muscle cells (BWM) are functionally and structurally similar to mammalian striated muscle [1]. In BWM, the sarcoplasmic reticulum (SR) forms a network of thin tubules localized in close association with the contractile apparatus. It is regularly localized around the dense body (DB, the actin attachment structures corresponding to both vertebrates *Z* discs and costameres) and below the contractile filaments adjacent to the apical membrane between DB and M-line (Fig. 1). The SR has to maintain a strong association with the DB and the muscle apical membrane, which resists the mechanical stress associated with the repeated contractile cycles. However, mechanisms and proteins involved both in establishing and maintaining the SR network are largely unknown.

Emmanuel Culetto and Renaud Legouis (eds.), *The ESCRT Complexes: Methods and Protocols*, Methods in Molecular Biology, vol. 1998, https://doi.org/10.1007/978-1-4939-9492-2_4, © Springer Science+Business Media, LLC, part of Springer Nature 2019

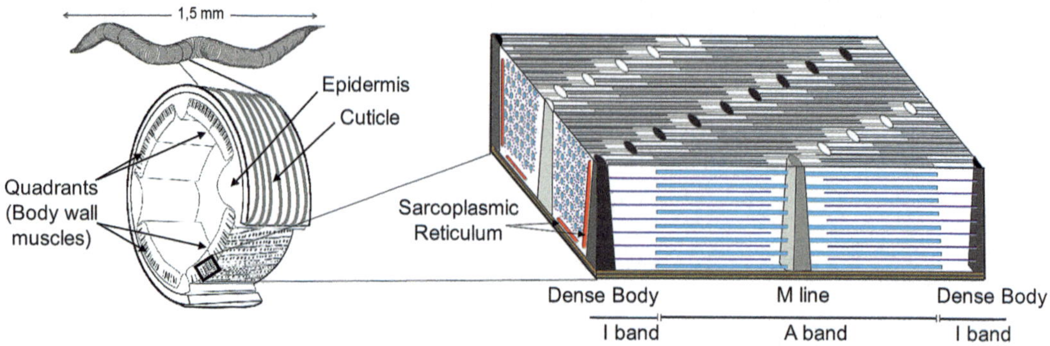

Fig. 1 Schematic representation of *C. elegans* body-wall muscles. Schematic representation of an adult *C. elegans* worm, and the organization of the Body Wall Muscles organized in quadrants (left). Schematic subcellular organization of a ventral Body Wall Muscle cell (right), showing the actin (purple) and myosin (blue) filaments, the contractile filaments anchoring dense body structures (black), and M-lines (gray). The BWM cell sarcolemma is in dark gray and the sarcoplasmic reticulum is shown as red tubules. Adapted from [8]

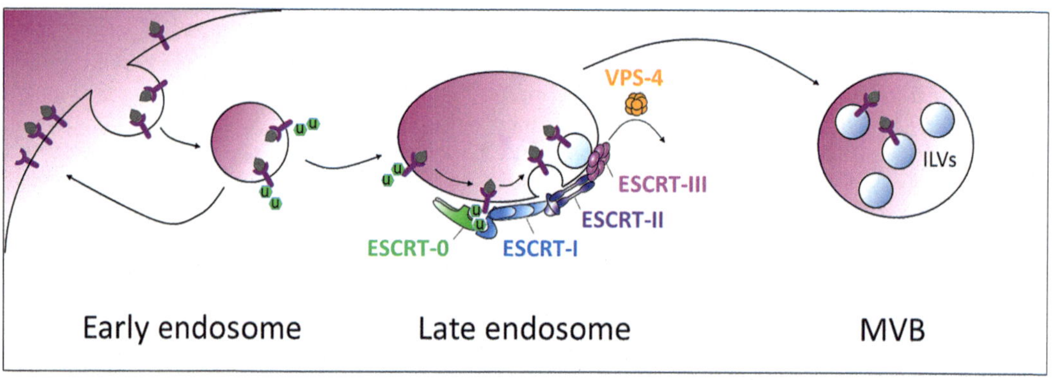

Fig. 2 ESCRT-mediated endocytosis pathways. During endocytosis, the endosomal Sorting Complexes Required for Transport (ESCRT 0 to III) are involved in endosome maturation, for sorting ubiquitylated proteins (u) into intraluminal vesicles (ILV) to form multivesicular bodies (MVB)

We investigated the role of ESCRT-II proteins on *C. elegans* sarcoplasmic reticulum structure and integrity. The endosomal sorting complexes required for transport (ESCRT 0 to III) have been initially discovered as key protein complexes involved in endosome maturation. During this process, the four ESCRT complexes, ESCRT-0, -I, -II, and -III, are sequentially recruited to the endosome membrane, allowing the sorting of ubiquitylated membrane proteins into intraluminal vesicles during multivesicular bodies maturation (MVB) (Fig. 2). Beyond this role in MVB formation, ESCRTs complexes are also involved in numerous cellular processes that involve membrane remodelling [2–4]. We have been investigating the ESCRT functions during *Caenorhabditis elegans* development and reported that ESCRT mutant phenotypes are heterogeneous [5, 6]. The observation that ESCRT-II mutants have late developmental defects, compared to that of other ESCRTs in *C. elegans* [7], supports a hypothesis of a specific role for ESCRT-II components. ESCRT-II is a hetero-tetramer made of VPS-22, VPS-36, and two VPS-25 subunits.

Using a combination of imaging and genetic approaches, we discovered that ESCRT-II, but neither ESCRT-0, ESCRT-I, nor ESCRT-III complexes, localizes to the sarcoplasmic reticulum, and that the depletion of ESCRT-II components results in alterations of sarcoplasmic reticulum shape. We also demonstrated that ESCRT-II mutant worms present a progressive alteration in locomotion during larval development [8].

We use Correlative Light and Electron Microscopy (CLEM) on *C. elegans* adults to characterize the subcellular localization of GFP-tagged VPS-36 protein in *C. elegans* muscle cells and showed it is closely associated to the DB and the apical membrane, which corresponds to the SR network organization. We report here the detailed protocol used to perform CLEM, which is subdivided in four main steps: (1) cryofixation of animal specimen by high pressure freezing to preserve structures; (2) freeze substitution and inclusion in GMA resin to preserve GFP fluorescence; (3) ultramicrotomy and successive observations of the sections by light and electron microscopies, and (4) correlation of fluorescent and electron microscopy images.

This protocol allows performing an easy and rapid correlation between the electron and the light microscopy, based on the dual imaging of thin sections. Although the CLEM pictures have a lower contrast than classical transmission electron microscopy [9], this approach circumvents the difficulty of the retrospective location of the regions of interest, which necessitates very specific approaches [10].

2 Materials

2.1 Caenorhabditis elegans *Strains and Culture*

The conditions for culturing nematode strains are very classical and they will not be detailed here [11].

1. Strains: RD112 *(N2; Ex[vps-36::gfp;rol-6(su1006)])*.
2. Nematode Growth Medium (NGM): dissolve 1.5 g of NaCl, 1.5 g of bactopeptone, and 0.5 ml of cholesterol (5 mg/ml) in 485 ml of water and autoclave. Cool down the medium to 55 °C and supplement it with 0.5 ml $CaCl_2$ (1 mol/l), 0.5 ml $MgSO_4$ (1 mol/l), 10 ml KH_2PO_4 (1 mol/l), and 1.65 ml KH_2PO_4 (1 mol/l). Pour the medium in Petri dishes (either 90 mm or 50 mm diameter plates).
3. *Escherichia coli* OP50 strain (available at the *Caenorhabditis* Genetic Center https://cgc.umn.edu/). Seed NGM plates with 100 μl of OP50 culture in exponential growth phase.

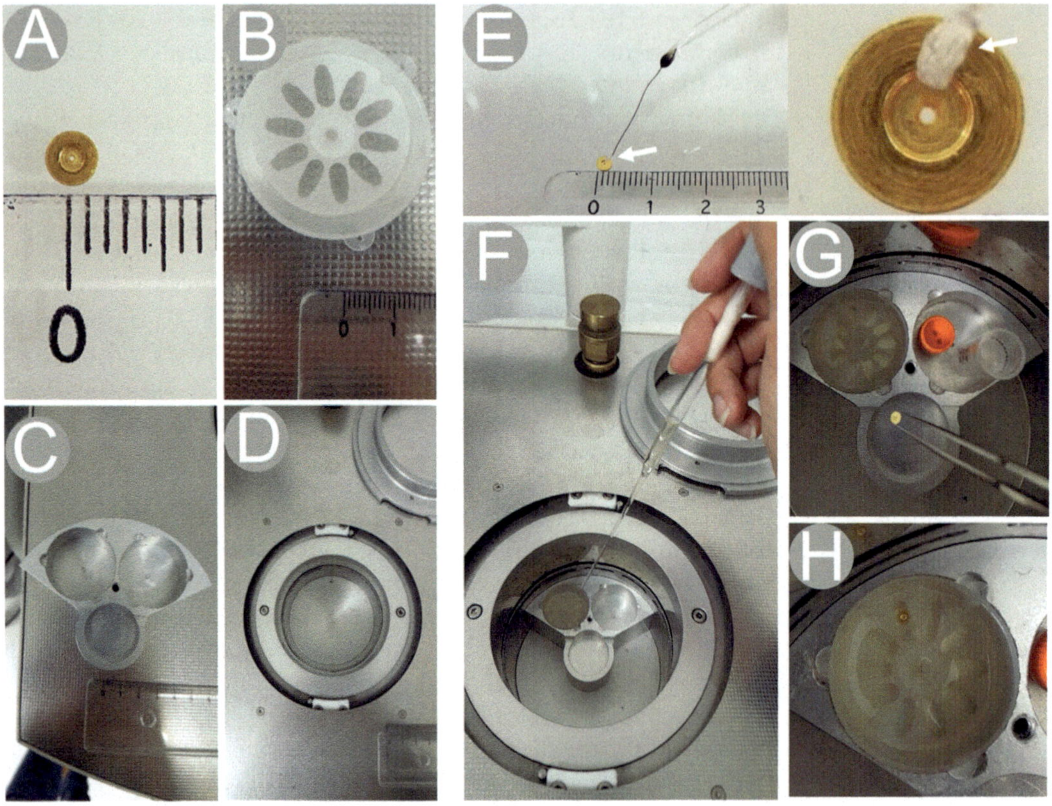

Fig. 3 Cryofixation and substitution materials. (**a**) 200 μm deep flat carrier. (**b**) Flat embedding insert. (**c**) Specimen processing metal block. (**d**) AFS incubating chamber. (**e**) Adults are transferred in the flat carrier with a "worm pick" (white arrow). (**f**) A Pasteur pipette is used for loading and changing the buffers inside the flat embedding insert. (**g–h**) 200 μm deep flat carrier is transferred into the flat embedding insert with tweezers. Ruler graduations are cm

2.2 Cryofixation

1. HPF apparatus EM PACT2 (Leica Microsystem) using liquid nitrogen for cryogenic conditions and methyl cyclohexane to generate pressure.
2. M9 buffer: Dissolve 6 g of Na2HPO4, 3 g of KH2PO4, and 5 g of NaCl in 1 L of water, autoclave and aliquot by 200 ml. Supplement 200 ml aliquots with 200 μl of 1 M MgSO4.
3. M9 buffer containing 20% (w/v) Bovine Serum Albumin (BSA), store at 4 °C after complete dissolution.
4. 1% phosphatidylcholine (sigma-aldrich) pre-coated 200 μm deep flat carriers (Leica Microsystem) (Fig. 3a). Prepare extemporaneously.
5. 4% (w/v) low gelling agarose in 1 ml aliquots in small glass tubes, store at 4 °C.

2.3 Freeze Substitution and Resin Inclusion

1. Freeze substitution is performed with Automated Freeze substitution System (AFS2) with integrated binocular lens and incubating chamber (Leica Microsystem) (Fig. 3d).
2. Pasteur pipettes.
3. Flat embedding insert (sample container containing a reagent tray and a continuous flow ring) (Leica Microsystem) (Fig. 3b).
4. Aclar sheet (Leica Microsystem).
5. Glass bottle of 10 ml.
6. Tweezers and « worm pick » with platinum wire (Fig. 3e, g).
7. Specimen processing metal block (Leica Microsystem) (Fig. 3c).
8. Freeze substitution buffer: 0.1% (w/v) KMnO4; 5% (v/v) water; 95% (v/v) anhydride acetone; 0.01% (v/v) osmium tetroxide (stock 4% w/v in water).
9. 95% (v/v) acetone in water.
10. 95% (v/v) ethanol in water.
11. 5% uranyl acetate (w/v) in anhydride ethanol. Prepare extemporaneously, vortex, and filter (0.2 μm) the solution.
12. 0.1% (v/v) uranyl acetate in 95% acetone. Prepare extemporaneously.
13. Glycol methalacrylate (GMA) acrylic resin: 68.1 ml glycol methacrylate supplemented with 30 ml of butyl methacrylate, 3 ml of H_2O, and 600 mg of benzoyl peroxide. Place the tube on a rotator and rotate slowly for 1 h at 4 °C and store at −20 °C in aliquot of 10 ml.
14. 30% (v/v) GMA in 95% ethanol. Prepare extemporaneously.
15. 70% (v/v) GMA in 95% ethanol. Prepare extemporaneously.
16. 0.32% (v/v) *N,N*-dimethyl-p-toluidine (accelerator polymerization agent) (Merck) in GMA acrylic resin. Prepare extemporaneously.

2.4 Ultramicrotomy

1. Ultramicrotome EM UC7 (Leica Microsystems).
2. Razor blade.
3. Glass knife.
4. Diamond knife for histology (Diatome, Histo, 2 mm, 45°).
5. Toluidine blue: 1% (w/v) toluidine blue and 1% (w/v) borate sodium in water. Filter (0.2 μm).
6. Standard light stereomicroscope.
7. 200 mesh copper formvar carbon-coated finder grid.

2.5 Light Microscopy

We used either upright or inverted microscopes.

1. Inverted microscope: AxioOberver Z1 microscope (Zeiss) equipped with Evolve EMCCD camera (Roper Scientific).
2. Upright microscope: Axioskop 2 Plus microscope (Zeiss) equipped with Nomarski optics coupled to a CoolSNAP camera (Roper Scientific).
3. 35 mm cell view, a cell culture dish with glass bottom (Greiner Bio-One n°627861).
4. Glass slides and coverslips (22 mm × 22 mm, thickness 0.13–0.17 mm).

2.6 Transmission Electron Microscopy

JEOL1400 transmission electron microscope operating at 80 kV coupled with a Gatan 11 Mpixels SC1000 Orius CCD camera.

2.7 Correlation Analysis

Treatment and superimposition of light and electron microscopy pictures are performed with ImageJ, PowerPoint, and Photoshop softwares.

3 Methods

3.1 Cryofixation

The current protocol is designed for the analysis of adults, but it can be used for larvae and embryos upon adaptation of the nematode culture.

1. Grow worms on NGM plate seeded with OP50 at 20 °C.
2. Transfer 20–30 adults with the worm pick, into 1% phosphatidylcholine pre-coated 200 μm deep flat carrier (*see* **Note 1**) containing 5 μl of 20% BSA in M9 (Fig. 3e). Eventually, before freezing, add a drop of agarose low gelling 4% to seal the worm and the cups (*see* **Note 2**).
3. Put 200 μm deep flat carriers containing adults in High Pressure Freezing apparatus (HPF) and cryofix according to manual protocol using automatic mode.
4. Transfer flat carriers to cryogenics tubes and store in liquid nitrogen container, where they can be kept several months. During the whole process, keep flat carriers in liquid nitrogen.

3.2 Freeze Substitution and Resin Inclusion

Freeze substitution process has been adapted from previously published protocols in *C. elegans* or plants [12–14].

1. Day 1, put the specimen processing metal block containing the flat embedding insert in the incubating chamber of AFS2 preset at −90 °C. Using a Pasteur pipette, fill delicately the flat embedding insert with about 1.5 ml of freeze substitution buffer (Fig. 3f). Perform all subsequent steps of buffer removal or washes this way.

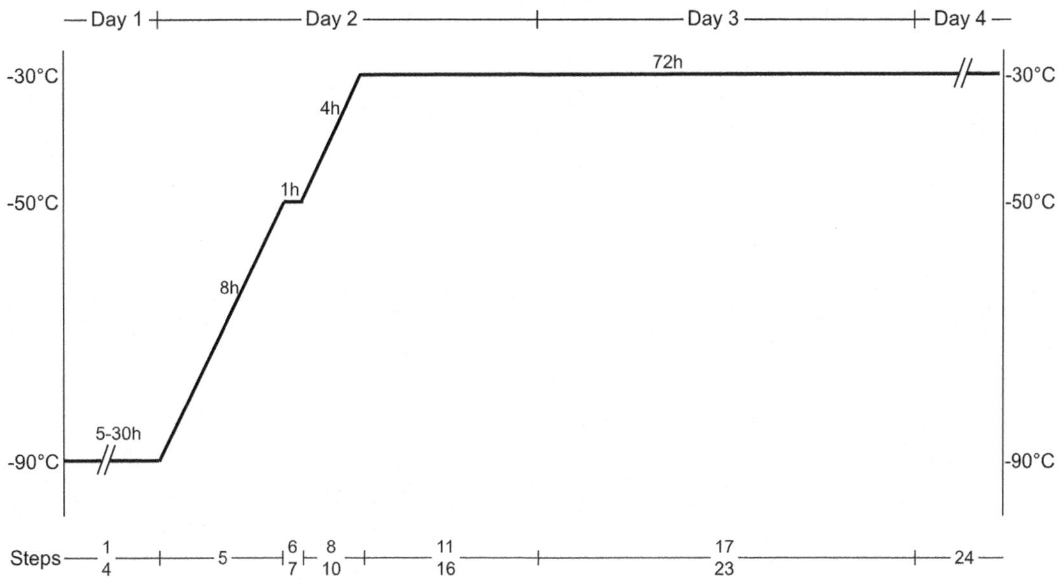

Fig. 4 Graphical summary of the freeze substitution process. The cryosubstitution program temperature is set between −90 °C and − 30 °C. Temperature slope is 5 °C per hour. Steps are detailed in the Subheading 3.2

2. After precooling the AFS2 for 2 h at −90 °C, remove flat carrier from cryogenic tubes using tweezers and put them in the specimen processing metal block without defrosting the samples (*see* **Note 3**) (Fig. 3g).
3. Transfer a single flat carrier from cryogenics tubes in each compartment of the flat embedding insert (Fig. 3h).
4. Close AFS and start the AFS program (Fig. 4). Keep samples at −90 °C for at least 5 h and up to 30 h, if convenient.
5. Set a temperature slope from −90 °C to −50 °C with a gradient of 5 °C per hour (Fig. 4).
6. Day 2, prepare 95% acetone and put in the AFS2 incubating chamber to allow temperature equilibrium for at least 15 min.
7. Remove the freeze substitution medium from the flat embedding insert and add 95% acetone. Incubate for 10 min and repeat the operation four times. Maintain AFS2 at −50 °C for 1 h (Fig. 4).
8. Prepare the 5% acetate uranyl and the 0.1% acetate uranyl in acetone and put it in the AFS2 incubating chamber to allow temperature equilibrium for at least 15 min (−50 °C).
9. Remove the 95% acetone solution and add 0.1% acetate uranyl in acetone solution.
10. Set a temperature slope from −50 °C to −30 °C with a gradient of 5 °C per hour and program AFS to maintain temperature

at −30 °C for the next 72 h (Fig. 4), but proceed to **step 11** when temperature has reached −45 °C.

11. Remove acetone 0.1% acetate uranyl and add acetone 95%. Incubate for 10 min. Repeat the acetone washes four times. Then keep samples in the AFS incubating room for 2 h until it reaches −30 °C.
12. Prepare the 95% ethanol and put in the AFS2 incubating chamber for temperature equilibrium for at least 15 min (−30 °C). Remove 95% acetone and add 95% ethanol for 15 min. Repeat ethanol washes four times.
13. Prepare 30% and 70% GMA in 95% ethanol and store in a −20 °C freezer until using.
14. Remove 95% ethanol from the flat embedding insert and add 30% GMA for a 2 h incubation.
15. Remove 30% GMA, add 70% GMA for a 3 h incubation.
16. Remove 70% GMA and add 100% GMA for an overnight incubation.
17. Day 3, remove 100% GMA and add fresh 100% GMA for 2.5 h.
18. Repeat **step 17**.
19. Remove sample from the deep flat carrier and the deep flat carrier from the flat embedding insert, using integrated binocular lens and curved metallic spike and tweezers. Remove very carefully 100% GMA and incubate in fresh 100% GMA for 15 min.
20. Prepare 100% GMA with accelerator.
21. Put an aclar ring on the flat embedding insert.
22. Remove 100% GMA and wash with 100% GMA containing the accelerator polymerization agent.
23. Remove immediately the solution and add 100% GMA with the accelerator polymerization agent for an overnight incubation. Avoid overflow of each compartment.
24. Day 4, switch off the AFS, take off the flat embedding insert, and unmold the blocks of resin that have polymerized in each compartment. Conserve blocks at −20 °C in darkness in a vacuum-sealed freezer bag.

3.3 Ultramicrotomy

Thin sectioning follows standard protocol. The main steps are indicated below without details (Leica Microsystems EM UC7 user manual) (*see* **Note 4**).

1. Make a pyramid shape structure in the block around the sample with razor blade.
2. Smooth pyramid surface with a glass knife (speed 10 feed 500 nm) (*see* **Note 5**).

3. Make semi-thin sections using histological diamond knife (speed 5 feed 200 nm).
4. Collect semi-thin sections on a glass slide and stain them with toluidine blue.
5. Look at stained semi-thin sections under a classical microscope.
6. If the sample is correct, make semi-thin sections at slower speed (speed 2 feed 200 nm) and collect them on 200 mesh copper formvar carbon-coated finder grid. Keep the grid in darkness and at 4 °C until microscopic observations (*see* **Note 6**).

3.4 Fluorescent and Electronic Microscopy

3.4.1 Fluorescent Microscopy Using an Upright Microscope (Fig. 5a and b)

1. Put the grid on a glass slide with semi-thin sections on the top side.
2. Add 10–20 μl of water on grid and add coverslip (*see* **Note 7**).
3. After bright field and fluorescent observations, remove the coverslip and collect gently the grid with tweezers (*see* **Note 8**).

3.4.2 Fluorescent Microscopy Using Inversed Microscope (Fig. 5c and d)

1. Take a 35 mm cell view.
2. In the middle of the glass bottom put a drop of water (5 μl).
3. Put the grid upside-down on the water drop using tweezers. The water should just make a link between grid and glass slide, and the grid should not move (*see* **Note 9**).
4. Take bright field and fluorescent images.
5. Remove the grid after observation, by gently adding 1 ml of water in the dish and using tweezers.

3.4.3 Electron Microscopy

1. Sections are observed with a JEOL 1400 TEM instrument operating at 80 kV, using a standard procedure.
2. Fill the anticontamination device with liquid nitrogen, Switch on the high tension and increase it to 80 kV, with the condenser aperture inserted.
3. Mount grids on the specimen holder and insert it in the microscope. Turn the filament on.
4. For each grid, localize the areas of interest using the letter and the number identified during light microscopy on 200 mesh copper formvar carbon-coated finder grid.
5. Image areas of interest at low magnification and then go to higher magnification.

3.5 Correlation Between Fluorescent and Electronic Images

1. Perform composite images of bright field and fluorescent images with NIH ImageJ Software (https://imagej.nih.gov/ij/) (Fig. 6a).

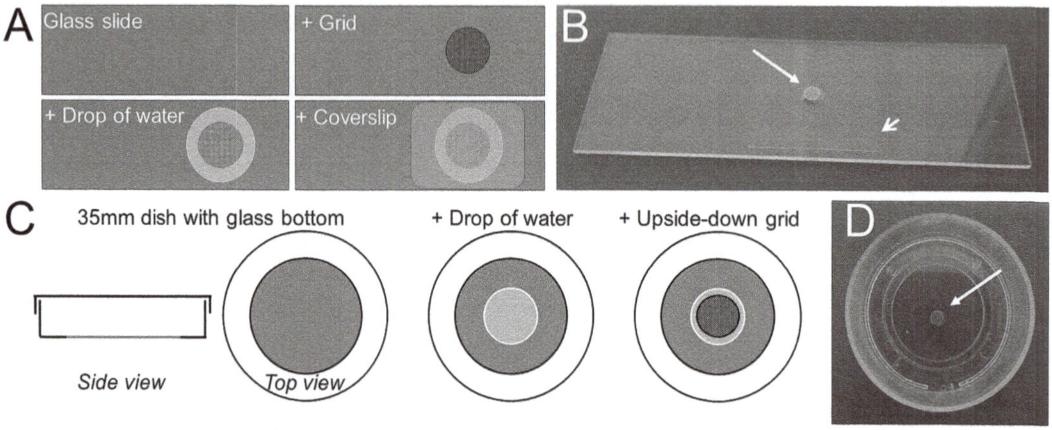

Fig. 5 Mounting grids for light microscopy. (**a–b**) When using an upright light microscope, the grid is deposited, upright, on a glass slide and successively covered with a drop of water and a coverslip. (**c–d**) When using an inverted microscope, the grid is deposited upside-down on a dish with glass bottom above a drop of water. The long arrows point to the grid and the short to the coverslip

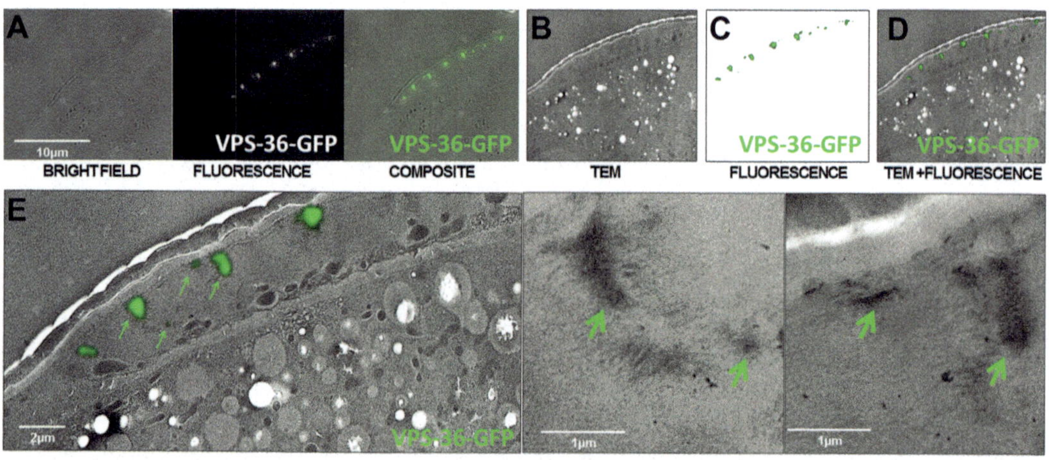

Fig. 6 Correlation between light and electron pictures. (**a**) Bright field and fluorescent images are merged using ImageJ software to obtain a composite image. (**b**) The corresponding transmission electron microscopy image is cropped and sized to fit the light images. (**c**) Using PhotoShop software, a thresholding and removal of background noise deletion is performed on the fluorescent image. (**d**) The fluorescent image is then superimposed with the TEM picture. (**e**) Using CLEM, we observed in muscle cells that GFP::VPS-36 localizes close to the dense bodies and to the apical membrane, which corresponds to the Sarcomeric Reticulum localization

2. Import the composite image in a PowerPoint file. Outline the area of interest manually (*see* **Note 10**) and report this draw on transmission electron microscopy images to put them at the same size (Fig. 6b).
3. Open the composite image, of bright field and fluorescent images, in Photoshop (64 Bit).

4. Select the image or use a selection tool to specifically choose one region of interest.
5. Copy the image or the selected area; create a new file with a transparent background.
6. Paste the image in the new file, select the magic wand tool, and click on to select the background of the image. Use the delete command to suppress the background and to obtain an image with a transparent background (Fig. 6c).
7. Save the file in Portable Network Graphics format (PNG) and superimpose manually using powerpoint, the PNG image, and the composite image to obtain images with identical size. Then, superimpose in a similar way the corrected-size PNG image and the electron microscopy image (Fig. 6d) (*see* **Note 11**).
8. Repeat **steps 1–7** on higher magnification electron microscopy images to correlate ultrastructure and fluorescent signals (Fig. 6e).

4 Notes

1. Under a fume hood, put the 200 μm deep flat carriers in crystallizer containing 1% phosphatidylcholine in chloroform and wait for 5 min. Keep working under fume hood, remove the carriers, and transfer them in an empty crystallize and wait 5 min. Store them at room temperature in a closed box.
2. Heat an agarose low gelling 4% glass tube with a Bunsen burner to melt agarose. Keep it in a heating block at 55 °C. Transfer a glass tube containing water in the heating block. After putting the adults in the 200 μm deep flat carriers, preheat a Pasteur pipette by pipetting up and down 55 °C water. Pipet, immediately, a few microliters of agarose low gelling and quickly pour a small drop on the 200 μm deep flat carriers containing the adults.
3. During the whole time of manipulation, keep working in the cold compartment of the AFS2, with the minimal lighting in the room. Remember to put a cache on the porthole when you do not manipulate in the AFS2. Check regularly the level of liquid nitrogen in the AFS.
4. To limit photobleaching of GFP, set the light of the ultramicrotome and the room at minimum level.
5. This process is generally called trimming and is essential to prepare the sample for optimal thin sectioning.
6. It is not necessary to contrast grid with uranyl acetate and lead citrate because there is enough contrast with freeze

substitution medium. Additionally, the section integrity is better conserved.

7. Add enough water to avoid the grid to be compressed, but not too much to avoid the grid to move during the observation.
8. Thin sections on the grids are fragile, therefore the fluorescence observation should be performed for a very short duration.
9. Centre the grid as much as possible; otherwise the correct focussing of the image could be difficult.
10. To superimpose the area of interest manually on the composite image, use different landmarks like the outline of the worm or the cuticle, small holes, or dirt.
11. A free open-source software called eC-CLEM, which could facilitate the correlation process, has been recently proposed [15]. It is available as a plugin in the Icy platform.

Acknowledgments

The authors would like to thank the *C. elegans* community and the Caenorhabditis Genetic Center which is funded by the NIH National Center for Research Resources (NCRR) for sharing C. elegans strains. We are grateful to Christophe Lefebvre for Fig. 2 and Claire Boulogne for critically reading the manuscript. The present work has benefited from the expertise of Béatrice Satiat-Jeunemaitre, Cynthia Gillet, Jessica Marion, and Claire Boulogne and the core facilities of Imagerie-Gif (http://www.i2bc.paris-saclay.fr) member of IBiSA (http://www.ibisa.net), supported by "France-BioImaging" (ANR-10-INBS-04-01) and the Labex "Saclay Plant Science" (ANR-11-IDEX-0003-02). This work was supported by the Agence Nationale de la Recherche (project EAT, ANR-12-BSV2-018) and the Association pour la Recherche contre le Cancer (SFI20111203826).

References

1. Waterston R (1988) Muscle. In: W. B. Wood). (ed) The nematode *C. elegans*. Cold Spring Harbor Laboratory, Cold Spring Harbor, NY, pp 281–335
2. Hurley JH (2015) ESCRTs are everywhere. EMBO J 34(19):2398–2407
3. Christ L, Raiborg C, Wenzel EM, Campsteijn C, Stenmark H (2017) Cellular functions and molecular mechanisms of the ESCRT membrane-scission machinery. Trends Biochem Sci 42(1):42–56
4. Lefebvre C, Legouis R, Culetto E (2017) ESCRT and autophagies: endosomal functions and beyond. Semin Cell Dev Biol 74:21–28
5. Roudier N, Lefebvre C, Legouis R (2005) CeVPS-27 is an endosomal protein required for the molting and the endocytic trafficking of the low-density lipoprotein receptor-related protein 1 in *Caenorhabditis elegans*. Traffic 6:695–705
6. Michelet X, Alberti A, Benkemoun L, Roudier N, Lefebvre C, Legouis R (2009) The ESCRT-III protein CeVPS-32 is enriched in domains distinct from CeVPS-27 and CeVPS-23 at the endosomal membrane of epithelial cells. Biol Cell 101:599–615
7. Djeddi A, Michelet X, Culetto E, Alberti A, Barois N, Legouis R (2012) Induction of autophagy in ESCRT mutants is an adaptive

response for cell survival in *C. elegans*. J Cell Sci 125:685–694

8. Lefebvre C, Largeau C, Michelet X, Fourrage C, Maniere X, Matic I, Legouis R, Culetto E (2016) The ESCRT-II proteins are involved in shaping the sarcoplasmic reticulum in *C. elegans*. J Cell Sci 129(7):1490–1499
9. Jenzer C, Simionato E, Legouis R (2015) Tools and methods to analyze autophagy in *C. elegans*. Methods 75:162–171
10. Kolotuev I, Schwab Y, Labouesse M (2009) A precise and rapid mapping protocol for correlative light and electron microscopy of small invertebrate organisms. Biol Cell 102 (2):121–132
11. Brenner S (1974) The genetics of *Caenorhabditis elegans*. Genetics 77:71–94
12. Watanabe S, Punge A, Hollopeter G, Willig KI, Hobson RJ, Davis MW, Hell SW, Jorgensen EM (2011) Protein localization in electron micrographs using fluorescence nanoscopy. Nat Methods 8(1):80–84
13. Watanabe S, Richards J, Hollopeter G, Hobson RJ, Davis WM, Jorgensen EM (2012) Nano-fEM: protein localization using photo-activated localization microscopy and electron microscopy. J Vis Exp 70:e3995
14. Marion J, Le Bars R, Satiat-Jeunemaitre B, Boulogne C (2017) Optimizing CLEM protocols for plants cells: GMA embedding and cryosections as alternatives for preservation of GFP fluorescence in Arabidopsis roots. J Struct Biol 198(3):196–202
15. Paul-Gilloteaux P, Heiligenstein X, Belle M, Domart MC, Larijani B, Collinson L, Raposo G, Salamero J (2017) eC-CLEM: flexible multidimensional registration software for correlative microscopies. Nat Methods 14 (2):102–103

Chapter 5

Proximity Ligation Assay (PLA) to Determine the Endosomal Localization of ESCRT Subunit in Virus-Infected Cells

Binod Kumar, Mohanan Valiya Veettil, Arunava Roy, and Bala Chandran

Abstract

Proximity ligation assay (PLA) is a newly developed technique that outperforms the traditional immunoassays for visualizing the in situ endogenous protein-protein interactions and localizations and the activation of proteins in cell culture systems as well as in tissue sections. PLA, when combined with cellular marker staining, becomes a powerful approach to identify differential interaction of the proteins of endosomal sorting complex required for transport (ESCRT) at distinct stages of virus infection. In this chapter, we describe a PLA protocol to study the localization and interaction between the ESCRT protein TSG101 and endosomal markers in early stages of viral endocytosis in in vitro infected cells.

Key words ESCRT, Virus infection, Proximity ligation assay, Colocalization, Endocytosis, Trafficking

1 Introduction

Advancements in life sciences and medicine research have surged tremendously in the last 4 decades due to the development of the polymerase chain reaction (PCR) for detecting nucleic acids and the antibody-based techniques for specific detection of proteins. The constant efforts to improve and develop nucleic-acid detection methods have given rise to the development of immuno-PCR (iPCR) and, more recently, the proximity ligation assays (PLA) [1]. PLA, since its first description in 2002 [2], has been extensively used to study individual proteins, protein-protein interactions, localizations, and quantification under natural biological conditions by detecting amplification of DNA linked to the PLA probes [3–5]. It relies on the principle that, if two epitopes/proteins are within the proximity of ~40 nm or below, the PLA oligo probes linked to two secondary antibodies bound to primary antibody-antigen complexes will be amplified to give a PLA signal visualized as a fluorescent dot. This method detects protein interactions at ~40 nm resolution compared to the 200 nm resolution of immunofluorescence assay (IFA), as well as weak or transient interactions

Emmanuel Culetto and Renaud Legouis (eds.), *The ESCRT Complexes: Methods and Protocols*, Methods in Molecular Biology, vol. 1998, https://doi.org/10.1007/978-1-4939-9492-2_5,

that might not survive the co-immunoprecipitation manipulations. PLA detects endogenous levels and gives spatial and subcellular resolution, while Fluorescence Resonance Energy Transfer (FRET) requires high expression levels of target proteins.

Viral entry is a complex and multistep process accompanied by coordinated events of several viral and host protein interactions, leading to cell signaling events and protein modifications [6–8]. Detection of specific proteins during any viral infection often provides valuable information about the key events involved in viral endocytosis, transport, and replication of the virus in the host cells. A virus may enter the host cells utilizing different pathways, but mainly traffics through the cytoplasm in a specialized compartment called endosomes before delivering the genetic material to the cytoplasm or to the nucleus. Studies from our laboratory have shown that Kaposi's sarcoma-associated herpesvirus (KSHV) utilizes cellular macropinocytic pathway to gain entry into its natural target cells, human microvascular dermal endothelial cells (HMVEC-d) [9–11]. Macropinocytosis is initiated upon binding of the virus with the receptors, which is followed by a set of sequential stages such as the endocytosis of virus, fusion of viral envelope with the endosomal membrane, release of viral capsid and transport of the capsid to the nuclear periphery, and delivery of viral nucleic acid into the nucleus of host cells [9, 10] (Fig. 1).

Additionally, we have shown that, during KSHV infection, the ESCRT-0 (Endosomal Sorting Complexes Required for Transport) protein HRS migrates to the plasma membrane to facilitate the macropinocytosis of the virus, and subsequently, ESCRT-I protein TSG101 interacts with endosomal and other ESCRT proteins to facilitate trafficking of the virus through the cytoplasm [10, 11].

In this chapter, we describe a PLA procedure to determine the endosomal localization and interaction between the ESCRT-I subunit TSG101 protein and the early endosomal maker Rab5 and the late endosomal marker Rab7 in KSHV-infected cells. The PLA protocol, applied in virus-infected cells, follows several steps which include: (1) Incubation of fixed virus-infected cells with the primary antibodies against TSG101 and endosomal markers (Rab5 or Rab7) raised in two different species; (2) Incubation with a pair of PLA probes from different species (one PLUS and one MINUS); (3) Circularization and ligation of the probes which bind specifically to primary antibodies; (4) Amplification of the circularized DNA via rolling-circle model; (5) Detection of the amplified products using a detection reagent (Fig. 2).

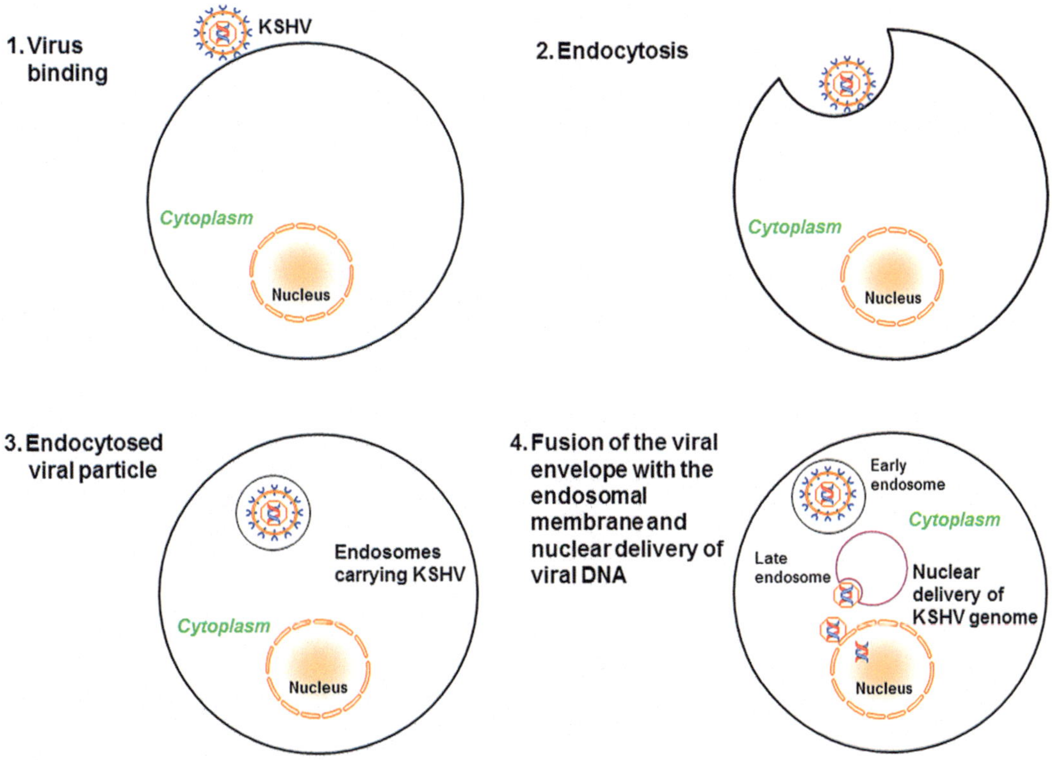

Fig. 1 Distinct stages of KSHV endocytosis and infection. After KSHV infection, endocytosis and nuclear delivery of the viral genome occurs nonsynchronously within 1–30 min post-infection

2 Materials

2.1 Cells and Virus

1. Human microvascular dermal endothelial cells (HMVEC-d) (*see* **Note 1**).
2. Endothelial cell growth medium 2 (EGM-2) (*see* **Note 2**).
3. Eight-well chambered glass slides.
4. Purified Kaposi's sarcoma-associated herpesvirus (KSHV) (*see* **Note 3**).

2.2 PLA- Reagents and Materials

Prepare all the solutions using either commercially available ultrapure RNAse/DNAse-free water or purified deionized water (sensitivity of 18 MΩ-cm at 25 °C) and analytical grade reagents. The prepared reagents should be stored at the appropriate temperature indicated throughout this protocol.

1. HMVEC-d cells in 8-well chambered slide.
2. A PLA kit which is commercially available (Duolink PLA; Sigma-Aldrich) and contains the following buffers and reagents: Wash buffer A (Prepare buffer by dissolving the contents of one pouch in sterile water to obtain 1× solution.),

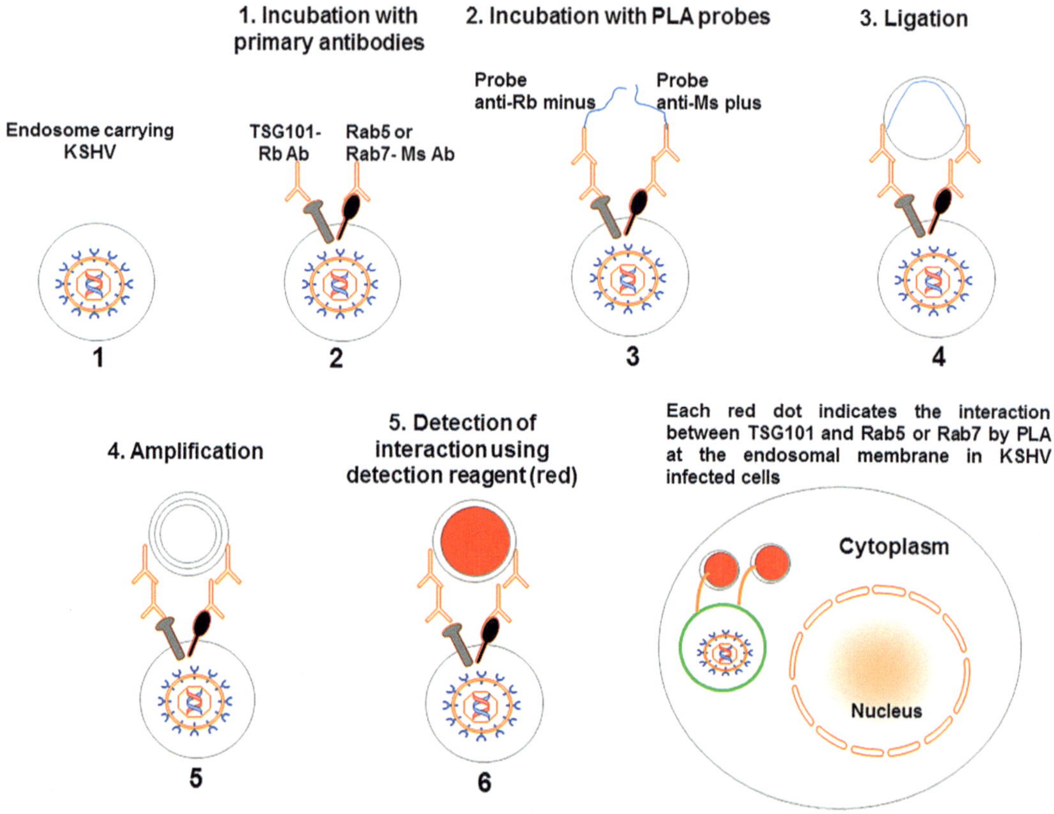

Fig. 2 Main steps involved in the detection of in situ interaction between the ESCRT protein TSG101 and the endosomal markers in KSHV-infected cells by PLA

wash buffer B (Prepare buffer by dissolving the contents of one pouch in sterile water to obtain 1× solution.), blocking buffer, primary antibody diluent, secondary antibody diluent, amplification and detection reagent (red/green), PLA probes (PLUS and MINUS), ligase, DNA-polymerase, and mounting medium with DAPI (*see* **Note 4**).

3. Primary antibodies raised in two different species [Mouse-anti-TSG101 (Santa Cruz Biotechnology, CA) and rabbit-anti-Rab5 (Cell Signaling Technology, MA) and rabbit-anti-Rab7 (Cell Signaling Technology, MA)] to study protein-protein interaction.
4. Phosphate buffered saline (PBS), 1× (Corning, NY) (*see* **Note 5**).
5. TritonX-100: Prepare 0.2% TritonX-100 in 1× PBS (*see* **Note 5**).
6. Paraformaldehyde (PFA): Prepare 4% PFA in 1× PBS from the 16% stock PFA (Electron Microscopy Sciences, PA).

7. Fluorescence microscope, staining jar, freeze block for enzymes, shaker, humidity chamber, incubator for 37 °C, cover slips (24 × 60 mm size, No.1; 0.13–0.17 mm thickness; VWR catalogue # 48393 106), and high purity water.

3 Methods

The PLA protocol in virus-infected cells has two stages. In the first stage, we describe cell culture and preparation of virus-infected cells for PLA. In the second stage, a detailed step-by-step protocol for analyzing the interaction between ESCRT protein TSG101 and endosomal markers Rab5/Rab7 using PLA technique is described.

3.1 Cell Culture and Preparation of Virus-Infected Cells

1. Seed approximately 15,000 HMVEC-d cells per well in 8-well chambered slides.
2. Culture the cells to 60–70% confluency (*see* **Note 6**) before proceeding to viral infection and PLA.
3. Dilute virus stock to get a multiplicity of infection (MOI) of 30 DNA copies/cell.
4. Remove medium from 8-well chambered slides (**step 2** above) and add 250 μl of the virus to each well to infect cells for 5, 10, and 30 min (in duplicate) (*see* **Note 7**). Leave two control wells uninfected as shown in Fig. 3.
5. Incubate the cells at 37 °C, 5% CO_2 incubator for different time points—5, 10, and 30 min.
6. After incubation, remove the virus containing media and wash the cells twice with 1× PBS.
7. Fix the uninfected and infected cells by incubating with 200 μl of 4% PFA per well for 20 min at room temperature.

1	2	3	4
Uninfected	KSHV infection- 5 minutes	KSHV infection- 10 minutes	KSHV infection- 30 minutes
5	**6**	**7**	**8**
Uninfected	KSHV infection- 5 minutes	KSHV infection- 10 minutes	KSHV infection- 30 minutes

Fig. 3 Typical position of samples within the 8-well chambered slides

3.2 Proximity Ligation Assay

1. Remove PFA and wash the cells gently three times with 1× PBS.
2. Add 200 μl of 0.2% of TritonX-100 to each well and incubate for 5 min at room temperature.
3. Wash the cells gently with 1× PBS three times.
4. Add 80–100 μl of blocking solution (PLA kit reagent) and incubate the slide in preheated humidity chamber (*see* **Note 8**) for 30 min at 37 °C.
5. Dilute TSG101 (1:100) and Rab5 (1:200) or TSG101 (1:100) and Rab7 (1:200) primary antibodies (from two different species) in two separate tubes using the antibody diluent from PLA kit.
6. Gently tap off/pipette out the blocking solution from the slides (*see* **Note 9**).
7. Add the TSG101-Rab5 primary antibody mix to the well numbers 1–4 and TSG101-Rab7 antibodies (200 μl) to well numbers 5–8 and incubate in a humidity chamber for 60 min at 37 °C.
8. Dilute the two PLA probes (*see* **Note 10**) in the ratio of 1:5 in the secondary antibody diluent (For an 80 μl reaction, take 16 μl of PLA probe MINUS stock, 16 μl of PLA probe PLUS stock, and 48 μl of antibody diluent).
9. Gently tap off/ pipette out the primary antibody solution from the slides.
10. Wash the slides twice with 1× wash buffer A for 5 min each (*see* **Note 11**).
11. Add the PLA probe solution (80 μl) and incubate the slides in a preheated humidity chamber for 1 h at 37 °C.
12. Dilute the ligation stock in the ratio of 1:5 in high purity water and mix gently. (For an 80 μl reaction, take 16 μl of the 5× Ligation stock and 62 μl of high purity water).
13. Gently tap off/ pipette out the PLA probe solution from the slides.
14. Wash the slides in 1× Wash Buffer A twice for 5 min each under gentle agitation.
15. Take out the ligase from the freezer and add ligase to the ligation solution at a 1:40 dilution and vortex briefly. (For an 80 μl reaction, add 2 μl of ligase to 78 μl of ligation solution).
16. Add the Ligation-Ligase solution to each sample and incubate the slides in a preheated humidity chamber for 30 min at 37 °C.
17. Dilute the Amplification stock in the ratio of 1:5 in high purity water and mix gently. (For an 80 μl reaction, take 16 μl of the 5× Amplification stock and 63 μl of high purity water). The

amplification stock consists of components needed for rolling-circle amplification (RCA) and fluorescently labelled oligonucleotides that hybridize to the RCA product.

18. Gently tap off/ pipette out the Ligation-ligase solution from the slides.
19. Wash the slides in 1× Wash Buffer A twice for 2 min each under gentle agitation.
20. Gently tap off/ pipette out all wash solution after the last washing.
21. Add Polymerase to the Amplification solution prepared above, at a 1:80 dilution and vortex. (For an 80 μl reaction, add 1 μl of Polymerase to 78 μl of Amplification solution).
22. Add the Amplification-Polymerase solution to each well and incubate the slides in a preheated humidity chamber for 100 min at 37 °C.
23. Gently tap off/ pipette out the Amplification-Polymerase solution from the slides.
24. Prepare 1× Wash Buffer B solution and wash the slides twice for 10 min each.
25. Wash the slides in 0.01× Wash Buffer B for 1 min.
26. Let the slides dry at room temperature in the dark.
27. Mount the slides with a cover slip using a minimal volume of in situ mounting medium with DAPI (*see* **Note 12**).
28. The slide is now ready for observation under an immunofluorescent microscope. The amplified PLA signals have an excitation wavelength of 594 nm and appear as red dots that are detected in the Texas red channel as shown in Fig. 4.
29. The slides can also be stored at −20 °C in the dark for several months.

4 Notes

1. The HMVEC-d cells should be at the lowest possible passage number.
2. The EGM medium supplemented with EGM-2 bullet kit is optimized for cultivation of primary endothelial cells. It contains hEGF, hydrocortisone, GA-1000 (Gentamicin, Amphotericin-B), FBS (Fetal Bovine Serum), VEGF, hFGF-B, R3-IGF-1, ascorbic acid, and heparin and does not contain BBE (bovine brain extract).
3. Prepare purified KSHV stock by ultracentrifugation. Remove debris by filtration before use.

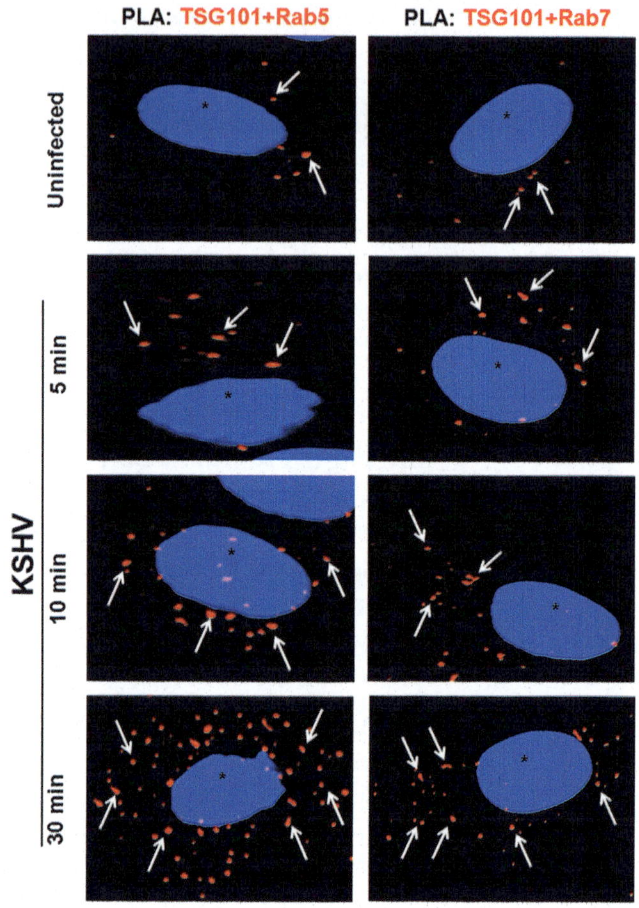

Fig. 4 Proximity ligation assay result for interactions between the ESCRT protein TSG101 and the early endosomal marker Rab5 or TSG101 and late endosomal marker Rab7 in uninfected and KSHV-infected endothelial cells at 5, 10, and 30 min after infection. Amplified PLA signals appearing as red dots (white arrows) indicate the interaction between TSG101 and endosomal markers, which increases with increase in time of infection. The blue structure is the nucleus (labelled with asterix). (Reproduced from ref. 10 with permission from authors/PLOS Pathogens)

4. All the components of PLA kits must be stored at the temperature indicated by the manufacturer. The wash buffer A and B can be stored at 4 °C for 1–2 weeks upon reconstitution in sterile water.
5. The solution can be stored at room temperature for several months.
6. The HMVEC-d cells seeded at 15,000 cells/ well should reach the confluency of 60–70% on second day of culture. This may need further standardization based on cell type chosen and their rate of growth.

7. All viral infections must be performed inside a recommended biosafety level 2 hood with full precautions following standard laboratory protocols. The fixed cells can be taken out of hood now after fixation and processed for next step outside hood on a work bench.
8. Any incubator maintained at 37 °C with a source of humidity (water) inside it can be used as humidity chamber for incubation.
9. Make sure that the samples do not dry before adding the primary antibodies as this may cause background staining.
10. If you have used a mix of primary antibodies raised in rabbit and mouse, use anti-rabbit PLUS and anti-mouse MINUS probes or vice versa.
11. Bring the wash buffers to room temperature before use. For best results, washing should be performed in a staining jar with sufficient volume (around 70–80 ml) on a shaker with gentle orbital shaking.
12. Insure that no air bubbles get trapped under the cover slip.

Acknowledgments

This study was supported in part by the Public Health Service grants CA 168472 and CA 180758, RFUMS H. M. Bligh Cancer Research Fund, and USF start up fund to Bala Chandran.

References

1. Hansen MC, Nederby L, Henriksen MO, Hansen M, Nyvold CG (2014) Sensitive ligand-based protein quantification using immuno-PCR: a critical review of single-probe and proximity ligation assays. BioTechniques 56(5):217–228
2. Fredriksson S, Gullberg M, Jarvius J, Olsson C, Pietras K, Gustafsdottir SM, Ostman A, Landegren U (2002) Protein detection using proximity-dependent DNA ligation assays. Nat Biotechnol 20(5):473–477. https://doi.org/10.1038/nbt0502-473
3. Soderberg O, Gullberg M, Jarvius M, Ridderstrale K, Leuchowius KJ, Jarvius J, Wester K, Hydbring P, Bahram F, Larsson LG, Landegren U (2006) Direct observation of individual endogenous protein complexes in situ by proximity ligation. Nat Methods 3 (12):995–1000. https://doi.org/10.1038/nmeth947
4. tom Dieck S, Kochen L, Hanus C, Heumuller M, Bartnik I, Nassim-Assir B, Merk K, Mosler T, Garg S, Bunse S, Tirrell DA, Schuman EM (2015) Direct visualization of newly synthesized target proteins in situ. Nat Methods 12(5):411–414. https://doi.org/10.1038/nmeth.3319
5. Nong RY, Wu D, Yan J, Hammond M, Gu GJ, Kamali-Moghaddam M, Landegren U, Darmanis S (2013) Solid-phase proximity ligation assays for individual or parallel protein analyses with readout via real-time PCR or sequencing. Nat Protoc 8(6):1234–1248. https://doi.org/10.1038/nprot.2013.070
6. Kumar B, Chandran B (2016) KSHV entry and trafficking in target cells-hijacking of cell signal pathways, actin and membrane dynamics. Viruses 8(11):E305. https://doi.org/10.3390/v8110305
7. Chakraborty S, Veettil MV, Chandran B (2012) Kaposi's sarcoma associated herpesvirus entry into target cells. Front Microbiol 3:6. https://doi.org/10.3389/fmicb.2012.00006

8. Veettil MV, Bandyopadhyay C, Dutta D, Chandran B (2014) Interaction of KSHV with host cell surface receptors and cell entry. Viruses 6 (10):4024–4046. https://doi.org/10.3390/v6104024
9. Bandyopadhyay C, Valiya-Veettil M, Dutta D, Chakraborty S, Chandran B (2014) CIB1 synergizes with EphrinA2 to regulate Kaposi's sarcoma-associated herpesvirus macropinocytic entry in human microvascular dermal endothelial cells. PLoS Pathog 10(2):e1003941. https://doi.org/10.1371/journal.ppat.1003941
10. Kumar B, Dutta D, Iqbal J, Ansari MA, Roy A, Chikoti L, Pisano G, Veettil MV, Chandran B (2016) ESCRT-I protein Tsg101 plays a role in the post-macropinocytic trafficking and infection of endothelial cells by kaposi's sarcoma-associated herpesvirus. PLoS Pathog 12(10): e1005960. https://doi.org/10.1371/journal.ppat.1005960
11. Veettil MV, Kumar B, Ansari MA, Dutta D, Iqbal J, Gjyshi O, Bottero V, Chandran B (2016) ESCRT-0 component hrs promotes macropinocytosis of kaposi's sarcoma-associated herpesvirus in human dermal microvascular endothelial cells. J Virol 90 (8):3860–3872. https://doi.org/10.1128/JVI.02704-15

Chapter 6

Immuno-localization of ESCRT Proteins in Virus-Infected Cells by Fluorescence and Electron Microscopy

Keisuke Tabata, Atsuki Nara, Hiroko Omori, and Eiji Morita

Abstract

Many enveloped viruses utilize the cellular ESCRT pathway for budding, even flaviviruses, which form viral particles inside replication organelles derived from the endoplasmic reticulum (ER). In this section, we introduce methods for detecting several ESCRT subunit proteins in virus-infected cells by immunofluorescence microscopy and immunoelectron microscopy (immuno-EM). We also introduce a new method; correlative light microscopy and electron microscopy (CLEM), which allows the observation of target structures with both high-resolution EM and fluorescence labeling.

Key words Fluorescence light microscopy, Conventional transmission EM (TEM), Immunoelectron microscopy, Correlative light microscopy and electron microscopy (CLEM), Positive strand RNA virus, Dengue virus (DENV), Japanese encephalitis virus (JEV)

1 Introduction

The ESCRT pathway was originally identified in yeast through a genetic analysis of factors required for the generation of intraluminal vesicles of multivesicular bodies (MVB). After finding that TSG101, a subunit of the ESCRT-I complex, directly binds to the Gag structural protein of human immunodeficiency virus (HIV) in infected cells, it was then shown that all retroviruses as well as the majority of enveloped negative-stranded RNA viruses utilize the ESCRT pathway for budding to enable virus escape from the infected cell [1, 2]. This process of small vesicle formation, in which the membrane buds from the cytoplasm, is topologically equivalent to the intraluminal vesicle formation of MVB. Therefore, it was not surprising that many viruses have been evolved to use the ESCRT pathway to obtain their envelope [1, 2].

Recently, our group and others reported that the positive-stranded RNA (ssRNA(+))viruses, such as flaviviruses, also utilize the ESCRT pathway for budding [3, 4]. Flaviviruses rearrange the intracellular membranes to create a specific compartment called the

Emmanuel Culetto and Renaud Legouis (eds.), *The ESCRT Complexes: Methods and Protocols*, Methods in Molecular Biology, vol. 1998, https://doi.org/10.1007/978-1-4939-9492-2_6, © Springer Science+Business Media, LLC, part of Springer Nature 2019

"replication organelle" at the ER, which is known to be involved in viral genome replication and virion assembly [5, 6]. Previous EM and 3D-tomography studies showed that flaviviruses, such as Dengue virus (DENV), Japanese encephalitis virus (JEV), and Zika virus (ZIKV), formed replication organelles comprising vesicle packets, convoluted membrane, and membranes associated with progeny virus [7–10]. Convoluted membrane is a highly electron-dense structure with a stacked membrane that has been proposed to be the site of viral protein synthesis. Vesicle packets that look like "snow peas" contain several inner vesicles (~90 nm in diameter) with a large outer membrane and are considered to be the sites of viral RNA genome replication because double-stranded RNA was detected in this compartment. Subsequently, the newly synthesized viral proteins and RNA genome are assembled on the ER membrane and bud into the ER lumen. We recently demonstrated that specific ESCRT factors (CHMP2B and CHMP4A) are recruited near the vesicle packet at the replication organelle by immunofluorescence microscopy and immunoelectron microscopy (immuno-EM) [4]. We also found that knockdown of those ESCRT subunits affects virus budding in the replication organelle by correlative light microscopy and electron microscopy (CLEM) analysis. These data suggest that the ER membrane rearrangement by the ESCRT pathway requires a unique, specific subset of ESCRT factors.

In the field of cell biology, fluorescence microscopy and EM are indispensable for determining the localization of target factors or membranous structures in virus-infected cells. In this chapter, we introduce specific methods, including fluorescence light microscopy, Immuno-EM, and CLEM, for detecting ESCRT factors in virus-infected cells. Fluorescence microscopy allows visualization and quantification of target factors and events, with high spatial and temporal resolution. In the past, organelle morphology and their detailed membranous structures were usually visualized only by EM. However, CLEM, which is a combination of fluorescence microscopy and high-resolution electron microscopy, allows high-resolution observation of target structures along with fluorescence labeling. These methods are excellent tools for studying the complex relationship between membrane structure and ESCRT function in virus-infected cells.

2 Materials

2.1 Virus Amplification and Stock (See Note 1)

1. JEV: AT31.
2. DENV2 New Guinea C strain (ATCC VR-1584).
3. Cell culture medium.
4. 293A, BHK-21 cells.

5. Transfection reagent: Lipofectamine 2000.
6. Filter (0.45 μm pore size).

2.2 In Vitro Transcription

1. Complete JEV clone: rJEV (Mie/41/2002)/pMW119 (*see* **Note 1**).
2. mMESSAGE mMACHINE T7 kit.
3. Restriction enzyme: NdeI.
4. Spin column: Wizard SV Gel and PCR Clean-Up System.
5. Lithium chloride precipitation solution: 7.5 M lithium chloride and 50 mM ethylenediaminetetraacetic acid (EDTA).
6. 70% (v/v) ethanol in nuclease-free water
7. Nuclease-free water.

2.3 Virus Infection and Virus Titer (Focus Forming Assay)

1. 96 well plate.
2. Cell culture medium.
3. 1.0% (w/v) methylcellulose in cell culture medium: Put methylcellulose and magnetic stirring bar in glass bottle. After autoclave, add culture medium to glass bottle and mix it at 4 °C until methylcellulose is completely solved.
4. 1× Phosphate-buffered saline (PBS): 137 mM NaCl, 2.7 mM KCl, 10 mM Na_2HPO_4, 1.8 mM KH_2PO_4, pH of 7.4.
5. 4% (w/v) paraformaldehyde in PBS: Dissolve 40 g of paraformaldehyde powder in 800 mL of PBS and heat to 60 °C. Add small amounts of 1.0 M NaOH until the solution is clear. Adjust the volume to 1 L with 1× PBS (*see* **Note 2**).
6. 0.2% (v/v) Triton X-100 in PBS.
7. 10% (v/v) Fetal Calf Serum (FCS) in PBS.
8. Primary antibodies: rabbit anti-JEV NS3 antibody, rabbit anti-DENV capsid antibody in PBS containing 10% FCS.
9. Secondary antibodies: biotinylated goat anti-rabbit IgG antibody in PBS containing 10% FCS.
10. Streptavidin- Horseradish peroxidase (HRP).
11. VIP peroxidase substrate kit.

2.4 Immunofluorescence Microscopy

1. Coverslips for Microscope slide: circle cover slips, thickness 1, 12 mm.
2. Transfection reagent: Lipofectamine 2000.
3. 1× PBS: *see* Subheading 2.3, **item 4**.
4. 4% (w/v) paraformaldehyde solution in PBS: *see* Subheading 2.3, **item 5**.
5. 50 mM NH_4Cl in PBS: Add 0.27 g of ammonium chloride to 100 mL of 1× PBS.

6. Cell permeabilization solution: 0.2% (v/v) TritonX-100 in PBS.
7. Blocking solution: 10% (v/v) FCS in PBS.
8. Primary antibodies (*see* **Note 3**): CHMP2B antibody: rabbit polyclonal CHMP2B antibody (ab33174; Abcam), CHMP4A antibody: rabbit polyclonal CHMP4A antibody (H-52; Santa Cruz, *see* **Note 4**), JEV E antibody: mouse monoclonal JEV E antibody (6B4A-10; Millipore), rabbit polyclonal JEV NS3 antibody (*see* **Note 5**), rabbit polyclonal DENV capsid antibody (*see* **Note 6**), and dsRNA antibody (mouse monoclonal dsRNA J2 antibody, English & Scientific Consulting).
9. Secondary antibodies (e.g., Alexa Fluor 488-conjugated goat anti-mouse antibody or Alexa Fluor 568-conjugated goat anti-rabbit antibody).
10. SlowFade Gold antifade reagent.

2.5 Immuno-EM Using Pre-embedding Nanogold Intensification

1. Cell desk LF1 low autofluorescence 13.5-mm-diameter plastic coverslips (*see* **Note 7**).
2. 1× PBS: *see* Subheading 2.3, **item 4**.
3. 0.3 M Sodium phosphate buffer, pH 7.4 (PB): Dissolve 4.5 g of NaH_2PO_4 $2H_2O$ and 43.5 g of Na_2HPO_4 $12H_2O$ in 500 mL of deionized water. Autoclave the solution, and store it at room temperature.
4. 0.1 M PB: Mix 100 mL of 0.3 M PB with 200 mL of deionized water.
5. 10% (w/v) stock solution of paraformaldehyde: Dissolve 5.0 g of paraformaldehyde in 45 mL of deionized water and heat to 60 °C. Add small amounts of 1.0 M NaOH until the solution is clear. Adjust the volume to 50 mL with deionized water (*see* **Note 2**).
6. 4% paraformaldehyde in 0.1 M PB: Dilute 3.6 mL of 10% paraformaldehyde solution with 3 mL of 0.3 M PB and 2.4 mL of deionized water (*see* **Note 2**).
7. 30, 50, 70, and 90% ethanol in deionized water
8. Nickel EM grids: Coat Veco 200 mesh grids (Electron Microscopy Sciences) with polyvinyl formal films.
9. 2% (w/v) uranyl acetate: Dissolve 0.6 g uranyl acetate in 30 mL of distilled water and filter the solution. The solution is stored at 4 °C and it is stable.
10. Lead citrate (Sigma-Aldrich): We commercially purchased Lead citrate EM-grade solution from Sigma-Aldrich.
11. Light microscope.

12. Cell permeabilization solution: 0.25% (w/v) saponin in 0.1 M PB.
13. Blocking solution A: 0.005% saponin, 10% (w/v) BSA, and 10% (v/v) serum in 0.1 M PB.
14. Secondary antibodies: Nanogold Fab′ fragments from goat anti-mouse IgG or anti-rabbit IgG covalently conjugated with 1.4 nm colloidal gold particles.
15. Rinse solution: 0.005% (w/v) saponin in 0.1 M PB.
16. 25% stock solution of glutaraldehyde (Electron Microscopy Sciences): This stock solution is stored in closed vials at −80 °C. When it's thawed to 4 °C, the solution can be used for a few weeks.
17. 50 mM Glycine in 0.1 M PB.
18. 1% (w/v) BSA in 0.1 M PB.
19. GoldEnhance EM kit (Nanoprobes).
20. 1.0% (w/v) sodium thiosulfate: Dissolve 0.5 g of sodium thiosulfate in 50 mL of deionized water.
21. 2.0% (v/v) osmium tetroxide stock solution: It is purchased in closed, 1 g glass ampules. Break the ampule carefully and add 5 mL of deionized water. Further, place the ampule in a closed flask and add 45 mL of deionized water. Incubate the flasks overnight with gentle shaking. Store the solution in a refrigerator.
22. 1.0% osmium oxide in PB supplemented with 1.5% (w/v) potassium hexacyanoferrate: Dissolve 0.045 g of potassium hexacyanoferrate in a mixture containing 1.5 mL of 2.0% (v/v) osmium tetroxide and 1.0 mL of 0.3 M PB in a fume hood. Adjust the volume to 3.0 mL with deionized water.
23. Epoxy resin kit (TAAB Laboratories Equipment): Combine 24.5 mL of Epon 812, 15 mL of dodecenyl succinic anhydride (DDSA), 13.5 mL of methyl nadic anhydride (MNA), and 0.75 mL of 2,4,6-tri(dimethylaminomethyl)phenol (DNP30). Mix continuously for 5 min. Store the mixed resin at −20 °C (*see* **Notes 8–10**).
24. Flat embedding mold (Nisshin EM Co., Ltd.).
25. Hitachi H-7600 transmission electron microscope equipped with a CCD camera (HT-7650; Gatan).

2.6 Double-Labeled Immuno-EM

See Subheading 2.5 to prepare reagents for 1–11. Additional reagents are required for double-labeled immuno-EM as following.

1. LR-white resin supplemented with 0.5% (w/w) benzoin methyl ether (LR-White resin mixture): Dissolve 0.15 g of benzoin methyl ether (Tokyo Kasei Kogyo) in 30 g of

LR-white resin (Electron Microscopy Sciences), and store at −30 °C (*see* **Note 11**).

2. 8-mm Flat embedding capsule (TAAB).
3. UV polymerization apparatus (TUV200; Dosaka EM).
4. Blocking solution B: 2% (w/v) BSA and 2% FCS in 0.1 M PB.
5. Primary antibody solution: Dilute rabbit polyclonal anti-CHMP2B and mouse monoclonal anti-JEV E antibodies with blocking solution B at 1: 50, respectively.
6. Secondary antibody solution: Dilute anti-rabbit IgG conjugated with 10 nm and anti-mouse IgG conjugated with 5 nm gold particles, with blocking solution B at 1: 100, respectively.
7. 0.1 M sodium cacodylate buffer solution, pH 7.4.
8. 2.5% (v/v) glutaraldehyde in 0.1 M cacodylate buffer, pH 7.4 (Wako).

2.7 CLEM

1. Gridded 35-mm glass bottom dishes (P35G-2-14-CGRD; Mat-Tek).
2. 4% formaldehyde in PBS: *see* Subheading 2.3, **item 4**.
3. 0.1 M phosphate buffer (PB), pH 7.4. *See* Subheading 2.5, **item 4**.
4. 4% (w/v) sucrose in 0.1 M PB, pH 7.4.
5. 2.5% glutaraldehyde and 2% formaldehyde in 0.1 M PB, pH 7.4.
6. 0.25% (w/v) saponin in 0.1 M PB, pH 7.4.
7. 0.005% saponin in 0.1 M PB, pH 7.4.
8. Blocking solution; 0.1 M PB, pH 7.4 containing 0.005% saponin, 10% (w/v) BSA, 10% normal serum that is derived from immune animal species of secondary antibody, and 0.01% (w/v) cold water fish skin gelatin.
9. Primary antibody.
10. Secondary antibody, fluorescently labeled.
11. Confocal laser scanning microscope.
12. Image-processing software.
13. 1% (v/v) osmium tetroxide and 1.5% (w/v) potassium ferrocyanide in 0.1 M PB.
14. 50, 70, and 90% (v/v) ethanol in deionized water.
15. 100% ethanol (99.5%, special grade).
16. Embedding mix [11] (*see* **Notes 8–10**).
17. 50% (v/v) epoxy resin mixture dissolved in 100% ethanol.
18. Embedding capsules for epoxy resin mixture.
19. Oven set at 60 °C.

20. Diamond knife (Ultra 45°; Diatome AG).
21. Ultra-microtome (EM UC7; Leica microsystems, DE).
22. Cu grid, 200 hexagonal mesh thin-bar grid, G200HH (Gilder, UK).
23. 8% (w/v) uranyl acetate in deionized water.
24. Saturated uranyl acetate.
25. Lead citrate (Sigma-Aldrich): We commercially purchased Lead citrate EM-grade solution by Sigma-Aldrich.
26. JEOL JEM-1011 transmission electron microscope 80 kV (JEOL, Ltd., JP) equipped with a CCD camera (OLYMPUS VELETA 2K × 2K).

3 Methods

3.1 Cell Culture and Virus Amplification

Maintain mammalian cells in growth medium under 5% CO_2 at 37 °C. Amplify virus strains (JEV strain AT31 and DENV 2 strain NGC/ATCC VR-1584) in 293A and BHK-21 cells, respectively.

3.2 In Vitro Transcription of the JEV RNA Genome

1. Prepare plasmid DNA. The DNA template should be relatively free of contaminating proteins and RNA, as clean DNA should lead to high yields.
2. Linearize rJEV (Mie/41/2002)/pMW119 plasmid by digestion with NdeI.
3. Cut the band from the gel and purify the linearized DNA with a spin column kit.
4. Mix the following in a tube for a single 20 μL reaction at room temperature; 10 μL 2× NTP/CAP, 2 μL 10× reaction buffer, 1 μg linearized template DNA, 2 μL enzyme mix, and nuclease-free water. You can also refer user guide for mMESSAGE mMACHINE Kit.
5. Incubate at 37 °C for 1–2 h.
6. Stop the reaction by adding 30 μL of nuclease-free water and 30 μL of lithium chloride precipitation solution to precipitate RNA.
7. Mix by inverting the tube and incubate at −20 °C for 30 min.
8. Centrifuge at 4 °C for 15 min at top speed (20,000 × *g*) to pellet the RNA.
9. Carefully remove the supernatant and rinse the pellet once with 1 mL of ice-cold 70% ethanol.
10. After centrifugation (20,000 × *g*), remove the ethanol, and resuspend the RNA in nuclease-free water.
11. Determine the RNA concentration, and store at −20 °C or −70 °C.

3.3 Preparation of JEV Stock (See Note 12)

1. Transfect BHK-21 cells or 293A cells with in vitro transcribed RNA by using Lipofectamine 2000, according to the manufacturer's protocol.
2. Collect supernatant after 3 days (*see* **Note 13**), filter (0.45 μm pore size), and store at −70 °C.

3.4 Virus Infection and Virus Titer (Focus Forming Assay)

1. Seed target cells in a 96-well plate 1 day before infection. The total volume of medium is 200 μL per well.
2. Serially dilute the virus inoculum (1:10 or 1:5) with growth medium and use at least 6 wells per dilution. Replace the medium with this virus suspension medium. For a 1:1 dilution, aspirate medium and add 200 μL of virus suspension in the first row.
3. Incubate the cells and virus for 4–6 h.
4. Add 1.0% methylcellulose to the growth medium (*see* **Note 14**).
5. Incubate the cells for 36 h for JEV or 72 h for Dengue Virus.
6. Fix the cells with 4% paraformaldehyde in PBS for at least 20 min.
7. Wash with PBS three times.
8. Permeabilize the cells with 0.2% (v/v) Triton X-100 in PBS.
9. Block the cells with 10% (v/v) FCS in PBS.
10. Incubate the cells with diluted primary antibodies (rabbit anti-JEV NS3 antibody (1:500) for JEV or rabbit anti-DENV capsid antibody (1:500) in PBS containing 10% FCS) for 60 min at room temperature without agitation.
11. Wash with PBS three times.
12. Incubate with the secondary antibodies (biotinylated goat anti-rabbit IgG antibody in PBS containing 10% FCS) for 30 min at room temperature without agitation.
13. Wash with PBS three times.
14. Incubate with streptavidin- Horseradish peroxidase (HRP) for 30 min at room temperature without agitation.
15. Wash with PBS three times.
16. Mix reagents in VIP peroxidase substrate kit before use: three drops (approximately 90 μL) of reagent 1, three drops (approximately 90 μL) of reagent 2, three drops (approximately 90 μL) of reagent 3, and three drops (approximately 120 μL) of hydrogen peroxide solution in 5 mL of PBS.

17. Incubate with the substrate solution prepared in **step 16** for 5–30 min. Optimal development times should be determined by investigator.
18. Stop reaction by washing with water.
19. Count the foci (*see* **Note 15**).
20. Calculate the infectious virus titer.

3.5 Immunostaining of Virus-Infected Cells

1. Seed target cells on coverslips in a multi-well plate 1 day before infection.
2. Remove the supernatant and replace with growth medium containing virus. It should be clear in advance what volume of virus suspension solution is necessary for infection (*see* **Note 16**).
3. Change the medium after 4–6 h.
4. After 24–72 h for JEV or 24–96 h for DENV (*see* **Note 17**), fix the cells by adding 1 mL of 4% paraformaldehyde in PBS to each well and incubating for 20 min at room temperature.
5. Wash the cells three times with PBS.
6. Add 1 mL of 50 mM NH_4Cl in PBS and incubate for 10 min at room temperature to quench-free aldehyde groups.
7. Wash the cells three times with PBS.
8. Permeabilize the cell with 0.2% Triton X-100 in PBS.
9. Wash the cells twice with PBS.
10. Block the cells with 10% FCS in PBS.
11. Incubate cells on coverslips with 40 μL of primary antibody diluted in 10% (v/v) FCS in PBS (e.g., 200-fold dilution of CHMP2B antibody) for at least 60 min at room temperature.
12. Wash the cells three times with PBS.
13. Incubate the cells with 40 μL of the appropriate secondary antibody diluted in 10% FCS/PBS (1:500–1:1000) for 60 min at room temperature.
14. Wash the cells three times with PBS.
15. Mount coverslips on glass slides using 5 μL of SlowFade Gold Antifade Reagent. Remove the excess reagent with filter paper.
16. Seal the coverslip to the glass slide using nail polish.
17. Store at 4 °C in a black box.
18. Confocal microscopy. Observe the samples mounted on the glass slide with a confocal laser scanning fluorescent microscope equipped with appropriate lenses, e.g., 63×/1.40 and 100×/1.34 or 1.40 (Fig. 1).

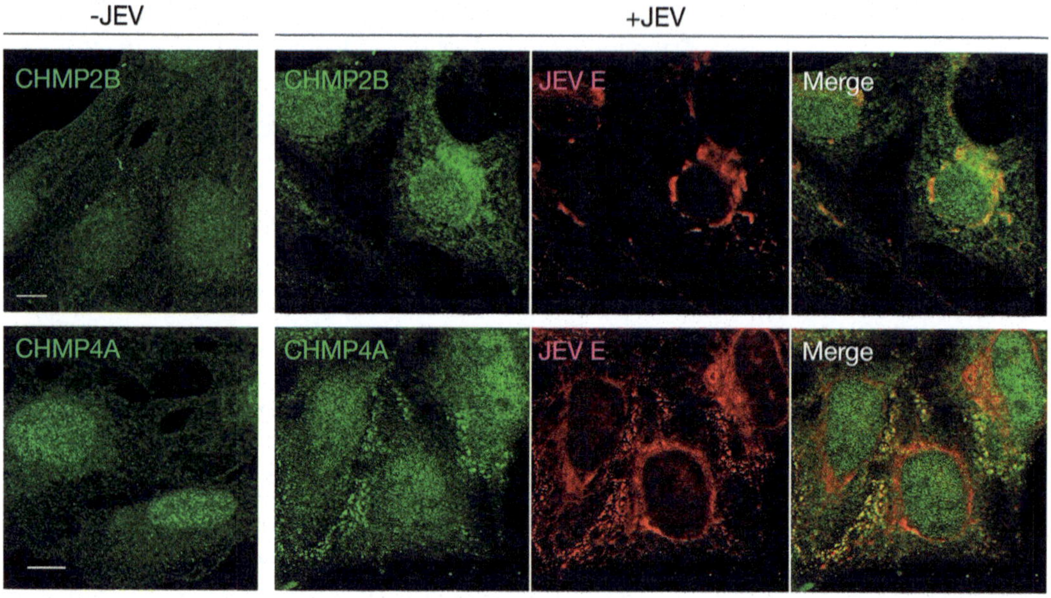

Fig. 1 Immunostaining of virus-infected cells. Vero cells infected with JEV (+JEV) for 48 h or uninfected (−JEV) were stained with an anti-CHMP2B (upper panels, green) or anti-CHMP4A (lower panels, green) antibody along with an anti-JEV E antibody (red). Images were acquired with a confocal laser microscope. Scale bars are 10 μm

3.6 Single-Labeled Immuno-EM

Cell culture and fixation

1. Culture cells on LF1 plastic coverslips in a 24-well plate containing 0.5 mL of culture medium per well.
2. Remove the culture medium, wash the cells with 1 mL of PBS for a few minutes, and then fix the cells with 1 mL of 4% paraformaldehyde in PB for 30 min.
3. Wash the cells with 1 mL of 0.1 M PB three times.
4. Permeabilize the fixed cells with 0.5 mL of 0.25% (w/v) saponin in 0.1 M PB for 30 min at room temperature (*see* **Note 18**).
5. Wash once with 1 mL of PB for 5 min.

Immunogold labeling

6. Incubate the fixed and permeabilized cells with 0.5 mL of blocking solution A for 10 min.
7. Remove the blocking solution, and transfer the coverslips on Parafilm in a moist chamber to prevent drying.
8. Incubate the cells with 50 μL of ice-cold primary antibody in blocking solution A. Samples can be incubated overnight at 4 °C.
9. Remove the primary antibody solution by quickly touching the cover slip with a filter paper. Subsequently, transfer the coverslips to a 24-well plate containing 500 μL of rinse solution per well.

10. Wash the cells with 500 μL of rinse solution six times for 10 min each.
11. Transfer the coverslips to a new piece of Parafilm in a moist chamber.
12. Incubate the cells on the Parafilm with 50 μL of nanogold Fab′ conjugates to goat anti-rabbit IgG (1:250 dilution) in blocking solution A as a secondary antibody for 2 h at room temperature.
13. Remove the secondary antibody, and transfer the coverslips to the 24-well plate again.
14. Wash the cells with 500 μL of the 0.005% (w/v) saponin in 0.1 M PB six times for 10 min each.
15. Fix the immunogold-labeled cells with 500 μL of 1% (w/v) glutaraldehyde in PB for 10 min.

Nanogold intensification procedure

16. Wash the immunogold-labeled cells with 1 mL of 50 mM glycine three times for 5 min each.
17. Wash the cells with 1 mL of 1% (w/v) BSA in PB three times for 5 min each.
18. Wash the cells with 1 mL of distilled water three times for 5 min each.
19. Prepare the gold intensification solution just before use by mixing the reagents from the Nanoprobes Gold Enhancement kit.
20. Transfer the coverslips to a new piece of Parafilm, and incubate the cells with 50 μL of gold intensification mixture at room temperature for 5 min.
21. Remove the gold intensification mixture, and transfer the coverslips to the 24-well plate.
22. Wash the immunogold-labeled cells once with 1 mL of 1% (w/v) sodium thiosulfate for 1 min.
23. Wash the cells with 1 mL of distilled water three times.
24. Incubate the immunogold-labeled cells with 300 μL of 1% (w/v) OsO_4 and 1.5% (w/v) potassium ferrocyanide in PB for 60 min at room temperature in a fume food.

Embedding in epoxy resin and ultrathin sectioning

25. Dehydrate the immunogold-labeled cells in a graded series of ethanol (30, 50, 70, and 90%, v/v; 1 mL each), and then twice in 100% ethanol for 10 min each.
26. Incubate the cells on the coverslips with 500 μL of 50% (v/v) epoxy resin in ethanol for 30 min.

27. Place the coverslips on small drops of epoxy resin and incubate twice for 1 h each. This procedure can be done on the Parafilm where the drops are placed (*see* **Note 19**).
28. Fill the wells in the flat embedding mold with epoxy resin.
29. Cover the wells with the cover slips, cell side down. If necessary, cut the coverslips with scissors to adjust them to fit the wells.
30. Polymerize the samples for 2 days at 60 °C (*see* **Note 20**).
31. Remove the polymerized epoxy-containing and coverslip-attached blocks from the mold. The coverslips should be easy to remove from the epoxy blocks.
32. Observe the cells on the surface of the blocks by light microscopy. Mark the areas of interest with a marker.
33. To generate ultrathin sections of areas of interest, cut the epoxy blocks horizontally (85 nm thick) with an ultra-microtome equipped with a diamond knife. Set the cutting speed at 1.2 mm S^{-1}.
34. Place the ultrathin sections on the polyvinyl formal-coated nickel EM grids.
35. Stain the sections on the grids with 2% (w/v) uranyl acetate for 1 h, and then wash three times in distilled water.
36. Stain the sections with lead citrate for 2 min, and then wash three times in distilled water.
37. Observe the sections with a Hitachi H-7600 transmission electron microscope.

3.7 Double-Labeled Immuno-EM

Cell culture, fixation, and embedding in LR-White resin

1. Culture cells on LF1 plastic coverslips in a 24-well plate containing 0.5 mL of culture medium per well.
2. Remove the culture medium, wash with 1 mL of PBS for a few minutes, and then fix the cells with 1 mL of 4% paraformaldehyde in PB for 30 min.
3. Wash the cells in 1 mL of PB three times.
4. Dehydrate the cells in a graded series of ethanol (30, 50, 70, and 90%, v/v; 1 mL each) and then twice in 100% ethanol for 10 min at −20 °C.
5. Incubate the cells with 500 μL of 50% (v/v) LR-White resin mixture in ethanol for 30 min at −20 °C.
6. Remove the solution, replace it with 500 μL of LR-White resin mixture, and incubate for 30 min at −20 °C.
7. Cut the coverslips with scissors, and place the cut fragments on the bottom of the capsules in a fume hood. Fill the capsules with LR-White resin mixture.

8. Polymerize the resin in the sample with UV irradiation using a UV polymerization apparatus at −20 °C overnight.
9. Continue to UV irradiate the samples at room temperature for 24 h.
10. Remove the polymerized LR-White resin and coverslip-attached blocks from the capsules. The coverslips should be easy to remove from the blocks.
11. Observe the cells on the surface of the blocks by light microscopy. Mark areas of interest with a marker.
12. To generate ultrathin sections of areas of interest, cut the blocks horizontally (80 nm thick) using an ultra-microtome equipped with a diamond knife. Set the cutting speed at 1.2 mm s^{-1}.
13. Place the sections on the polyvinyl formal-coated nickel EM grids.

Immunogold labeling

14. Place 100 μL of blocking solution B on a piece of Parafilm.
15. Place grids containing the sections on the blocking solution and incubate for 30 min at room temperature.
16. Place 25–40 μL aliquots of primary antibody on Parafilm in a moist chamber (*see* **Note 21**).
17. Place the sections on the primary antibody solution and incubate overnight in a refrigerator.
18. To wash the sections, place 100 μL of blocking solution B on a piece of Parafilm and place the sections in the solution. Incubate for 1 min at room temperature.
19. Transfer the section to another 100-μL drop of blocking solution B and repeat this six times.
20. Place 50 μL of the secondary antibody solution on a piece of Parafilm in a moist chamber.
21. Place the sections on the grids in the second antibody solution and incubate for 2 h at room temperature (*see* **Note 22**).
22. To wash the sections with cacodylate buffer, place 100 μL aliquots of 0.1 M cacodylate buffer, pH 7.4, on a piece of Parafilm. Place the sections in the drops of blocking solution B to wash the sections six times for 1 min at a room temperature as described in **steps 5** and **6** (*see* **Note 23**).
23. Place 100 μL aliquots of 2.5% (v/v) glutaraldehyde in 0.1 M cacodylate buffer, pH 7.4, on a piece of Parafilm. Place the sections in the buffer and incubate for 10 min at room temperature to fix the cells.
24. Wash the sections with 100 μL of distilled water on a piece of Parafilm at room temperature three times for 1 min each.

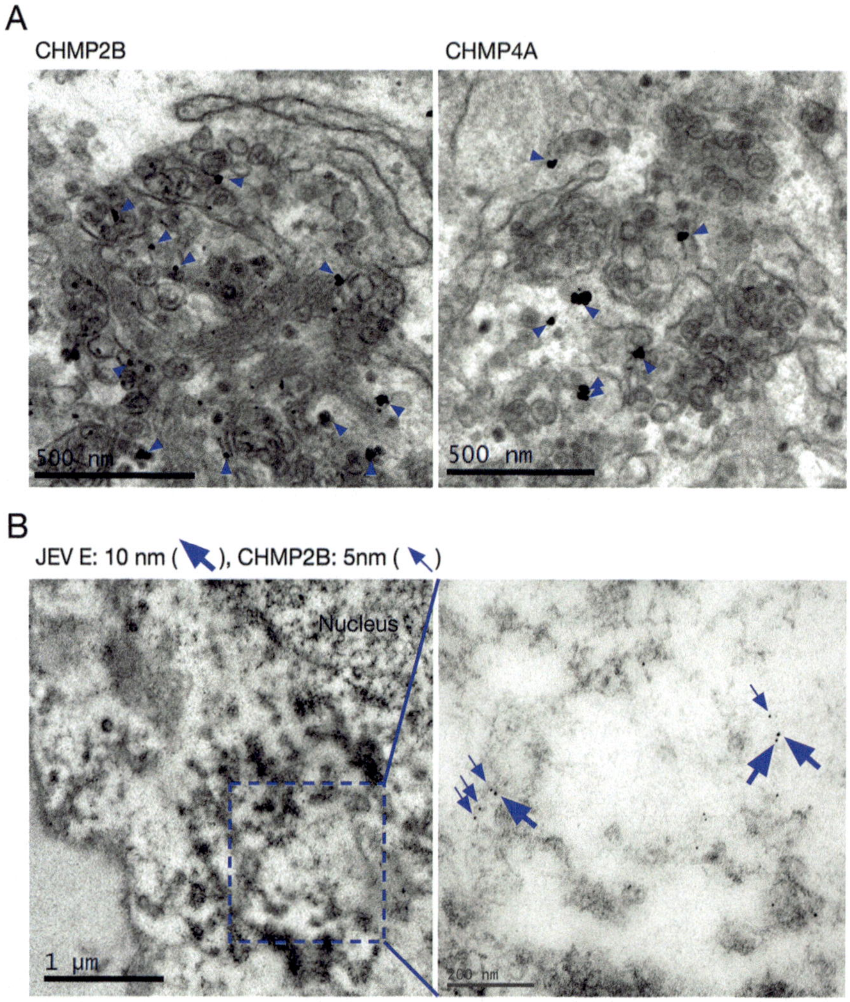

Fig. 2 Single- and double-labeled immunoelectron microscopy. (**a**) Vero cells infected with JEV for 48 h were fixed and stained with antibodies specific for CHMP2B (left) or CHMP4A (right). The enhanced signals for CHMP2B and CHMP4A are indicated with blue arrowheads. (**b**) Vero cells infected with JEV for 48 h were co-stained with anti-CHMP2B (5 nm gold particles) and anti-JEV E antibodies (10 nm gold particles). The area in the square is shown on the right

25. Remove excess water from the grids by touching them to a filter paper.
26. Stain the sections on the grids with 2% (w/v) uranyl acetate for 10 min, and then wash three times in distilled water.
27. Stain the sections with lead-staining solution for 2 min, and then wash three times in distilled water.
28. Observe the sections with an electron microscope (Fig. 2).

3.8 CLEM

1. Culture cells on gridded 35 mm glass bottom dishes (P35G-2-14-CGRD; Mat-Tek, USA) and then transfect the cells with JEV RNA using Lipofectamine 2000 or infect with virus (*see* **Note 24**).
2. Incubate for 48 h. For virus-infected cells, change medium after 4–6 h.
3. Remove the culture medium, and fix the cells with 4% formaldehyde in 0.1 M PB, pH 7.4, for the optimal time at room temperature (*see* **Note 25**).
4. Wash the cells in 0.1 M PB, pH 7.4, containing 4% (w/v) sucrose three times for 5 min each.
5. Permeabilize the cells with 0.1 M PB, pH 7.4, containing 0.25% saponin for 30 min at room temperature (*see* **Note 26**).
6. Incubate the cells with blocking solution for 30 min at room temperature.
7. Remove the blocking solution and replace it with the solution containing primary antibodies, the specific antibodies for JEV NS3 and dsRNA, diluted to the optimal concentration in blocking solution (*see* **Note 27**).
8. Incubate the sample dish overnight at 4 °C.
9. Remove the solution containing the primary antibodies, and wash the cells with 0.005% (w/v) saponin, 0.1 M PB, pH 7.4, six times for 5 min each.
10. Incubate the cells in a dark chamber with the fluorescence-labeled secondary antibodies (at the optimal dilution) in blocking solution for the optimal time at room temperature.
11. Remove the solution containing secondary antibodies, wash the cells with 0.005% saponin, 0.1 M PB, pH 7.4, six times for 5 min each, and then with 0.1 M PB, pH 7.4, for 5 min at room temperature (*see* **Note 28**).
12. Search for a qualified cell as a target and take a fluorescent image of the cell at 60× or 100× magnification with a confocal laser scanning microscope. In addition, take a differential interference contrast (DIC) image of the same target cell with the letters around it on the glass bottom at 100× magnification to confirm its location in the dish. The 100× magnification image will be used as a cell map during observation by electron microscopy (*see* **Note 29**).
13. Store the high-resolution image files of the target cell in a JPEG or TIF format that can be opened by most image-processing software.
14. Incubate the cells with 2.5% (v/v) glutaraldehyde and 2% (v/v) formaldehyde, 0.1 M PB pH 7.4, at 4 °C overnight. The next morning, wash the cells three times with ice-cold 0.1 M PB.

15. Postfix the cells with 1% (v/v) osmium tetroxide and 1.5% (v/v) potassium ferrocyanide, 0.1 M PB, for 1 h at room temperature, and then wash the cells with distilled water three times.
16. Check the location of the target cell under a phase contrast microscope and obtain a DIC (Differential Interference Contrast) image at 100× magnification to use as a map and mark the backside of the target cell with an oily marking pen.
17. Dehydrate the cells in a graded series of ethanol (50, 70, and 90%, v/v) on ice for 5 min each, and then wash in 100% ethanol at room temperature three times for 5 min each (*see* **Note 30**).
18. Incubate the samples in 50% (v/v) epoxy resin mixture dissolved in 100% ethanol for 1 h at room temperature.
19. Incubate the samples in epoxy resin mixture at room temperature twice for 1 h each.
20. Fill an embedding capsule with the epoxy resin mixture and place it upside-down on the target cell.
21. Place the sample dish with embedded resin into an oven at 60 °C and incubate for 2 days.
22. After the epoxy resin has been sufficiently polymerized, remove the glass bottom dish from the epoxy resin block in the embedding capsule by quick cooling using the cold air from liquid nitrogen, which makes removal possible due to the difference in the shrinkage rates of the resin and glass. Specifically, cool the epoxy resin block stuck on the glass with cold air from liquid nitrogen for 10–20 s, then twist the embedding capsule and pull it out (*see* **Note 31**). The target cell with the grid line and letter from the glass will be transferred to the resin surface and will appear as a mirror-reversed image. If any glass pieces remain on the resin after pulling the resin from the glass, place the resin block into cold air from liquid nitrogen or put the surface of the resin directly into liquid nitrogen for 10 s and remove the glass pieces with tweezers, taking care not to scar the target cell. Do not immerse the entire resin block into liquid nitrogen because this can cause cracking.
23. Trim the epoxy resin block around the target cell with a fresh sharp-edged razor under a stereoscopic microscope while confirming its location by referencing the printed low-magnification DIC image, which is a small-scale map of the cell (*see* **Note 32**). Before printing the map, flip the image horizontally with processing software because the target cell and the letter transcribed to the resin surface from the glass are mirror-reversed with respect to the DIC image.
24. Prepare ultrathin serial sections of the target cell (80 nm thick) with a diamond knife in an ultra-microtome, and place as many

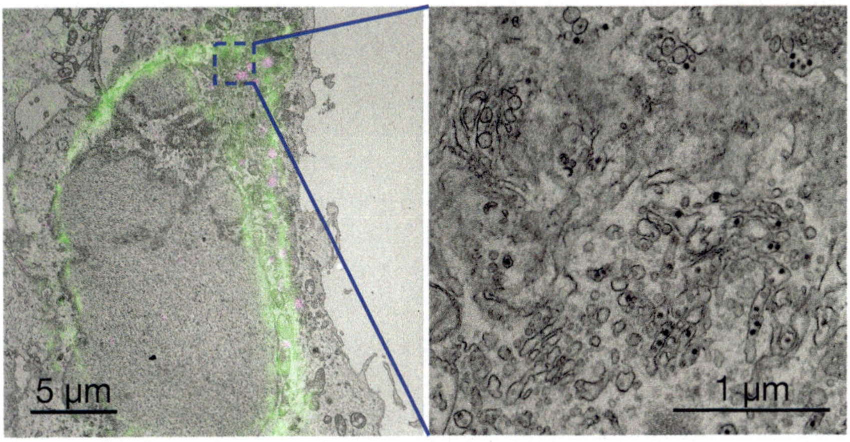

Fig. 3 CLEM of virus-infected cells. JEV-infected cells were stained with anti-JEV NS3 (green) and anti-dsRNA (magenta) antibodies and observed by confocal fluorescent microscopy. An ultrathin-sectioned image of the same cell was obtained by electron microscopy (left, merged with fluorescent images). Magnified images are shown on the right. (With permission from Tabata K, Arimoto M, Arakawa M, et al. (2016) Unique Requirement for ESCRT Factors in Flavivirus Particle Formation on the Endoplasmic Reticulum. Cell Rep 16:2339–2347. doi: 10.1016/j.celrep.2016.07.068)

sections as possible on a 200 hexagonal mesh thin-bar grid (Cu, G200HH; Gilder, UK) and stain with 8% (w/v) uranyl acetate and the lead citrate EM-staining solution.

25. Search the target cell under a transmission electron microscope at 80 kV while referencing the DIC and fluorescent images, which were printed ahead, and take electron micrographs with a side-mounted or bottom-mounted CCD camera that has at least 2K × 2K pixels. EM images of almost all serial sections must be taken to compare with the light microscope images, because each section will have a different image depending on its location (bottom to top) within the target cell.
26. Correlate the images with image-processing software (Fig. 3).

4 Notes

1. Before setting up the virus study, you must check whether the biosafety containment level of your laboratory is high enough to handle flaviviruses and check the regulations of your facility for the proper handling of viral plasmids.
2. Prepare reagent in a fume hood.
3. The available antibodies against ESCRTs for use in immunofluorescence microscopy are limited. Before starting any experiment, check the availability of the antibody for use in several experiments, such as immunofluorescence microscopy and western blotting, if you want to use another antibody.

4. This CHMP4A antibody is no longer available. Thus, alternative antibodies must be tested.
5. JEV NS3 antibody was raised against the N-terminal amino acids of NS3. Further, the rabbit antisera were affinity-purified using the same recombinant proteins.
6. Rabbit polyclonal DENV capsid antibody was raised against the N-terminal 100 residues of capsid protein. Furthermore, the rabbit antisera were affinity-purified using the same recombinant proteins.
7. The LF1 coverslips have two advantages. First, the coverslips have a bent edge and can be easily removed from a 24-well plate. Second, the coverslips can be removed from an epoxy resin block after embedding.
8. There are several recipes for epoxy resin. With respect to block hardness after polymerization, stability in the electron beam, ultrastructural observations, and so on, they are quite comparable.
9. The blocks need to be sufficiently hard. Use of an insufficient mixture produces soft resin blocks and eventually failed sections.
10. Epoxy resin mixture can be stored in syringes sealed with Parafilm in the freezer.
11. LR-White resin is capable of polymerization under low temperature; therefore, it is suitable for electron microscopic immunocytochemistry.
12. Virus-replicating cells must be cultured under proper biosafety containment and according to the regulations at your facility. To protect yourself, handle virus-infected cells and supernatants in a safety cabinet and wear a lab coat, protective glasses, and gloves.
13. Optimize the incubation time after transfection.
14. A semisolid overlay medium such as 1% methylcellulose in cell culture medium restricts the spread of infectious virus, creating localized clusters (foci) of infected cells.
15. Clusters (foci) of infected cells are visualized as red after incubation with VIP substrate solution. Count these red foci by microscopy and quantify the number of foci.
16. A multiplicity of infection (MOI) of 1.0 is preferable for IF. At this MOI, theoretically all cells can be infected.
17. Optimize the incubation time after virus infection.
18. Weak gold labeling is probably due to antigen destruction during fixation. As possible remedies to weak labeling, try decreasing the concentration of the fixation solution and shortening the fixation time. However, please note that weak fixation can lead to poor ultrastructural preservation.

19. Incomplete removal of the ethanol will produce soft resin blocks.
20. A polymerization time that is too short will yield soft resin blocks.
21. The primary antibodies can be mixed together if the antibodies were raised in different animal species.
22. The secondary antibody can be mixed if the antibodies are conjugated to different diameter nanoparticles.
23. Cacodylate buffer is more stain-resistant than some PB.
24. When the cells are placed, calculate the cell density and adjust it to 50% at fixation, because if the density of the cells is too high, they will hide the letters and the grid on the bottom. In addition, be aware that the 14 mm-diameter gridded circle in the glass bottom in the dish is recessed, and cells easily become concentrated in this recess. The recommended gridded 35 mm glass bottom dishes (P35G-2-14-CGRD; Mat-Tek, USA) are more expensive than normal 35 mm glass bottom dishes. Therefore, preliminary experiments using normal glass bottom dishes are recommended before using the gridded glass bottom dishes.
25. The cells can be washed in PBS for a few seconds before fixing if required.
26. Never use Triton X-100 to permeabilize the cells because it is too strong as a detergent to preserve the fine structure of cytoplasmic organelles. If the cells express fluorescent proteins, the immunostaining **steps 5–11** can be skipped.
27. Optimize the immunostaining conditions, such as the antibody dilution and the duration of incubation in saponin/PB, before starting the CLEM experiment.
28. From the washing process to observation, shade the sample dish from room lighting and store at 4 °C.
29. The approximately 400 nm squares on the glass bottom can become an ultrathin section. If it is necessary to observe multiple cells under the same experimental conditions, prepare as many dishes as required. Do not choose a cell that is directly over the letters and grid lines. Because the letters and lines are etched, they are transcribed as a raised impression onto the surface of the epoxy resin block in which the cells are embedded. Any cells on the raised letters and lines will be erased first and lost in the sectioning process.
30. Prevent the backside of the glass from coming into contact with ethanol, because it will remove the marking on the glass. Never use propylene oxide or any other substitutive solution after dehydration, because they are organic solvents and can melt the plastic parts of the glass bottom dishes.

31. There is another way to cool down and remove the glass, which does not require liquid nitrogen. After taking the dish containing the resin out of the 60 °C oven, place it immediately on ice via a copper or aluminum plate, which has good thermal conductivity. It is important to put the dish on ice immediately after taking it from the oven, because once the resin cools to room temperature, it cannot be removed from the glass well.
32. The diamond or glass trimming knife in an ultra-microtome can be used for block trimming.

Acknowledgments

This work was supported by grants from *JSPS KAKENHI (grant numbers: JP17K08849, JP17H06415, JP16H01188, and 26460555)*, a Health Labor Sciences Research Grant for Research on Emerging and Reemerging Infectious Diseases (grant number 12103320), and by *the Naito Foundation and the Uehara Memorial Foundation.*

References

1. Schöneberg J, Lee I-H, Iwasa JH, Hurley JH (2017) Reverse-topology membrane scission by the ESCRT proteins. Nat Rev Mol Cell Biol 18:5–17. https://doi.org/10.1038/nrm.2016.121
2. Votteler J, Sundquist WI (2013) Virus budding and the ESCRT pathway. Cell Host Microbe 14:232–241. https://doi.org/10.1016/j.chom.2013.08.012
3. Feng Z, Hensley L, McKnight KL et al (2013) A pathogenic picornavirus acquires an envelope by hijacking cellular membranes. Nature 496:367–371. https://doi.org/10.1038/nature12029
4. Tabata K, Arimoto M, Arakawa M et al (2016) Unique requirement for ESCRT factors in flavivirus particle formation on the endoplasmic reticulum. Cell Rep 16:2339–2347. https://doi.org/10.1016/j.celrep.2016.07.068
5. Chatel-Chaix L, Bartenschlager R (2014) Dengue virus- and hepatitis C virus-induced replication and assembly compartments: the enemy inside–caught in the web. J Virol 88:5907–5911. https://doi.org/10.1128/JVI.03404-13
6. Romero-Brey I, Bartenschlager R (2014) Membranous replication factories induced by plus-strand RNA viruses. Viruses 6:2826–2857. https://doi.org/10.3390/v6072826
7. Cortese M, Goellner S, Acosta EG et al (2017) Ultrastructural characterization of zika virus replication factories. Cell Rep 18:2113–2123. https://doi.org/10.1016/j.celrep.2017.02.014
8. Hase T, Summers PL, Dubois DR (1990a) Ultrastructural changes of mouse brain neurons infected with Japanese encephalitis virus. Int J Exp Pathol 71:493–505
9. Hase T, Summers PL, Ray P (1990b) Entry and replication of Japanese encephalitis virus in cultured neurogenic cells. J Virol Methods 30:205–214
10. Welsch S, Miller S, Romero-Brey I et al (2009) Composition and three-dimensional architecture of the dengue virus replication and assembly sites. Cell Host Microbe 5:365–375. https://doi.org/10.1016/j.chom.2009.03.007
11. Luft JH (1961) Improvements in epoxy resin embedding methods. J Biophys Biochem Cytol 9:409–414

Chapter 7

Single Cell Fluorescence Ratio Image Analysis for Studying ESCRT Function in Receptor Trafficking

Jalal M. Kazan, Gergely L. Lukacs, Pirjo M. Apaja, and Arnim Pause

Abstract

The endosomal sorting complexes required for transport (ESCRT) comprise a major trafficking pathway for plasma membrane proteins and are fundamental for ubiquitin-dependent cargo endocytosis. Here, we describe a method for studying the effect of ESCRT complexes on endo-lysosomal membrane trafficking and their role in receptor integrin α5β1 downregulation. Single cell fluorescence ratio image analysis (FRIA), using appropriate fluorescence probes, enables the measurement of dynamics of integrin α5β1 containing vesicles and represents a live cell-based method for studying the role of ESCRTs.

Key words ESCRT, Integrin α5β1, FRIA, HD-PTP, UBAP1, Vesicular pH, Recycling, Lysosomes

1 Introduction

The endosomal sorting complex required for transport (ESCRT) is a cellular machinery consisting of several complexes (ESCRT-0, I, II, III and Vps4-Vta1 ATPase complex) that are responsible for discrete steps in membrane scission events during cytokinesis, viral budding, and in formation of late endosomal multivesicular bodies (MVBs) [1]. ESCRT-mediated regulation is mandatory for intraluminal vesicle (ILV) invagination within late endosomes to form MVBs and is a key step in the delivery of ubiquitinated cargo proteins destined to lysosomes. Ubiquitinated cargo proteins targeted to late endosomes are recognized by early ESCRT complexes (ESCRT-0, I), which eventually recruit ESCRT-II complex. The regulation of lipid bilayer invagination resulting in the formation of ILVs is the main role of ESCRT-III complex, which at the end of the process will be recycled back to the cytosol through the Vps4-Vta1 ATPase complex.

ESCRT complexes regulate ubiquitin-dependent endosomal trafficking dynamics of extracellular ligand-stimulated cell surface receptors. Shortly following internalization, receptors are delivered to early endosomes that constitute the primary post-endocytic

Emmanuel Culetto and Renaud Legouis (eds.), *The ESCRT Complexes: Methods and Protocols*, Methods in Molecular Biology, vol. 1998, https://doi.org/10.1007/978-1-4939-9492-2_7,

sorting station. Sorting determines the fate of receptors, directing them either for recycling to the plasma membrane, to the Golgi, or through late endosomal MVBs to degradation in lysosomes [2]. Several ESCRT complexes have ubiquitin-binding motifs that recognize ubiquitinated receptors facilitating the recruitment of other ESCRTs and sorting toward MVBs. Perturbation of ESCRT-mediated cargo sorting may lead to a decrease in receptor degradation and/or their recycling back to the plasma membrane. For instance, several studies have shown that epidermal growth factor receptor (EGFR) accumulated at endosomes and its degradation was attenuated upon alterations in ESCRT levels [3, 4]. The same effect was also observed on the chemokine receptor CXCR4 [5] and the melanosomal receptor OA1 [6]. In *Drosophila*, the role of ESCRT in sorting the Notch receptor was demonstrated [7, 8]. Mutations in the fly ESCRT components triggered tissue overgrowth due to an increase in Notch signaling, hence showing a potential tumor suppressor activity of certain ESCRT subunits [9, 10]. Moreover, due to ESCRT loss of function and attenuation of receptor degradation, downstream signaling cascades such as the MAPK pathway were enhanced [11].

Integrin α5β1, the fibronectin (FN) receptor, is ubiquitinated and internalized upon ligand FN-binding [12, 13]. Early endosomal pH regulates dissociation of FN promoting α5β1 receptor complex deubiquitination and recycling [13]. Depletion of ESCRT-0 and -I subunits Hrs and TSG101 attenuated the lysosomal trafficking of integrin α5β1 [12, 13]. The ESCRT-I member histidine domain-containing protein tyrosine phosphatase (HD-PTP/PTPN23) is a scaffold for binding of ESCRT-0-components Stam1-Hrs to ESCRT-I component UBAP1 [14], and it may act as an alternative noncanonical ESCRT pathway between ESCRT-I to -III replacing ESCRT-II [15]. Kharitidi et al. have shown that both HD-PTP and UBAP1 proteins are required for the subsequent lysosomal targeting of integrin α5β1 upon FN stimulation. Depletion of HD-PTP and UBAP1 enhanced the recycling of integrin α5β1 back to the plasma membrane and increased migratory/invasive cellular phenotype [13].

The acidic luminal pH at the endosomes and lysosomes is tightly regulated, and progressive acidification along this pathway is required to correctly sort and target the cargoes [16, 17]. We have shown that single cell fluorescence ratio image analysis (FRIA) can be used as an effective tool to determine the vesicular pH (pHv) of different cargo containing vesicles [18, 19]. FRIA is based on synchronized internalization of the receptor-pH-sensitive fluorescein-isothiocyanate (FITC) complex, which permits the tracking and measurement of luminal pHv to the specific endo-lysosomal compartment. For instance, cells are alternatively excited with 495 nm and 440 nm light and the emitted fluorescence is detected at 535 nm. Since only the FITC emission following

495 nm excitation is sensitive to pH, the fluorescence intensity ratio (495 nm/440 nm) is largely pH-dependent and independent of the dye concentration. Here, we summarize the developed FRIA technique to determine how ESCRT complexes regulate the sorting kinetics of integrin α5β1 receptor at the endo-lysosomal compartment. This method can further be used to address the effect of other regulatory proteins on integrin α5β1 receptor trafficking.

2 Materials

2.1 HeLa Cell Culture

1. HeLa cells are maintained in Dulbecco's Modified Eagle's Medium (DMEM) containing: 10% (v/v) Fetal Bovine Serum (FBS), 1 μg/ml puromycin for cells expressing shRNA for HD-PTP or UBAP1 (DMEM-FBS-puromycin medium), in a tissue culture incubator (37 °C, 5% CO_2). Store the medium at 4 °C. For cell seeding, 6-well culture plates and 25 mm glass coverslip with standard #1 or #1.5 thickness are used.

2.2 Labeling of Integrin α5β1 Receptor

1. Anti-α5 (CD49e) (BD Pharmingen) or α5β1-blocking antibody (Millipore) at 1:500 dilution equivalent to 10 μg/ml, prepared in PBS^{++}-BSA (*see* **step 3** and **Note 1**).
2. FITC-conjugated goat anti-mouse secondary fluorescent antibody (Fab) (Jackson ImmunoResearch Laboratories) at 1:1500 dilution, prepared in PBS^{++}-BSA (*see* **Note 1**).
3. PBS^{++} for washing: phosphate-buffered saline supplemented with 0.1 mM $CaCl_2$, 1 mM $MgCl_2$.
4. For labeling, prepare 0.5% (w/v) bovine serum albumin (BSA) in PBS^{++}. Store solutions at 4 °C.

2.3 Detecting the pH of Recycling Endosomes and Lysosomes

1. FITC-Transferrin (Tf) prepared as 5 μg/ml in serum-free DMEM.
2. Albumin Chicken (ovalbumin).
3. Holotransferrin prepared as 0.1 mg/ml in DMEM-FBS containing 1 mg/ml ovalbumin.
4. FITC-Dextran (M.W: 10 kDa anionic) prepared as 50 μg/ml in DMEM-FBS.
5. Dextran (M.W: 68.8 kDa) prepared as 100 μg/ml in DMEM-FBS.

Experiments are performed on inverted fluorescence microscope equipped for ratio imaging and programs for advanced image analysis (*see* **Note 2**).

2.4 Single or Multi-point In Situ pH Calibration

1. Nigericin and monensin 10 mg/ml in ethanol. Store at −20 °C (*see* **Note 3**).
2. K^+-rich buffer for in situ pH calibration: 10 mM NaCl, 135 mM KCl, 10 mM glucose, 1 mM $CaCl_2$, 0.1 mM $MgCl_2$, 20 mM MES for pH $\leq$ 5.5, 20 mM HEPES for pH > 5.5. Adjust pH with KOH and store at 4 °C.
3. NaKH solution: 140 mM NaCl, 5 mM KCl, 10 mM HEPES, 10 mM glucose, 1 mM $CaCl_2$, 0.1 mM $MgCl_2$. Adjust pH to 7.3 and store at 4 °C.
4. Low pH medium: 0.2 M acetic acid, 0.2 M NaCl, adjust pH to 2.5 with NaOH.
5. Recovery medium: 150 mM NaCl, 20 mM HEPES, 1 mM $CaCl_2$, 5 mM KCl, 0.1 mM $MgCl_2$, adjust pH to 7.4 with NaOH.
6. First, prepare 50 ml aliquots of K^+-rich medium and adjust the pH to 4.5, 5.0, 5.5, 6.0, 6.3, 6.6, 6.9, 7.2, and 7.4. Then add 10 μg/ml of nigericin and monensin (*see* **Note 3**). The medium pH should be adjusted with 1 M KOH at 25 °C and sterilized by filtration. Do not store buffers longer than 4 weeks at 4 °C and check the medium pH before usage at 25 °C.

3 Methods

3.1 Depleting HD-PTP or UBAP1 Using shRNA

HeLa cells were stably transfected with constitutive Mission short hairpin RNA (shRNA) lentiviral plasmids pLKO.1 to achieve depletion of ESCRT components HD-PTP or UBAP1. Scrambled sequence was used as a control. Cells were selected in the presence of 1 μg/ml puromycin. Depletion was verified using immunoblotting [13].

3.2 Preparation of Glass Cover Slips

1. Prepare the sterile glass cover slips in the tissue culture hood on 6-well plates.
2. Seed the cells on sterile glass cover slips at least 24 h before the experiment. Add 2 ml of DMEM-FBS-puromycin medium. Grow the cells to ~70% confluence.

3.3 Labeling Integrin α5β1 Receptor

1. Wash the cells gently twice with 2 ml of ice-cold PBS^{++}-BSA.
2. Incubate the cells with 1 ml primary anti-α5 (CD49e, 1:500 dilution) or α5β1-blocking antibody, prepared in PBS^{++}-BSA for 1 h on ice (Figs. 1 and 2a).
3. Wash the cells three times with 2 ml of PBS^{++}-BSA.
4. Incubate the cells with 1 ml FITC-conjugated goat anti-mouse Fab (1:1500 dilution), prepared in PBS^{++}-BSA for 1 h on ice.
5. Wash the cells three times with 2 ml of PBS^{++}-BSA.

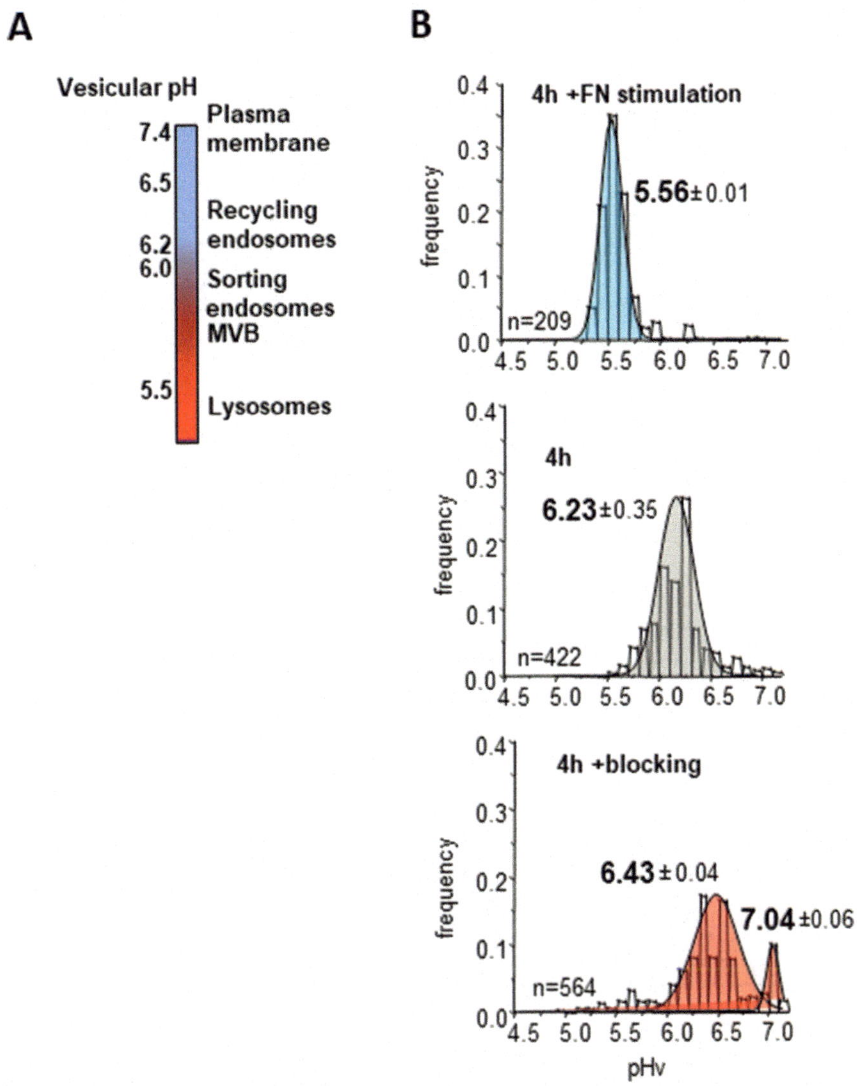

Fig. 1 Evaluation of integrin α5β1 receptor post-endocytic sorting using FRIA. (**a**) The endo-lysosomal pathway has a characteristic decreasing pH gradient reaching a pH 6.3–6.1 in early and recycling endosomes, pH 6–5.5 in late endosomes and MVB, and an acidic (<pH 5.5) environment in lysosomes. (**b**) Representative histograms of integrin α5β1 receptor endocytosis pathway after 4 h chase. The anti-integrin α5 antibody and FITC-conjugated Fab were bound to HeLa cell surface, after which the receptor was endocytosed. Ubiquitin-mediated α5β1 receptor trafficking was stimulated by its ligand FN (upper panel), without stimulation (middle panel) or alternatively blocked using α5β1-blocking antibody (lower panel). The pHv is measured by the ratio of emissions at 520 nm of excited FITC at 490 nm and 440 nm. A number of analyzed vesicles are indicated and data are expressed as frequency of pH_v of vesicle population. With permission from [13]

3.4 Endocytosis

1. Integrin α5β1 receptor is endocytosed by adding 2 ml of pre-warmed DMEM-FBS and incubated for the desired time at tissue culture incubator.

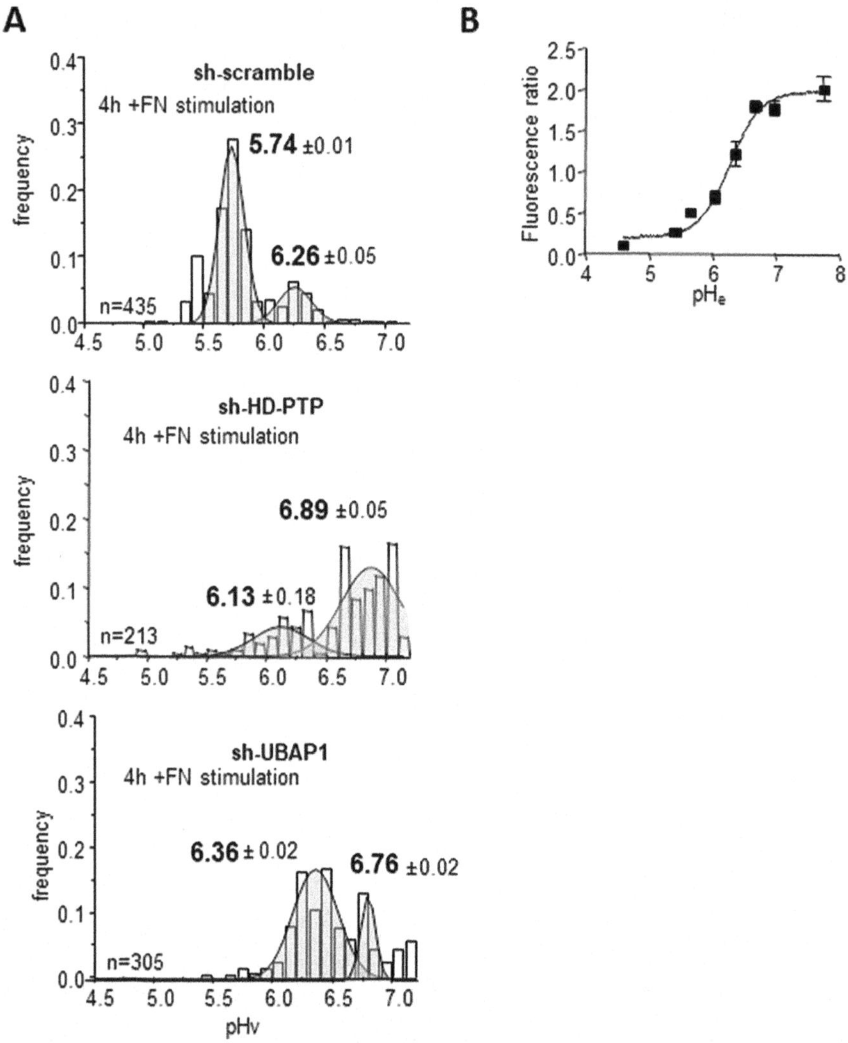

Fig. 2 Monitoring the effect of ESCRT complex component HD-PTP or UBAP1 depletion on post-endocytic sorting of integrin $\alpha5\beta1$ receptor. (**a**) Representative histograms of integrin $\alpha5\beta1$ receptor endocytosis pathway after 4 h chase. The anti-integrin $\alpha5$ antibody and FITC-conjugated Fab were bound to HeLa cell surface, after which the receptor was endocytosed in cells expressing sh-scrambled as a control or depleted for ESCRT-I components HD-PTP (sh-HD-PTP) or UBAP1 (sh-UBAP1). A number of analyzed vesicles are indicated and data are expressed as frequency of pH_v of vesicle population. (**b**) Multi-point calibration analyses for FRIA. In situ calibration curves of vesicular pH for HeLa cells expressing FITC-labeled integrin $\alpha5\beta1$ receptor. Distribution of fluorescence ratios of endosomal vesicles is presented as the ratio of emissions at 520 of excited FITC at 490/440 after clamping to indicate extracellular pH (pH_e). With permission from [13]

2. Endocytosis is attenuated by washing the cells with 2 ml of ice-cold PBS^{++} solution and placing the coverslip in the 35 mm chamber containing 2 ml of NaKH solution.

3.5 Acquisition of Fluorescent Ratio Images

1. Pre-warm the thermostated coverslip holder and microscope objective to 37 °C (*see* **Note 2**).
2. Acquire images at 495/440 nm excitation wavelengths by selecting at least 10–15 areas on the coverslip. This step should be performed in a timely manner to avoid significant cargo processing.

3.6 Single-Point In Situ pH Calibration

1. Perform the single-point calibration on each experimental coverslip after all images are acquired (*see* **Note 4**).
2. Aspirate the NaKH solution and add 0.8 ml of freshly prepared K+-rich medium containing 10 μg/ml nigericin and monensin (pH 6.5, 25 °C) to clamp the intracellular pH to 6.5.
3. Incubate till the fluorescence ratio reaches the steady state. This is monitored by acquiring images after 5 min incubation.
4. Record the fluorescence ratio. If this differs from the multi-point calibration value, the recalibration is needed (*see* Subheading 3.8).

3.7 Vesicular pH Analysis

1. Set up the MetaFluor program to obtain the average fluorescence intensity values of selected regions of interest (ROI) from images acquired at 495/440 nm excitation wavelengths and import them to Excel table. The average fluorescence intensity represents the integrated fluorescence intensity divided by the ROI area.
2. Select all the ROIs corresponding to the labeled vesicles in the first cell. Within the same cell, select three cytoplasmic ROIs free of labeled vesicles to calculate the background fluorescence. Select vesicles at least from three or more independent repeats (*see* **Note 5**).
3. Acquire the mean fluorescence intensity value of ROIs for all cells. The Excel file will contain the mean fluorescence intensity value for each ROI acquired at 495 and 440 nm excitation wavelengths and the background values for each cell.
4. Subtract the mean background fluorescence intensity acquired at both excitation wavelengths from the corresponding vesicle ROIs to obtain the specific fluorescent signal.
5. Calculate the mean fluorescence intensity ratio for each vesicle ROI by dividing the mean fluorescence intensity at 495 nm with 440 nm. Convert the fluorescence values to pHv using the multi-point calibration curve (*see* Subheading 3.8).
6. Frequency distributions of pHv values are plotted using the OriginPro software and single or multi-component Gaussian distribution. The mean pHv is calculated from the weighted mean of each Gaussian curve. Partitioning of cargo molecule into various organelles can be estimated based on the fractional distribution of vesicles in the individual peak relative to the total number of vesicles.

3.8 Multi-point In Situ pH Calibration

1. Prepare the cells (*see* Subheading 3.2–3.4).
2. Perform multi-point calibration using primary anti-α5 (CD49e) antibody and FITC-conjugated secondary Fab-labeled integrin α5β1 receptor as an endocytosed cargo. Oregon Green dextran, typically labeling lysosomes, can be used for the pHv values under 5.4.
3. Remove the plasma membrane-bound antibodies incubating the cells in 1 ml of ice-cold low pH medium for 2 min, wash twice with 1.5 ml ice-cold PBS^{++} and incubate the cells in 1 ml of cold recovery medium for 3 min. Repeat three times.
4. Clamp the luminal organelle pH using K + -rich buffers supplemented with 10 μg/ml of nigericin and monensin for in situ pH calibration with pH values ranging between 5.4 and 7.4 (*see* Subheading 3.6).
5. Acquire the fluorescence ratio images (*see* Subheading 3.5).
6. Perform the analysis steps (*see* Subheading 3.7) to acquire a multi-point calibration curve (Fig. 2b). The theoretical basis of this method has been previously described [20].

3.9 Measuring the pH of Recycling Endosomes and Lysosomes

The next steps describe the measurement of pHv in recycling endosomes and lysosomes using specific markers. This is done to verify integrin α5β1 localization and nonspecific effect of ESCRT depletion on membrane trafficking (*see* **Note 6**). Other specific organelle markers can be used such as markers for trans-Golgi network. Labeling recycling endosomes and lysosomes has been described in [21, 18, 19]. Transferrin is used, because its receptor is nearly 80% in recycling endosomes at the steady state. Uptake of fluid-phase marker dextran [22] is used to selectively label lysosomes. Colocalization of FITC-dextran loaded lysosomes with Lamp1 can validate the specificity of the dextran labeling [23].

3.9.1 Labeling Recycling Endosomes

1. Deplete endogenous Tf by incubating cells in 2 ml of serum-free DMEM for 45 min at 37 °C in a tissue culture incubator.
2. Add to the cells 1 ml of serum-free DMEM having 5 μg/ml FITC-Tf for 1 h at 37 °C in a tissue culture incubator.
3. Move the cells on ice and wash 4 times with 2 ml of DMEM-FBS containing 1 mg/ml ovalbumin and 0.1 mg/ml holotransferrin.
4. Perform the imaging and analysis steps (*see* Subheading 3.5–3.7) (Fig. 3).

3.9.2 Labeling Lysosomes

1. Incubate cells overnight with 2 ml of 50 μg/ml of FITC-dextran, prepared in DMEM-FBS, in a tissue culture incubator at 37 °C. Sterilize the dextran solution by filtration (0.22 μm disposable sterile filter unit) before use.

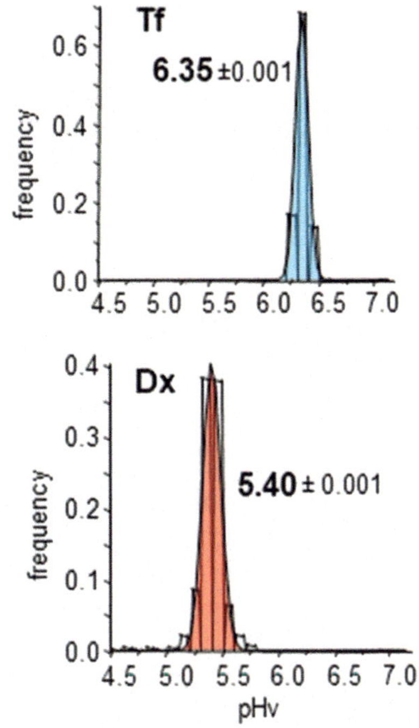

Fig. 3 Validation of pHv in recycling endosomes and in lysosomes. FITC-Transferrin (Tf) was used to monitor the pH of recycling endosomes and FITC-dextran (Dx) of lysosomes. Internalization and chase of markers is described in Subheading 3.9

2. Wash the cells three times with 2 ml of PBS^{++}.
3. Incubate the cells in 2 ml of DMEM-FBS supplemented with 100 μg/ml of dextran for 3 h in a tissue culture incubator at 37 °C to ensure that early and late endosomes were cleared of fluorophores.
4. Perform the imaging and analysis steps (*see* Subheading 3.5–3.7) (Fig. 3).

3.10 Calculations for In Situ Calibration

Calculation of the intracellular vesicle pH is based on acquired fluorescence ratio values established in the in situ multi-point calibration.

1. The fluorescence ratios of intracellular vesicles clamped to pH 4.5–7.4 are fitted using nonlinear regression analysis according to a modified version of the Henderson-Hasselbalch equation, as described originally in [20]:

$$\mathrm{pH_i} = \mathrm{pKa} - \log_{10}(R_{\max} - R_{\mathrm{i}}/R_{\mathrm{i}} - R_{\min})$$

2. R_i is the background subtracted fluorescence ratio (495 nm/ 440 nm); R_{max} and R_{min} are the asymptotes of minimum and maximum ratio values at highly acidic and alkaline pH,

respectively. The apparent dissociation constant of FITC is indicated by pKa, and pHi is the intracellular pH. Rearranging the equation for defining R yields [24]:

$$R_{\mathrm{i}} = R_{\min} + R_{\max} 10^{(\mathrm{pHi}-\mathrm{pKa})}/10^{(\mathrm{pHi}-\mathrm{pKa})} + 1$$

3. The R_{max} and R_{min} values were iterated using the Prism 6® (GraphPad software) software from measured ratios. The pHv of integrin α5β1 containing vesicles was calculated from extrapolated constants and the calculated R value using Excel.

4 Notes

1. Prepare all the solutions using ultrapure water and analytic grade reagents. Store the fluorescent probe in freezer at −20 °C and avoid light exposure. Endogenous integrin α5β1 receptor is labeled with antibody recognizing extracellular epitope.
2. Used instrumentation: a Zeiss Observer Z1 (Carl Zeiss Micro-Imaging) equipped with the use of X-Cite 120Q system (Lumen Dynamics Group) illumination system, an oil immersion objective (63× NA 1.4). Photometrics Evolve Delta EMCCD imaging camera recorded the fluorescence emission (535 ± 25) evoked by excitation through 495 ± 10/440 ± 20 band-pass filters housed by Lambda 10-2 (Sutter Instrument) filter wheel. The temperature 37 °C is maintained by a thermostated coverslip holder. Data acquisition and analysis was done with the use of MetaFluor® (Molecular Devices) and OriginPro® (OriginLab®) with a PC-based computer.
3. Luminal pH of intracellular compartments is clamped to the extracellular pH by using ionophores monensin and nigericin that mediate H^+/Na^+ and H^+/K^+ exchange.
4. The microscope optical characteristics and/or the experimental conditions can easily change during measurement or between the experimental repeats. The single-point calibration is done to ensure that the conditions match ones with the multi-point calibration (*see* Subheading 3.8).
5. Identify as many vesicles as possible acquiring 10–15 fields at the coverslip. One field typically contains five or more Hela cells.
6. Use of lysosomal cargoes and recycling markers is recommended to monitor the nonspecific effect of ESCRT depletion on membrane trafficking.

References

1. Schmidt O, Teis D (2012) The ESCRT machinery. Curr Biol 22(4):R116–R120
2. Scita G, Di Fiore PP (2010) The endocytic matrix. Nature 463(7280):464–473
3. Razi M, Futter C (2006) Distinct roles for Tsg101 and Hrs in multivesicular body formation and inward vesiculation. Mol Biol Cell 17 (8):3469–3483
4. Raiborg C, Bache KG, Mehlum A et al (2001) Hrs recruits clathrin to early endosomes. EMBO J 20(17):5008–5021
5. Verma R, Marchese A (2015) The endosomal sorting complex required for transport pathway mediates chemokine receptor CXCR4-promoted lysosomal degradation of the mammalian target of rapamycin antagonist DEPTOR. J Biol Chem 290(11):6810–6824
6. Giordano F, Simoes S, Raposo G (2011) The ocular albinism type 1 (OA1) GPCR is ubiquitinated and its traffic requires endosomal sorting complex responsible for transport (ESCRT) function. Proc Natl Acad Sci 108 (29):11906–11911
7. Le Borgne R (2006) Regulation of Notch signalling by endocytosis and endosomal sorting. Curr Opin Cell Biol 18(2):213–222
8. Baron M (2012) Endocytic routes to Notch activation. Semin Cell Dev Biol 23(4):437–442
9. Moberg KH, Schelble S, Burdick SK (2005) Mutations in erupted, the Drosophila ortholog of mammalian tumor susceptibility gene 101, elicit non-cell-autonomous overgrowth. Dev Cell 9(5):699–710
10. Rodahl LM, Haglund K, Sem-Jacobsen C et al (2009) Disruption of Vps4 and JNK function in Drosophila causes tumour growth. PLoS One 4(2):e4354
11. Malerod L, Stuffers S, Brech A et al (2007) Vps22/EAP30 in ESCRT-II mediates endosomal sorting of growth factor and chemokine receptors destined for lysosomal degradation. Traffic 8(11):1617–1629
12. Lobert VH, Brech A, Pedersen NM et al (2010) Ubiquitination of α5β1 integrin controls fibroblast migration through lysosomal degradation of fibronectin-integrin complexes. Dev Cell 19(1):148–159
13. Kharitidi D, Apaja PM, Manteghi S et al (2015) Interplay of endosomal pH and ligand occupancy in integrin α5β1 ubiquitination, endocytic sorting, and cell migration. Cell Rep 13(3):599–609
14. Ali N, Zhang L, Taylor S et al (2013) Recruitment of UBPY and ESCRT exchange drive HD-PTP-dependent sorting of EGFR to the MVB. Curr Biol 23(6):453–461
15. Parkinson MD, Piper SC, Bright NA et al (2015) A non-canonical ESCRT pathway, including histidine domain phosphotyrosine phosphatase (HD-PTP), is used for down-regulation of virally ubiquitinated MHC class I. Biochem J 471(1):79–88
16. Mellman I, Fuchs R, Helenius A (1986) Acidification of the endocytic and exocytic pathways. Annu Rev Biochem 55(1):663–700
17. Mukherjee S, Ghosh RN, Maxfield FR (1997) Endocytosis. Physiol Rev 77(3):759–803
18. Apaja PM, Xu H, Lukacs GL (2010) Quality control for unfolded proteins at the plasma membrane. J Cell Biol 191(3):553–570
19. Barrière H, Apaja P, Okiyoneda T et al (2011) Endocytic sorting of CFTR variants monitored by single-cell fluorescence ratiometric image analysis (FRIA) in living cells. Methods Mol Biol 741:301–317
20. Geisow MJ (1984) Fluorescein conjugates as indicators of subcellular pH: a critical evaluation. Exp Cell Res 150(1):29–35
21. Barriere H, Lukacs GL (2008) Analysis of endocytic trafficking by single-cell fluorescence ratio imaging. Curr Protoc Cell Biol Chapter 15:Unit 15.13
22. Lencer WI, Weyer P, Verkman AS et al (1990) FITC-dextran as a probe for endosome function and localization in kidney. Am J Phys Cell Phys 258(2):C309–CC17
23. Kumar KG, Barriere H, Carbone CJ et al (2007) Site-specific ubiquitination exposes a linear motif to promote interferon-alpha receptor endocytosis. J Cell Biol 179 (5):935–950
24. Erdahl WL, Chapman CJ, Taylor RW et al (1995) Effects of pH conditions on Ca2+ transport catalyzed by ionophores A23187, 4-BrA23187, and ionomycin suggest problems with common applications of these compounds in biological systems. Biophys J 69 (6):2350–2363

Chapter 8

Genetic and Biochemical Analyses of Yeast ESCRT

Sudeep Banjade, Shaogeng Tang, and Scott D. Emr

Abstract

Budding yeast *Saccharomyces cerevisiae* is an ideal model organism to study membrane trafficking pathways. The ESCRT (endosomal sorting complexes required for transport) pathway was first identified in this organism. Upon recognition of endocytosed ubiquitinated membrane proteins at endosomes, ESCRTs assemble at these organelles to catalyze the biogenesis of multivesicular bodies (MVBs). Formation of MVBs leads to the trafficking of these membrane proteins to vacuoles for degradation. Here, we describe genetic and biochemical approaches to study ESCRT function. We outline in vivo endocytosis assays using two model cargoes in *Saccharomyces cerevisiae* and also describe an in vitro approach to analyze ESCRT-III polymerization on lipid monolayers.

Key words ESCRT, Endocytosis, Multivesicular bodies (MVB), ESCRT-III polymerization, Snf7 spiral, Lipid monolayer, *Saccharomyces cerevisiae*

1 Introduction

The endosomal sorting complexes required for transport (ESCRT) pathway is critical for cell-surface receptor downregulation at endosomes by the creation of multivesicular bodies (MVBs) [1–3]. Since the initial discovery in MVB biogenesis, ESCRTs have been implicated in many other biological functions [1]. The ESCRT pathway consists of five distinct complexes, ESCRT-0, ESCRT-I, ESCRT-II, ESCRT-III, and the AAA-ATPase Vps4. ESCRT-0, -I, and -II recognize ubiquitinated transmembrane proteins at endosomes and recruit ESCRT-III. ESCRT-III forms spiraling polymers to sequester these transmembrane proteins. Together with Vps4, ESCRT-III induces membrane invagination and vesicle formation to create MVBs.

Saccharomyces cerevisiae is an ideal model organism for the study of ESCRTs in MVB biogenesis. Here, we describe methods for the study of ESCRTs' function in this model organism through analyses of two well-characterized endocytic cargoes, Mup1 and Can1. We also describe an in vitro strategy to analyze ESCRT-III

Emmanuel Culetto and Renaud Legouis (eds.), *The ESCRT Complexes: Methods and Protocols*, Methods in Molecular Biology, vol. 1998, https://doi.org/10.1007/978-1-4939-9492-2_8, © Springer Science+Business Media, LLC, part of Springer Nature 2019

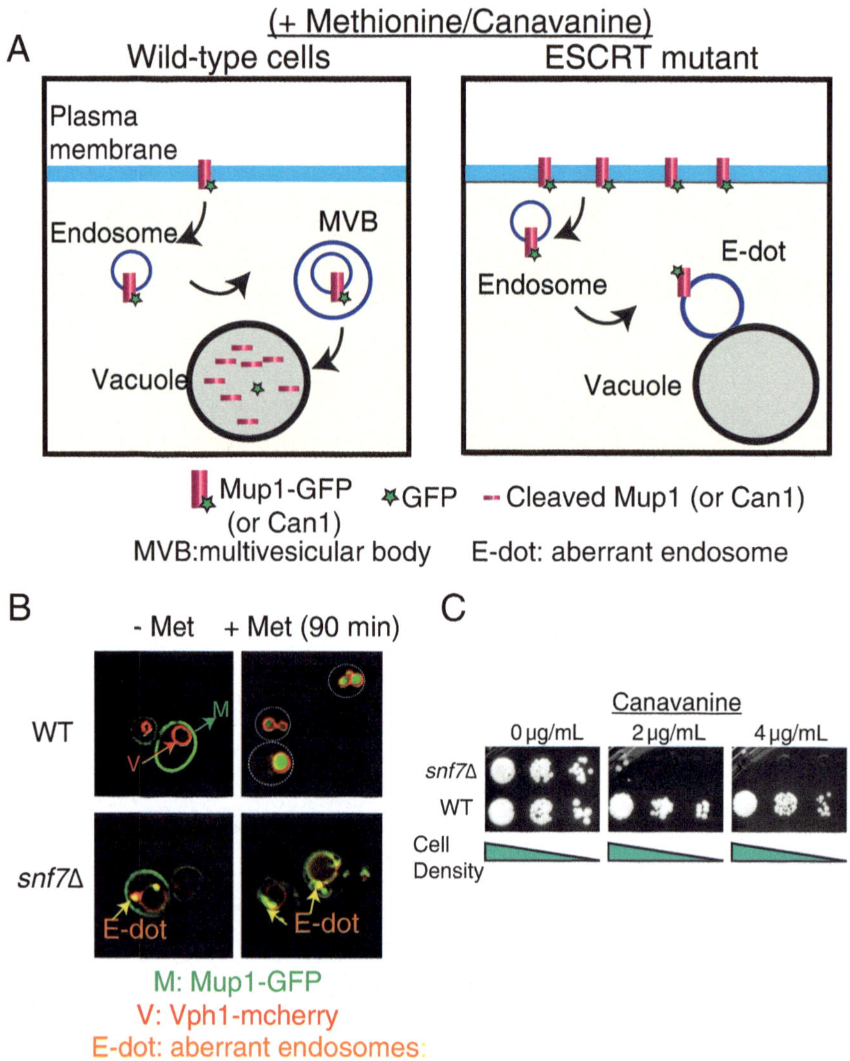

Fig. 1 (**a**) Model of endocytosis of plasma membrane proteins Mup1 and Can1 upon environmental stimuli. In ESCRT mutants, trafficking of plasma membrane proteins is trapped at the aberrant endosomes (E-dots). (**b**) Mup1-GFP trafficking experiments upon methionine (20 μg/mL) stimulation in wild-type and ESCRT mutant cells. The vacuolar membrane protein Vph1 is C-terminally tagged with mCherry to label vacuolar membranes. (**c**) Canavanine-sensitivity assay at day 5

(Snf7) polymerization. A combination of these in vivo and in vitro approaches is useful in delineating mechanisms of ESCRT assembly.

1.1 Analyze ESCRT Function in Endocytosis

Mup1 is a plasma membrane methionine permease, which undergoes endocytosis upon the addition of extracellular methionine (Fig. 1) [4, 5]. To follow its endocytic trafficking, Mup1 is C-terminally tagged with GFP (or the pH-sensitive GFP variant, pHluorin, [6]). In wild-type cells, Mup1-GFP is readily trafficked to the vacuole for degradation. However, in ESCRT mutants, Mup1-GFP is primarily trapped at aberrant endosomal compartments (Fig. 1).

Can1 is a plasma membrane arginine permease [7]. It can also transport canavanine, a toxic arginine analog. In the presence of extracellular canavanine, wild-type cells effectively endocytose Can1. However, ESCRT mutants that have Can1 trafficking defects accumulate more Can1 at the cell surface, and therefore, exhibit canavanine-sensitivity. This property enables us to dissect ESCRT function by varying extracellular canavanine concentration.

These assays in yeast provide in vivo relevance to hypotheses generated through structural and biochemical analyses of ESCRT members. For example, the crystal structure of Snf7 provided important clues on the Snf7:Snf7 interface during its polymerization [8]. The relevance of these interfaces was validated using these aforementioned in vivo assays. Furthermore, they have also been useful genetic selection tools to obtain mutant ESCRTs that are hyperactive. For example, using canavanine as a drug to select for mutants of the ESCRT-III protein Snf7 that bypass the nucleator Vps20, we identified several point mutations in Snf7 that shift the equilibrium of Snf7 to a more open conformation. These activating mutants of Snf7 also lower the critical concentration for polymerization [9, 10]. Similar selection approaches could be used to obtain suppressors of other family members, allowing us to generate new hypotheses on ESCRT properties and function.

1.2 In Vitro Approach to Study ESCRT-III (Snf7) Polymerization

We also describe the methods for the purification and polymerization assay of one of the core ESCRT-III subunits: Snf7. Snf7 has an intrinsic ability to form spiraling polymers on membrane surfaces. ESCRT-III spirals are predicted to be critical for ESCRT-mediated vesicle budding reactions during MVB biogenesis [8–11]. Membrane deformation and vesicle budding are probably direct consequences of the energy released by the dynamic reorganization of ESCRT-III polymerization. Since ESCRT function in MVB pathway is dependent upon the ability of ESCRT-III proteins to polymerize, understanding the mechanisms of ESCRT-III polymerization and how the polymerization is controlled remains a fundamental question in ESCRT biology.

Visual analyses of polymers through electron microscopy remain one of the best approaches to study ESCRT-III polymerization. Snf7 from *S. cerevisiae* expresses well in *Escherichia coli*, which can then be purified to homogeneity and studied for assembly on model membranes. Here, we describe an approach to form Snf7 polymers on lipid monolayers. While we describe this approach for Snf7 assembly, the same approach also applies for analyses of other ESCRT-III proteins. This in vitro method has been used to confirm Snf7 polymerization interfaces predicted by its crystal structure and to understand the roles of other ESCRT proteins (ESCRT-II complex, Vps20, Bro1, Vps24, Vps2) in nucleating or modifying the mesoscale structural reorganization of Snf7 polymers [8–10].

2 Materials

2.1 Endocytosis Assays

1. *Saccharomyces cerevisiae* strains expressing Mup1-GFP. The strains we use are derivatives of SEY6210 (*MATα leu2-3,112 ura3-52 his3-Δ200 trp1-Δ901 suc2-Δ9 lys2-801; GAL*).
2. Milli-Q water (deionized distilled water from Millipore).
3. YPD broth: 1%(w/v) yeast extract, 2%(w/v) bacto peptone made in Milli-Q water and autoclaved (40 min for 1 L). To this mix, sterile dextrose (made at 50% w/v and autoclaved) is added to a final concentration of 2% (w/v).
4. YPD plates: 1%(w/v) yeast extract, 2%(w/v) bacto peptone dissolved in Milli-Q water. Then agar is added to 2% (w/v). This mixture is autoclaved for 40 min (for 1 L), and then sterile dextrose (made at 50% w/v and autoclaved) is added to a final concentration of 2% (w/v). After the solution cools down to ~65 °C, 20–25 mL is poured to make one plate.
5. 1× YNB medium: 0.67% (w/v) yeast nitrogen base without amino acids, 2% (w/v) dextrose autoclaved (25 min for 1 L).
6. SCD (synthetic complete dextrose) plates: 0.67% (w/v) yeast nitrogen base without amino acids, 21.2 μg/mL adenine, 84.8 μg/mL histidine, 169.7 μg/mL leucine, 84.8 μg/mL lysine, 84.8 μg/mL methionine, 84.8 μg/mL tryptophan, 84.8 μg/mL tyrosine, 84.8 μg/mL uracil, and 2% (w/v) agar. This mixture is autoclaved for 25 min (1 L). Dextrose (50% w/v stock autoclaved) is added to 2%. After the solution cools down to ~65 °C, 20–25 mL is poured to make each plate.
7. SCD medium: 1× of YNB medium with 20 μg/mL adenine, 20 μg/mL arginine, 230 μg/mL lysine, 300 μg/mL threonine, 30 μg/mL tyrosine, 20 μg/mL histidine, 60 μg/mL leucine, 20 μg/mL tryptophan, and 20 μg/mL of uracil. The dropout mixtures are made separately—adenine, arginine, lysine, threonine, and tyrosine as a single 20× stock, dissolved in water, and sterile-filtered with a 0.22 μm filter. Histidine, leucine, tryptophan, and uracil are made individually as a 100× stock, dissolved in water, and sterile-filtered with a 0.22 μm filter.
8. Amino acid dropout plates: made similarly as SCD plates (*see* **item 6**), but excluding a certain amino acid. For example, to make -Leu plates, leucine is omitted from the SCD plate mixture in **item 5**. Since we use auxotrophic strains (*see* **item 1**), these dropout plates and media can be used to select for yeast transformants if using plasmids for the assays described here.
9. Methionine (20 mg/mL) stock solution made in water and sterile-filtered through a 0.22 μm filter.

10. L-Canavanine sulfate (5–10 mg/mL) stock solution made in water and sterile-filtered through a 0.22 μm filter.
11. SCD (synthetic complete dextrose) plate with canavanine: made similarly as SCD plates (*see* **item 6**), adding L-canavanine to 0.5 μg/mL–4.0 μg/mL when the autoclaved solution reaches ~65 °C.
12. Spot-plating replicators.
13. 96-well non-treated flat-bottomed plate.
14. Boiling buffer: 8 M Urea, 2% (w/v) SDS, 50 mM Tris-HCl pH 7.5, 1 mM EDTA.
15. Sample buffer: 7 M Urea, 10% (w/v) SDS, 24% (v/v) glycerol, 150 mM Tris-HCl pH 6.8, 1.4 M β-ME.
16. 0.425–0.600 mm acid-washed glass beads.
17. Torrey Pines Rabbit α-GFP antibody, mouse α-3-phosphoglycerate kinase antibody, mouse α-GFP antibody, rabbit anti–glucose-6-phosphate dehydrogenase (*see* **Note 1**).
18. Wide-field or confocal fluorescence microscope. We use a Deltavision RT system with the following filter sets: FITC (fluorescein isothiocyanate) for GFP and RD-TR-Cy3 for mCherry and take images with a 100× objective.

2.2 Snf7 Purification

1. *E. coli* strain Rosetta.
2. pET23d (+) vector, including a hexahistidine-*snf7*^{R52E} (or another ESCRT-III) construct.
3. 30 mg/mL chloramphenicol (1000×) made in ethanol and stored at −20 °C.
4. 50 mg/mL ampicillin (1000×), made in water, sterile-filtered through a 0.22 μm filter, and stored at −20 °C.
5. Luria-Bertani (LB) medium: 1% (w/v) tryptone, 1% (w/v) NaCl, 0.5% (w/v) yeast extract, pH adjusted to 7.0 with NaOH, autoclaved (20 min for 1 L). However, generally use mixed, commercially available LB powder.
6. LB-agar plates with 50 μg/mL ampicillin and overlaid with 30 μg/mL chloramphenicol.
7. Stock solutions of 1 M IPTG (2000×), dissolved in water, sterile-filtered through a 0.22 μm filter, and stored at −20 °C.
8. Protease inhibitors: Roche complete EDTA-free protease inhibitor cocktail, 200 mM PMSF in ethanol.
9. Lysis buffer: 25 mM HEPES pH 7.5, 500 mM NaCl, 20 mM Imidazole pH 8, 5 mM βME.
10. Lysis buffer with protease inhibitors: 25 mM HEPES pH 7.5, 500 mM NaCl, 20 mM Imidazole pH 8, 5 mM βME, 1×

Roche complete EDTA-free protease inhibitor cocktail, 1 mM PMSF.

11. Wash buffer: 25 mM HEPES-NaOH pH 7.5, 150 mM NaCl, 40 mM Imidazole pH 8, 5 mM βME.
12. Elution buffer: 25 mM HEPES-NaOH pH 7.5, 150 mM NaCl, 300 mM Imidazole pH 8, 5 mM βME.
13. Buffer QA: 25 mM HEPES-NaOH pH 7.5, 150 mM NaCl, 1 mM βME.
14. Buffer QB: 25 mM HEPES-NaOH pH 7.5, 500 mM NaCl, 1 mM βME.
15. EM buffer: 20 mM HEPES-NaOH pH 7.5, 150 mM NaCl.
16. Tip sonicator: Branson Sonifier 250.
17. Econo-Column columns.
18. Talon metal affinity (Co^{2+}) resin.
19. Hi-Trap Q HP anion exchange column (GE).
20. Superdex 200 Increase 10/300 GL gel-filtration column (GE).
21. Centrifuges: Sorvall RC6+, Sorvall RC-3B or similar.
22. FPLC (fast protein liquid chromatography) system from GE (AKTA FPLC), or similar.

2.3 Snf7 Polymerization Assay on Lipid Monolayers

1. Teflon-based lipid monolayer apparatus (home-made).
2. Chloroform.
3. Methanol.
4. Lipids: **POPC** (1-palmitoyl-2-oleoyl-*sn*-glycero-3-phosphocholine), **POPS** (1-palmitoyl-2-oleoyl-*sn*-glycero-3-phospho-L-serine), **PI3P** (1,2-dioleoyl-*sn*-glycero-3-phospho-(1′-myo-inositol-3′-phosphate)).
5. Glass syringes to handle lipids in organic solvents.
6. 2% (w/v) ammonium molybdate for negative staining, made in Milli-Q water, filtered through a 0.22 μm filter, and stored at room temperature.
7. Formvar-coated electron microscopy grids from Electron Microscopy Sciences.
8. Transmission electron microscope (TEM). (Our images are taken in an FEI Morgnani TEM at 80–100 kV).

3 Methods

3.1 Cell-Biological Analysis of ESCRT Function Using Model Cargoes Mup1 and Can1

3.1.1 Mup1-GFP Trafficking Assay

1. Streak out the appropriate *Saccharomyces cerevisiae* strains onto YPD or amino acid dropout (for plasmid selection) agar plates.
2. Inoculate single colonies into 5 mL of SCD or amino acid dropout cultures and grow overnight at 26 °C in a culture incubator-shaker.
3. Next day, cut back the cultures to an optical density ($OD_{600\ nm}$) of 0.005 and let them grow over 16–18 h (*see* **Note 2**).
4. Collect 5 optical density volumes (ODV) of sample for western blot analysis once the OD reaches a value of 0.5. This will be the time-zero sample to study methionine-induced Mup1 degradation. For microscopic analysis, collect 0.5 ODV of sample (*see* **Note 3**).
5. Add methionine to a final concentration of 20 μg/mL to the rest of the culture. Collect 5 ODV of cultures every 15–20 min for an h. For microscopic analysis, collect 0.5 ODV of sample.
6. Centrifuge the collected cultures at 3000 × *g* for 3 min. Resuspend in 1 mL cold distilled H_2O. For microscopic analysis, resuspend 0.5 ODs of cultures in 25 μL of YNB (or H_2O) and proceed immediately to image using a wide-field fluorescence microscope (*see* **Note 4**).
7. Centrifuge 2 min at 13,000 × *g*. Then add trichloroacetic acid (TCA) on ice to 10% (v/v).
8. Incubate on ice for 30 min. Then centrifuge 3 min at 13,000 × *g*.
9. Resuspend pellet in ice-cold acetone. Bath-sonicate at room temperature for ~1 min until the pellet comes into solution. Spin down the solution for 3 min at 13,000 × *g* and aspirate the acetone off.
10. Repeat acetone wash (*see* **step 9**).
11. Allow the pellet to dry (either on the bench top or using a speed-vac).
12. Resuspend the pellet in 100 μL of boiling buffer. Add glass beads to a volume corresponding to ~100 μL.
13. Incubate at 42 °C for 5 min. Then vortex for 5 min.
14. Add 100 μL of sample buffer. Spin down the solution for 2 min at 13,000 × *g*.
15. Load 6 μL on an 11% SDS-PAGE gel and proceed to perform western blot analysis (*see* **Note 1**).

3.1.2 Can1 Endocytosis Spot-Plating Assay

1. Collect mid-log cells (OD_{600} between 0.4 and 0.8) for spot-plating.
2. In a 96-well plate, make 200 μL of 6–8 tenfold serial dilutions in YNB medium, starting with an OD of 0.1–0.3 as the first dilution.
3. Incubate spot-plating replicators with the cultures and swirl the replicator head in the cultures for a few seconds.
4. Place the replicator on SCD agar plates without canavanine (or on amino acid dropout plates if working with plasmids) and hold for about 5 s, gently pressing the replicator on the plates (*see* **Note 5**).
5. Incubate the replicator again with the cultures, swirling gently. Repeat procedures 4–5 in plates containing different concentrations of canavanine (*see* **Note 6**).
6. Incubate at 26 °C for 3–7 days. Image at Day 3, Day 5, and Day 7 (*see* **Note** 7).

3.2 Purification of Recombinant Snf7

1. Transform pET23d (+) $snf7^{R52E}$ into Rosetta *E. coli* strain, and plate onto a LB-ampicillin-chloramphenicol plate. The R52E mutation is an activating mutation on Snf7, which allows the protein to form polymers readily in in vitro assays [9] (*see* **Note 8**).
2. Next day, pick a single colony into a 10 mL LB medium with 50 μg/mL ampicillin and with 30 μg/mL chloramphenicol. Let the culture grow overnight (~16 h) at 37 °C.
3. Use the overnight culture to inoculate a 1 L LB medium with ampicillin and chloramphenicol, to an OD_{600} of 0.05.
4. After the $OD_{600\ nm}$ reaches 0.8, induce with 1 mM IPTG. Let the culture grow for 4 h at 37 °C.
5. Spin the cells at 4000 × *g* for 25 min. Resuspend the pellets in ice-cold lysis buffer with protease inhibitors (Roche complete at 1× and PMSF at 1 mM).
6. Lyse the cells by tip sonication. In a Branson Sonifier 250, we use an amplitude of 50%, duty cycle of 40%, for 1 min, three times with an interval of 1 min between each of the three runs. Sonication must be performed on ice. Duty cycle of 40% (0.4 s on, 0.4 s off) allows reduction in the generation of bubbles or froth.
7. Clear the lysate by centrifugation, at 20,000 × *g* for 40 min.
8. Equilibrate 1 mL Talon metal affinity (Co^{2+}) resin (resin volume) with lysis buffer.
9. Apply the cleared lysate (supernatant) to the equilibrated resin and incubate by rotating at 4 °C for 30 min to 1 h.

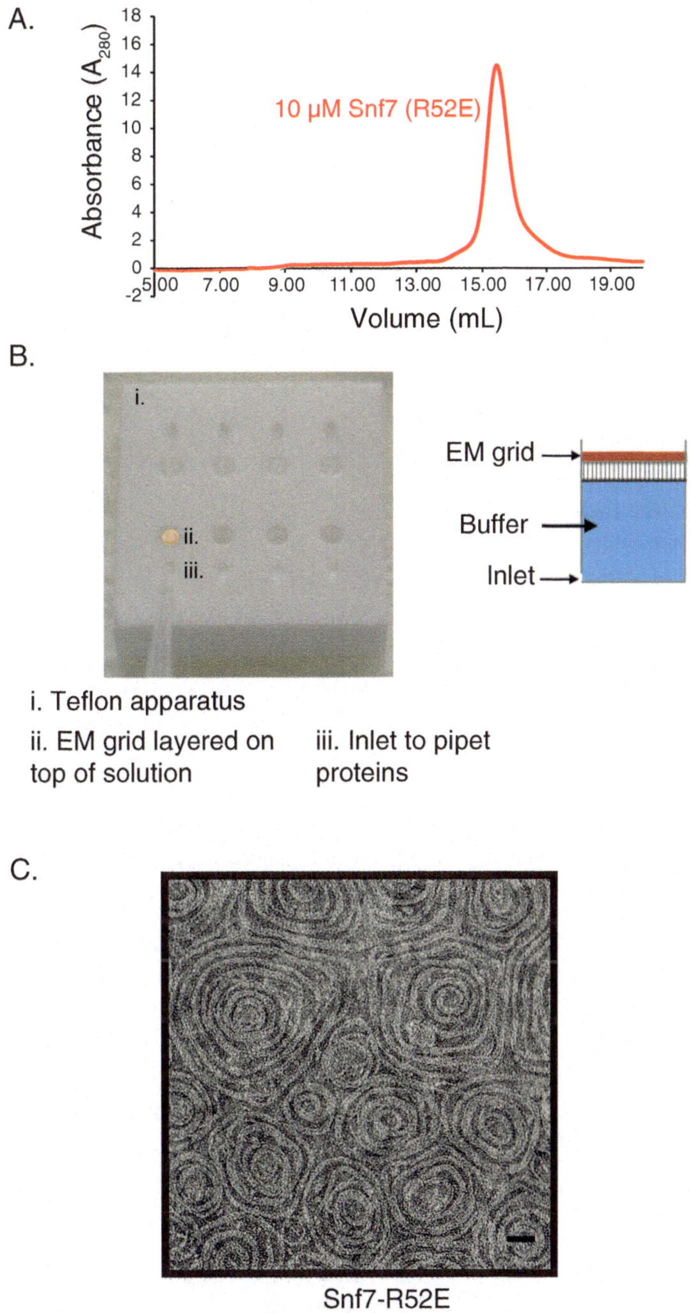

Fig. 2 (**a**) Size-exclusion chromatogram of His6-Snf7R52E protein. (**b**) Lipid monolayer setup with home-made Teflon apparatus. (**c**) Electron microscopy image of His6-Snf7R52E polymers formed at 500 nM. Scale bar, 50 nm

10. Pass through an Econo-column column and wash the resin with 10 column volumes (CV) of lysis buffer without protease inhibitors, 5 CV of wash buffer, and then again with 5 CV of lysis buffer. Elute with 3 CV of elution buffer.
11. Apply the protein through a Hi-Trap anion exchange column equilibrated in buffer QA. Wash with 2 column volumes of QA (*see* **Notes 9–11**).
12. Collect eluates in 1 mL volumes with buffer QB.
13. After running an SDS-PAGE to determine the eluate contents, run through a Superdex 200 increase gel-filtration column in buffer QB, and collect peak fractions (Fig. 2a).

3.3 Electron Microscopy Assay to Analyze ESCRT-III Polymerization on Lipid Monolayers

1. Create a lipid mixture of 60:30:10 (molar ratios) of POPC: POPS: PI3P in 95:5 CCl_4: CH_3OH at a total lipid concentration of 1.5 μg/μL.
2. Add 50 μL of buffer (150 mM NaCl, 20 mM HEPES pH 7.5) in the wells of a Teflon apparatus (home-made, Fig. 2b). Take out 25 μL from each well to generate a concave liquid surface in the wells.
3. Add 1.0 μL of lipid mixture to the wells at the surface of the aqueous buffer. Incubate for 1 h to allow monolayers to form. Perform in a moist environment to keep the solution from drying out.
4. Add Formvar-coated EM grids to the monolayers. Simultaneously, add your protein of interest through the pipetting inlet (Fig. 2b) and incubate for 1 h (*see* **Notes 12** and **13**).
5. At the end of the incubation, add ~20 μL EM buffer through the inlet (Fig. 2b) to raise the grid out of the wells. Invert the grids and blot with filter paper. Stain with 10 μL 2% filtered Ammonium Molybdate solution for 30 s. Blot again with filter paper and let the grids dry before proceeding to visualize using a transmission electron microscope (Fig. 2c).

4 Notes

1. If using pHluorin-conjugated Mup1 to analyze Mup1 degradation by immunoblot, using the rabbit-antibody from Torrey Pines is recommended.
2. In Mup1-endocytosis assays, letting the cultures grow in mid-log over 16–18 h is important to limit stress-induced Mup1 trafficking [12]. Cutting back to a particular OD600 (0.005) is determined empirically depending upon the specific strain. We cut back to an OD, such that in 16 or 18 h the OD will be around 0.5.

3. We also routinely use a flow-cytometry-based assay to study endocytosis of Mup1, which has been described before [13]. Mup1 tagged with a pH-sensitive GFP analog (pHluorin) is quenched upon sorting into the lumen of the vacuole. In ESCRT mutants, fluorescence quenching is avoided as the cargo stays in the cytoplasm. This property allows us to quantify the degree of quenching between wild-type and ESCRT mutant strains.
4. During preparation of samples for imaging assays, resuspension of pelleted cells in water allows for formation of larger, single vacuoles. Resuspending cells in SCD creates multilobed vacuoles. Mup1 endocytosis assays should be performed immediately after centrifugation as higher density of cells causes stress-induced Mup1 endocytosis.
5. For canavanine assays, arginine is excluded in the medium as Can1 is an arginine transporter, which undergoes arginine-induced internalization.
6. In the spot-plating assay, before spotting on to canavanine plates, ensuring that all the replicator pin-heads have visually equal volumes of cultures picked up from the wells will eliminate unequal spots on the plates.
7. Depending upon the specific mutant, the canavanine concentration to be used should be varied for canavanine-sensitivity assays. For example, *snf7*Δ is more sensitive to canavanine than *vps24*Δ or *vps2*Δ [14]. We routinely use canavanine concentrations from 0.5 to 6 μg/mL.
8. We use Rosetta (Novagen) *E. coli* strains for protein purification. This strain provides tRNAs for rare codons through a chloramphenicol-resistant plasmid. Therefore, it is important to select for chloramphenicol-resistant colonies in addition to selecting for the plasmid through which the ESCRT-III protein is expressed.
9. Because of the presence of basic surfaces in ESCRT-III proteins, they tend to co-purify nucleic acid contaminants. Using anion exchange columns is essential to get rid of these contaminants.
10. If possible, avoid concentrating the proteins as ESCRT-III proteins variably precipitate upon concentration. Also, limit the time the protein stays in lower concentration (150 mM) of NaCl.
11. Because ESCRT-III polymerization is salt-sensitive, we purify these proteins at high salt (500 mM NaCl) and perform the polymerization assays at 150 mM NaCl.
12. In the in vitro polymerization assay, time of incubation of proteins on monolayers will need to be varied under different

conditions. For example, to generate dense spirals of Snf7, incubation for 10 min at 1 μM will be sufficient.

13. To generate 3D helical structures with Vps24, Vps2, and Snf7, incubation for up to 1-h is necessary at similar concentrations. We also preincubate Snf7, Vps24, and Vps2 for 30 min at room temperature before adding the proteins to the monolayer solution.

Acknowledgments

Work in the Emr lab is supported by a Cornell University Research Grant CU3704. Sudeep Banjade is a HHMI fellow of the Damon Runyon Cancer Research Foundation (DRG-2273-16). Shaogeng Tang is a Merck fellow of the Damon Runyon Cancer Research Foundation (DRG-2301-17). We thank all members of the Emr lab for building these protocols in the lab over the years.

References

1. Henne WM, Buchkovich NJ, Emr SD (2011) The ESCRT pathway. Dev Cell 21:77–91
2. Christ L, Raiborg C, Wenzel EM, Campsteijn C, Stenmark H (2017) Cellular functions and molecular mechanisms of the ESCRT membrane-scission machinery. Trends Cell Biol 27:1–11
3. Schöneberg J, Lee IH, Iwasa JH, Hurley JH (2017) Reverse-topology membrane scission by the ESCRT proteins. Nat Rev Mol Cell Biol 18:5–17
4. Menant A, Barbey R, Thomas D (2006) Substrate-mediated remodeling of methionine transport by multiple ubiquitin-dependent mechanisms in yeast cells. EMBO J 25:4436–4447
5. Guiney EL, Klecker T, Emr SD (2016) Identification of the endocytic sorting signal recognized by the Art1-Rsp5 ubiquitin ligase complex. Mol Biol Cell 15:4043–4054
6. Miesenböck G, De Angelis DA, Rothman JE (1998) Visualizing secretion and synaptic transmission with pH-sensitive green fluorescent proteins. Nature 394:192–195
7. Lin CH, MacGurn JA, Chu T, Stefan CJ, Emr SD (2008) Arrestin-related ubiquitin-ligase adaptors regulate endocytosis and protein turnover at the cell surface. Cell 135:714–725
8. Tang S, Henne WM, Borbat PP, Buchkovich NJ, Freed JH, Mao Y, Fromme JC, Emr SD (2015) Structural basis for activation, assembly and membrane binding of ESCRT-III Snf7 filaments. elife:e12548
9. Henne WM, Buchkovich NJ, Zhao Y, Emr SD (2012) The endosomal sorting complex ESCRT-II mediates the assembly and architecture of ESCRT-III helices. Cell 151 (2):356–371
10. Tang S, Buchkovich NJ, Henne WM, Banjade S, Kim YJ, Emr SD (2016) ESCRT-III activation by parallel action of ESCRT-I/II and ESCRT-0/Bro1 during MVB biogenesis. elife 5:E15507
11. Chiaruttini N, Redondo-Morata L, Colom A, Humbert F, Lenz M, Scheuring S, Roux A (2015) Relaxation of loaded ESCRT-III spiral springs drives membrane deformation. Cell 163:866–879
12. MacDonald C, Payne JA, Aboian M, Smith W, Katzmann DJ, Piper RC (2015) A family of tetraspans organizes cargo for sorting into multivesicular bodies. Dev Cell 33:328–342
13. Prosser DC, Whitworth K, Wendland B (2010) Quantitative analysis of endocytosis with cytoplasmic pHluorin chimeras. Traffic 11:1141–1150
14. Teis D, Saksena S, Judson BL, Emr SD (2010) ESCRT-II coordinates the assembly of ESCRT-III filaments for cargo sorting and multivesicular body vesicle formation. EMBO J 29:871–883

Chapter 9

Live Imaging of ESCRT Proteins in Microfluidically Isolated Hippocampal Axons

Veronica Birdsall, Jose C. Martinez, Lisa Randolph, Ulrich Hengst, and Clarissa L. Waites

Abstract

Live imaging of microfluidically isolated axons permits study of the dynamic behavior of fluorescently tagged proteins and vesicles in these neuronal processes. We use this technique to study the motility and transport of ESCRT proteins in axons of primary hippocampal neurons. This chapter details the preparation of microfluidic chambers, as well as the seeding, fluidic isolation, and lentiviral transduction of hippocampal neurons in these chambers, optimized for the study of ESCRT protein dynamics.

Key words ESCRT, Microfluidic, Primary neuronal culture, Hippocampal neuron, Axonal transport, Axonal isolation, Live cell imaging

1 Introduction

The emerging roles of the ESCRT pathway in multiple aspects of nervous system function, from neuronal morphogenesis to synaptic plasticity and vesicle secretion [1, 2], have led to heightened interest in the study of this pathway in neurons. Such interest has only intensified as ESCRT pathway dysfunction is increasingly linked to neurodegenerative disease etiology. For instance, mutations in the ESCRT-III protein CHMP2B cause frontotemporal dementia and amyotrophic lateral sclerosis, and deletion of other ESCRT components (i.e., HRS, TSG101) in mouse models induces profound neurodegeneration [3–8]. It has been difficult to study the ESCRT pathway in axons and presynaptic terminals due to the low endogenous expression of ESCRT proteins in these compartments. However, expression of fluorescently tagged ESCRT proteins via lentiviral transduction, coupled with microfluidic isolation and live cell imaging in rat hippocampal neurons, has allowed us to study the localization and dynamic behavior of ESCRT proteins in axons of cultured neurons. Here we provide a detailed protocol for this process.

Emmanuel Culetto and Renaud Legouis (eds.), *The ESCRT Complexes: Methods and Protocols*, Methods in Molecular Biology, vol. 1998, https://doi.org/10.1007/978-1-4939-9492-2_9, © Springer Science+Business Media, LLC, part of Springer Nature 2019

1.1 Fabrication of Microfluidic Devices to Cultivate and Image Axons of Primary Hippocampal Cells

Microfluidic isolation of axons facilitates investigation of axonal ESCRT dynamics in three important ways. First, it allows for the easy identification of axons when fluorescently -tagged cargo exhibits low expression or axonal localization, enabling study of ESCRT proteins in this particular subcellular structure of the neuron. Second, it allows the experimenter to control which cells express lentivirus, and thus deduce the directionality of axonal transport. Third, it permits the environmental segmentation of axonal and somatodendritic compartments, allowing the experimenter to differentially manipulate these compartments. Preparing microfluidic devices to isolate axons of primary neurons is the first and most involved task of this protocol. Our methods are adapted from well-established protocols [9, 10] in order to maximize neuronal health and ensure high levels of tagged ESCRT protein expression. Our microfluidic chambers are fabricated using soft lithography in a poly (dimethylsiloxane) (PDMS)-based encapsulant, using silicon wafer molds that can be fabricated [9] or purchased commercially. PDMS-based chambers can also be purchased commercially from several different companies. Once chambers have been fabricated, they can be seeded with freshly isolated and dissociated primary neurons [11, 12] (Fig. 1), which can be cultured for 5–7 days before lentiviral transduction. Dissection and dissociation of embryonic rat hippocampal neurons will not be discussed in detail here, but a comprehensive protocol can be found in Pacifici and Peruzzi [13], and additional protocols in Kaech and Banker and in Waites et al. [11, 12].

1.2 Lentiviral Transfection and Imaging Axons

Lentivirus is produced from transfected HEK293T cells as described previously [14–16] with minor modifications. Given the large size of many ESCRT proteins, it is vital to produce virus of high titer to ensure optimal expression levels. After 5–7 days of viral expression, live imaging can be performed on isolated axons using an epifluorescence or confocal microscope with a high-magnification objective, and data analysis can be performed using image processing software such as ImageJ/FIJI. We have used these techniques to study the axonal dynamics of several ESCRT proteins, including HRS, STAM, and TSG101 (Fig. 2), and our preliminary data indicate that ESCRTs are bidirectionally motile in axons. Future studies will lead to a deeper understanding of the axonal dynamics of ESCRT proteins, and how this pathway is regulated both in axons and in the neuron as a whole.

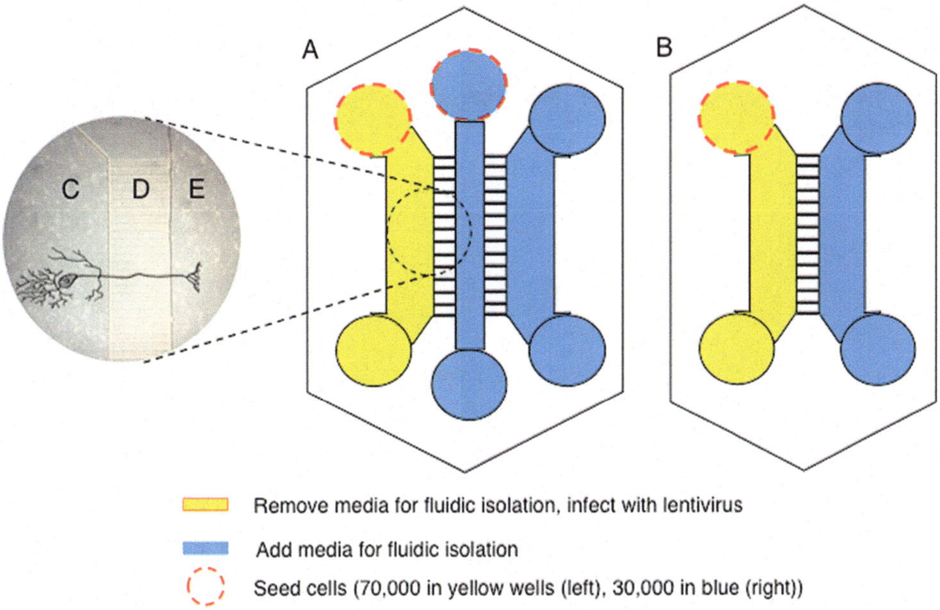

Fig. 1 Schematic diagram for seeding neurons and microfluidic isolation in tripartite (**a**) and bipartite (**b**) chambers. Medium must be removed from the side to be transduced, and placed into corresponding wells on the other side of the chamber to maintain a continuous flow across the microgrooves. This step is necessary to prevent diffusion of lentivirus into other compartments. The image to the left of (**a**) is a representative photograph of a tripartite chamber, showing the somatic compartment that is lentivirally transduced (**c**), the axonal compartment containing channels that are imaged (**d**), and the somatic compartment that is not transduced (**e**). The cartoon neuron illustrates how neurons grow in the chamber to isolate the different cellular domains

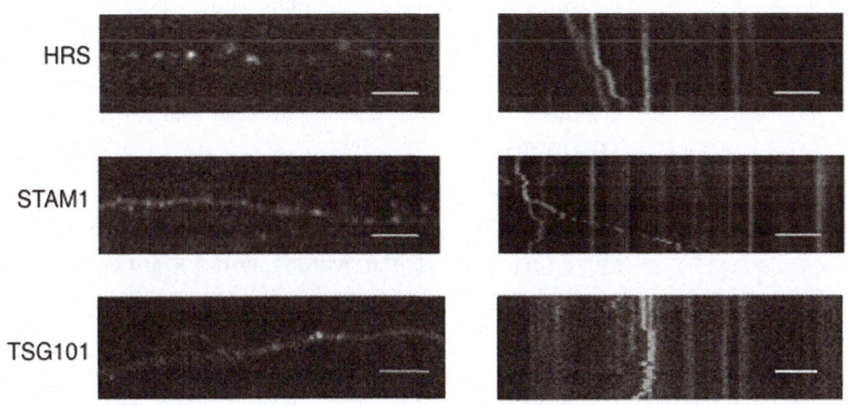

Fig. 2 Representative still images (left) and corresponding kymographs (right) of fluidically isolated axons from neurons expressing the indicated ESCRT proteins. Still images were taken before corresponding videos, and movement of each puncta over the course of the video is represented by the kymographs. HRS is very punctate, while STAM1 and TSG101 are punctate but also show low levels of diffuse protein. All three ESCRT proteins have motile and nonmotile pools in axons. Scale bar is 10 μm

2 Materials

2.1 Microfluidic Chamber Preparation, Reversible Bonding, and Seeding

1. Transparency mask.
2. Photoresists, SU-8 5 and SU-8 2050 (Microchem Corp., Westborough, MA).
3. Propylene glycol methyl ether acetate (PGMEA) photoresist developer.
4. Silicon wafer.
5. Isopropanol.
6. Deionized sterile water (dH_2O).
7. Pressurized nitrogen gas.
8. 100 mM boric acid solution: 6.2 g of boric acid in 1 L of dH_2O (pH 8.0). The pH is adjusted with NaOH.
9. Poly-L-lysine (PLL) solution: 0.25 mg/mL PLL in 100 mM boric acid (pH 8.0).
10. Commercially available two-component clear silicone encapsulant (PDMS based, e.g., Sylgard 184, Qsil216).
11. Dissociated hippocampal neurons (*see* Pacifici and Peruzzi [13] for protocol and additional reagents).
12. Neurobasal media.
13. B27 supplement.
14. 100× GlutaMAX.
15. 1× antibiotic-antimycotic: 10,000 units/mL of penicillin, 10,000 μg/mL of streptomycin, and 25 μg/mL of amphotericin B.
16. Neuronal culture media: 2% (v/v) B27, 1× GlutaMAX, 1× antibiotic-antimycotic in neurobasal media (all from Thermofisher).
17. Minimum essential media (MEM).
18. Plating media: 1 mM pyruvic acid, 0.6 mM glucose, 10% (v/v) horse serum (heat-inactivated), and 1× antibiotic-antimycotic in minimum essential media (MEM).
19. Disposable plastic cups for mixing PDMS solution.
20. Disposable stir rods (plastic knife).
21. Clean room tape.
22. 50 mm glass bottom dishes No. 1.5, uncoated and gamma irradiated (MatTek).
23. Razor blades.

24. Disposable biopsy punch, 5 mm diameter.
25. Spin coater.
26. Mask aligner.
27. Digital balance.
28. Hot plate.
29. Laboratory oven for curing PDMS polymer.
30. Vacuum desiccator for degassing PDMS polymer.
31. UV light box.
32. Dissection and tissue culture hood.

2.2 Lentiviral preparation

1. HEK293T cell line.
2. Dulbecco's modified Eagle medium (DMEM).
3. HEK293 cell media: 1× GlutaMAX, 10% (v/v) fetal bovine serum (heat inactivated), and 1× Pen/Strep in Dulbecco's modified Eagle medium (DMEM).
4. Lentiviral transfer plasmid (e.g., FUGW).
5. Packaging plasmid (e.g., delta P).
6. Envelope plasmid (e.g., VSV-G—plasmid information available from Addgene under "lentiviral guide").
7. Calfectin transfection reagent.
8. 10 cm cell culture-treated plates.
9. 0.45 μm syringe filter.
10. 10 mL syringe.
11. 2 mL cryotubes.
12. 1.5 mL Eppendorf tubes.
13. qPCR lentivirus titration kit.
14. Tissue culture hood.
15. Vortex mixer.
16. Microcentrifuge.

2.3 Imaging and Data Analysis

1. Computer with image acquisition software.
2. Epifluorescence or confocal microscope equipped for live cell imaging.
3. Stage warmer or incubation chamber capable of warming samples to 37 °C.
4. Image processing software, for example, ImageJ/FIJI.

3 Methods

3.1 Preparing Microfluidic Chambers

3.1.1 Fabrication of Master Mold (Steps 1–19 To Be Performed in a Nanofabrication Facility) (See Note 1)

1. Dehydrate a 4 in. silicon wafer at 200°C for 5 min on a hot plate. Let it cool to room temperature before proceeding (*see* **Note 2**).
2. Clean wafer with pressurized nitrogen.
3. Flood SU-8 5 on the silicon wafer (about 4 mL or enough to cover the surface).
4. Place the silicon wafer on a spin coaster and hold vacuum.
5. Spin on the spin coater at 2500–3500 rpm for 60 s. Use a faster speed for shorter height of the grooves. For height specifications, *see* **Note 1**.
6. Bake the wafer for 2 min at 95 °C on a hot plate.
7. Transfer the wafer to a mask aligner equipped with a UV lamp.
8. Expose the wafer to UV light through the transparency mask containing the microgroove pattern (100 mJ/cm^2, constant dose).
9. Bake the wafer for 2 min at 95°C on a hot plate.
10. Immerse wafer in developer (propylene glycol methyl ether acetate) for 2–3 min or until fully developed.
11. Wash with isopropanol and dry with pressurized nitrogen/inert gas. If the isopropanol wash creates a white or yellow residue, then develop for longer.
12. Bake the wafer for 5 min at 95 °C to completely dry the wafer.
13. Flood SU-82050 on the wafer (about 4 mL or enough to cover the surface) and spin at 1500–2000 rpm for 60 s. Use a faster speed for shorter height of the compartments.
14. Bake for 3 min at 65 °C, and then for 9 min at 95 °C on a hot plate.
15. Align the chamber mask to the grooves pattern. Expose to UV light (250 mJ/cm^2, constant dose).
16. Bake for 1–2 min at 65 °C, and then for 6–7 min at 95 °C on a hot plate.
17. Immerse wafer in SU-8 developer for 10 min or until fully developed. Rinse once with fresh SU-8 developer. If the isopropanol wash creates a white or yellow residue, then develop for longer.
18. Wash with isopropanol, and then with deionized water. Dry with pressurized nitrogen/inert gas.
19. Bake for 5 min at 150 °C.
20. Prepare about 130 g PDMS polymer per master mold.

21. Cover the bottom of a glass Petri dish with ~30 g PDMS polymer and bake for 15 min at ~70 °C.
22. Place the master mold on top, cover with remaining PDMS polymer, and bake at ~70 °C for at least 3 h or overnight.
23. Cut out chambers, and the mold is now ready to use.

3.1.2 Fabrication of Poly (dimethylsiloxane) Polymer-Based Chamber

1. Weigh silicone polymer and curing agent in the ratio of 10:1 in a disposable plastic cup and mix thoroughly for 2–5 min. Follow the manufacturer's recommendation for making and mixing the polymer solution.
2. Vacuum mix PDMS polymer solution in a vacuum desiccator for 1 h or until bubbles are mostly gone.
3. Clean dust and particles from the master mold using tape.
4. Carefully pour solution into the mold, and let it sit at room temperature for 20 min.
5. Remove any remaining bubbles from the surface by lightly blowing.
6. Place the mold into a level laboratory oven and cure for at least 3 h at 70 °C. The polymer should solidify when fully cured (*see* **Note 3**).
7. Carefully cut around the inner edge of the mold using a razor blade and slowly remove the cured polymer piece (*see* **Note 4**).
8. Keeping the microgrooved side up, cut out individual chambers using a razor blade and punch out channels using a steel punch. Chambers can be kept in a sealed container at room temperature.

3.1.3 Preparation of Chambers and Seeding of Cells (All Steps Performed in Laminar Flow Hood)

1. Coat glass bottom dishes with PLL solution for >12 h at 37 °C.
2. Wash 3× with deionized sterile water (dH_2O), 2 min per wash, and then let dry.
3. Sterilize chambers by immersing briefly in 70% ethanol, and dry fully in laminar flow hood before use.
4. Assemble sterile, dried chambers and dishes by placing the chamber lightly on glass, grooved side down. Ensure full seal by lightly tapping the corners of the chamber (*see* **Note 5**).
5. After assembly, fill channels with warmed neuron culture media. Assemble the chambers 3 h before seeding and place them in a well-humidified incubator at 37 °C with 5% CO_2 (*see* **Notes 6** and **7**).
6. Prepare E18 rat hippocampal dissociated cell suspension, according to Kaech and Banker or Pacifici and Peruzzi [11–13], with an approximate density of 10,000 cells/mL of plating media.

7. Aspirate media from chamber wells, being careful not to remove media from the channels (*see* **Note 8**).
8. Seed cells into the two top left wells of a tripartite chamber, 70,000 cells in the upper left well and 30,000 in the upper middle well (Fig. 1) (*see* **Note 9**).
9. Place chamber back in incubator for 30 min–1 h, then fill the remaining wells with plating media.
10. Remove plating media and replace with neurobasal containing antibiotic-antimycotic 24 h later.
11. To counteract media evaporation, add enough media to fill the wells every 3–4 days.

3.2 Lentiviral Preparation (All Steps Performed in Laminar Flow Hood)

1. Plate HEK293T cells in HEK cell media in a culture-treated 10 cm dish. Cells should be 60–70% confluent at transfection. (*see* **Note 10**).
2. Change media prior to transfection.
3. For each plate, set up transfection mixture in a 1.5 mL tube (*see* **Note 11**). To each tube, add: 1 mL DMEM (serum-free), 10 μg FUGW plasmid (containing the desired ESCRT construct), 7.5 μg delta P (packaging plasmid), and 5 μg VSV-G (envelope gene).
4. Vortex tubes to mix, then spin down briefly.
5. Add 35–40 μL of CalFectin reagent/tube and mix gently.
6. Incubate tubes for 15 min at room temperature.
7. Add the mixture drop-wise to a 10 cm plate of HEK cells; swirl gently to mix, and place cells back in the incubator.
8. After 12–18 h of transfection, remove HEK cell media and replace with 8 mL of Neurobasal media.
9. Collect virus 24 h after media change.
10. To collect, pipette media off the cells, filter into a 50 mL conical tube using a 10 mL syringe and a 0.45 μm filter.
11. Aliquot into labeled cryotubes and store at −80 °C.
12. Test viral titer before use (*see* **Note 12**).

3.3 Lentiviral Transduction (Microfluidic Isolation, Fig. 1)

1. Transduce neurons between day in vitro (DIV) 5–7.
2. Fill all four wells with the same volume (100 μL) of fresh media.
3. For a dipartite chamber, move 50 μL of media from a well on one side of the chamber (e.g., left) and add it to the well on the opposite side (right). For a tripartite chamber, add media to both wells across from the well to be transduced (*see* Fig. 1 for example).

4. Repeat for all wells of the chamber (remove from left well, add to corresponding right well).
5. Add 30–50 μL of thawed virus to the desired well(s) on the left side of the chamber (*see* **Note 13**).
6. Image chambers 5–7 days after viral transduction. Check for viral expression and fluidic isolation and adjust accordingly (*see* **Notes 14** and **15**).

3.4 Imaging and Data Analysis (see Note 16, Fig. 2)

1. Image fluidically isolated axons expressing fluorescently tagged ESCRT proteins using an epifluorescence or confocal microscope with 40× or 63× oil immersion objective (*see* **Note 17**).
2. Acquire images every 5 s for a total of 50 frames for imaging of early ESCRT proteins (e.g., HRS, Tsg101). Faster or slower image acquisition can be used based on the experimental question (*see* **Note 18**).
3. Analyze videos using the Manual Tracking ImageJ/FIJI plug in. Information about puncta movement and directionality (i.e., speed, anterograde/retrograde/bidirectional movement, moving puncta per axon) can be extracted from these raw data by writing a custom macro in ImageJ, or with custom Matlab code.

4 Notes

1. To determine proper spin conditions for desired thickness of mold, consult http://cleanroom.byu.edu/su8. The conditions described are what we generally use for 5-μm-tall microgrooves and 75–150-μm-tall compartments.
2. After all dehydration steps, the wafer should be allowed to cool to room temperature before proceeding. However, proceed to coating as soon as the wafer is cooled to avoid re-adsorption of water.
3. Chambers can be baked overnight. Make sure that the inside of the oven is clean and level.
4. Make sure to cut all the way around the mold, and that the tip of the razor blade goes all the way through the PDMS when cutting it out of the molds. If the PDMS is not separated completely from the mold, it can rip when pulled out.
5. The polymer will stick easily to the treated glass of the dishes; plasma bonding is not necessary for the protocol.
6. Warm up media before adding it to the chamber in order to prevent bubbles from being trapped in the channels. If bubbles do form, they can be removed by repeatedly pipetting media through the channels.

7. Given the small volume of liquid in the chambers, it is vital that they be stored in a well-humidified incubator after seeding to ensure cell survival. If media evaporation occurs, chambers can be placed in a container with a small amount of deionized sterile water (dH_2O), (chambers should be above the dH_2O, not in contact with it).
8. Leave a small amount of media in the wells to avoid introducing bubbles.
9. The use of a bipartite or tripartite chamber depends on the needs of the experiment to be performed.
10. Because open reading frames (ORFs) encoding ESCRTs tend to be long (~2.5 kb), the size of resulting FUGW plasmids can be >12 kb, often resulting in low-efficiency virus. Thus, it is essential to use healthy, low passage number (i.e., <15 passages) HEK293T cells for lentiviral production.
11. These are the specific transfer, packaging, and envelope plasmids that we use for our experiments, based on previous work [14]. Multiple lentiviral plasmids are available at Addgene (addgene.org) and may be optimal for expression of different proteins.
12. Viral titer can be tested directly using a qPCR Lentivirus Titration Kit (Applied Biological Materials, cat. # LV900; 10^6 virus/mL is excellent), and viral efficiency should be measured empirically by transducing test wells of neurons with varying amounts of lentivirus. On average, 10 μL of viral volume per 10 mm diameter coverslip should transduce ~20–30% of neurons.
13. Lentiviral transduction volume will depend on the quality and titer of the virus, the ESCRT protein being expressed, and the goals of the experiment. The ideal volume should be determined by test transductions, as described earlier.
14. Relatedly, as the volume of virus changes, the volume moved to ensure fluidic isolation will also have to be adjusted. The basic concept is that in order to isolate the virus in one side of the chamber (necessary for knowing the directionality of axon growth and puncta movement), it is important to maintain a hydrostatic pressure imbalance between the transduced and non-transduced chambers. This hydrostatic pressure difference is created by a small difference in volume between compartments. Determining the correct volume difference may take some trial and error to figure out, given the goals of the experiment, transduction volume, and size of the chamber, but a volume difference of ~50 μL is a good place to start.
15. Finally, the ideal DIV for imaging and transduction will also depend on the goals of the experiment. This protocol is

optimized to ensure that imaging is done in neurons that have formed synapses. Transducing and imaging as early as possible will make it easier to maintain neuronal health, as long-term culturing (i.e., longer than 12–14 DIV) in microfluidic chambers can be difficult.

16. ESCRT protein expression in neurons tends to be relatively sparse and punctate in axons (*see* Fig. 2). These features allow for the tracking of individual ESCRT-positive puncta.
17. The microscope should be equipped with a stage warmer or incubation chamber to allow for imaging at 37 °C.
18. Number of frames and exposure time must be calibrated to minimize bleaching of fluorescent tags. If a long exposure is necessary, videos may need to be shorter. Moreover, with a longer exposure time, it may not be possible to image a given axon for more than a few minutes.

Acknowledgments

This work was supported by NIH grants 5T32NS064928 to V.B.; F31GM11661701 to J.C.M.; 5T32HD007430-20 to L.R., R01NS080967 to C.L.W., and Hirschl Research Scientist Award to U.H. and C.L.W.

References

1. Sweeney NT, Brenman JE, Jan YN, Gao F-B (2006) The coiled-coil protein shrub controls neuronal morphogenesis in Drosophila. Curr Biol 16(10):1006–1011. https://doi.org/10.1016/j.cub.2006.03.067
2. Rodal AA, Blunk AD, Akbergenova Y, Jorquera RA, Buhl LK, Littleton JT (2011) A presynaptic endosomal trafficking pathway controls synaptic growth signaling. J Cell Biol 193(1):201
3. Talbot K, Ansorge O (2006) Recent advances in the genetics of amyotrophic lateral sclerosis and frontotemporal dementia: common pathways in neurodegenerative disease. Hum Mol Genet 15(suppl_2):R182–R187. https://doi.org/10.1093/hmg/ddl202
4. Lee J-A, Beigneux A, Ahmad ST, Young SG, Gao F-B (2007) ESCRT-III dysfunction causes autophagosome accumulation and neurodegeneration. Curr Biol 17(18):1561–1567. https://doi.org/10.1016/j.cub.2007.07.029
5. Cox LE, Ferraiuolo L, Goodall EF, Heath PR, Higginbottom A, Mortiboys H, Hollinger HC, Hartley JA, Brockington A, Burness CE, Morrison KE, Wharton SB, Grierson AJ, Ince PG, Kirby J, Shaw PJ (2010) Mutations in CHMP2B in lower motor neuron predominant amyotrophic lateral sclerosis (ALS). PLoS One 5(3):e9872. https://doi.org/10.1371/journal.pone.0009872
6. Vernay A, Therreau L, Blot B, Risson V, Dirrig-Grosch S, Waegaert R, Lequeu T, Sellal F, Schaeffer L, Sadoul R, Loeffler J-P, René F (2016) A transgenic mouse expressing CHMP2Bintron5 mutant in neurons develops histological and behavioural features of amyotrophic lateral sclerosis and frontotemporal dementia. Hum Mol Genet 25 (15):3341–3360. https://doi.org/10.1093/hmg/ddw182
7. Walker WP, Oehler A, Edinger AL, Wagner K-U, Gunn TM (2016) Oligodendroglial deletion of ESCRT-I component TSG101 causes spongiform encephalopathy. Biol Cell 108 (11):324–337. https://doi.org/10.1111/boc.201600014
8. Tamai K, Toyoshima M, Tanaka N, Yamamoto N, Owada Y, Kiyonari H, Murata K, Ueno Y, Ono M, Shimosegawa T, Yaegashi N, Watanabe M, Sugamura K (2008)

Loss of Hrs in the central nervous system causes accumulation of ubiquitinated proteins and neurodegeneration. Am J Pathol 173 (6):1806–1817. https://doi.org/10.2353/ajpath.2008.080684

9. Park JW, Vahidi B, Taylor AM, Rhee SW, Jeon NL (2006) Microfluidic culture platform for neuroscience research. Nat Protoc 1:2128. https://doi.org/10.1038/nprot.2006.316
10. Batista AFR, Martínez JC, Hengst U (2017) Intra-axonal synthesis of SNAP25 is required for the formation of presynaptic terminals. Cell Rep 20(13):3085–3098. https://doi.org/10.1016/j.celrep.2017.08.097
11. Kaech S, Banker G (2007) Culturing hippocampal neurons. Nat Protoc 1:2406. https://doi.org/10.1038/nprot.2006.356
12. Waites CL, Specht CG, Härtel K, Leal-Ortiz S, Genoux D, Li D, Drisdel RC, Jeyifous O, Cheyne JE, Green WN, Montgomery JM, Garner CC (2009) Synaptic SAP97 isoforms regulate AMPA receptor dynamics and access to presynaptic glutamate. J Neurosci 29 (14):4332
13. Pacifici M, Peruzzi F (2012) Isolation and culture of rat embryonic neural cells: a quick protocol. J Vis Exp (63):e3965. https://doi.org/10.3791/3965
14. Lois C, Hong EJ, Pease S, Brown EJ, Baltimore D (2002) Germline transmission and tissue-specific expression of transgenes delivered by Lentiviral vectors. Science 295 (5556):868
15. Leal-Ortiz S, Waites CL, Terry-Lorenzo R, Zamorano P, Gundelfinger ED, Garner CC (2008) Piccolo modulation of Synapsin1a dynamics regulates synaptic vesicle exocytosis. J Cell Biol 181(5):831
16. Sheehan P, Zhu M, Beskow A, Vollmer C, Waites CL (2016) Activity-dependent degradation of synaptic vesicle proteins requires Rab35 and the ESCRT pathway. J Neurosci 36 (33):8668

Chapter 10

Studying the Spatial Organization of ESCRTs in Cytokinetic Abscission Using the High-Resolution Imaging Techniques SIM and Cryo-SXT

Shai Adar-Levor, Inna Goliand, Michael Elbaum, and Natalie Elia

Abstract

The ESCRT machinery mediates scission of the intercellular bridge that connects two daughter cells at the end of cytokinesis. Structured illumination microscopy (SIM) and cryo-soft-X-ray tomography (cryo-SXT) have been used in recent years to study the topology of ESCRT-driven cytokinetic abscission. These studies revealed that the intercellular bridge is occupied by cortical rings and spiral-like filaments and that ESCRTs form ring-like structures in this region during abscission. In this chapter, we provide two protocols: a protocol for determining the spatial organization of specific ESCRT components at the intercellular bridge using SIM and a protocol for resolving the ultrastructural organization of cortical filaments at the intercellular bridge using cryo-SXT.

Key words Membrane fission, Soft-X-ray, Tomography, Cell division, Cytokinesis, ESCRT, ESCRT-III, Super-resolution microscopy, Fluorescence imaging, Cryo, Synchronization, Immunostaining, Correlative imaging

1 Introduction

The spatial organization of the ESCRT machinery during membrane fission has been studied in recent years by high-resolution imaging techniques including live cell imaging, electron microscopy, and super resolution (SR) microscopy [1–6]. In many of these studies, abscission, that is, the final stage of animal cytokinesis, was used as a model system. In abscission, the ESCRT complex drives constriction and fission of the 1 μm wide intercellular membrane bridge that connects the two nascent daughter cells (Fig. 1) [1, 7]. The bridge is cleaved on each side of the dark zone, a densely packed structure located at center of the intercellular bridge (Fig. 1). Final fission occurs at the abscission sites, located 1 μm away on each side of the dark zone. The relatively large dimensions of the cytokinetic intercellular bridge (>1 μm compared to ~100 nm of an HIV viral bud and ~50 nm in MVB formation)

Emmanuel Culetto and Renaud Legouis (eds.), *The ESCRT Complexes: Methods and Protocols*, Methods in Molecular Biology, vol. 1998, https://doi.org/10.1007/978-1-4939-9492-2_10,

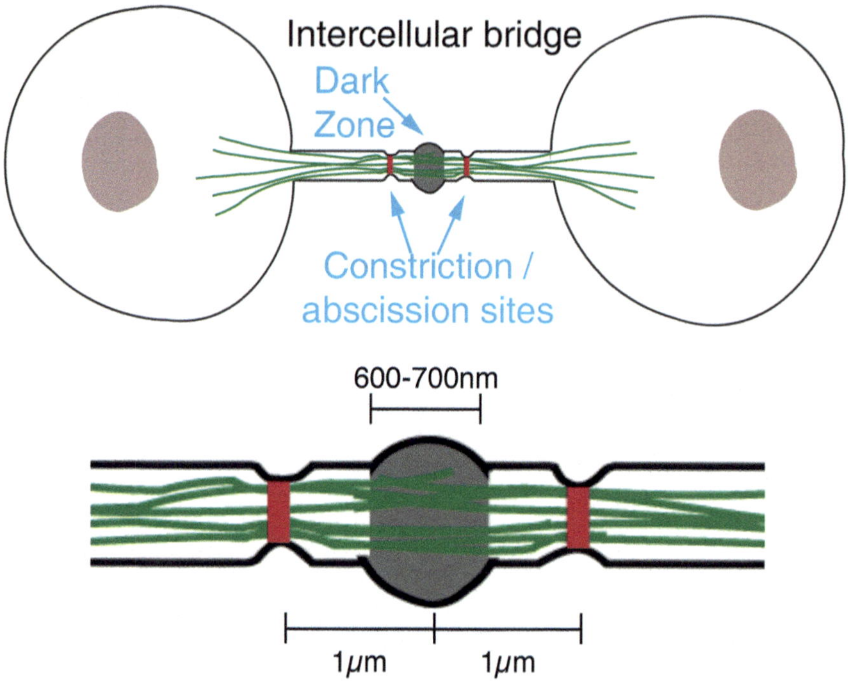

Fig. 1 Morphology of the intercellular bridge. Top panel: At the end of cytokinesis, cells are connected with a thin intercellular membrane bridge called the intercellular bridge. The bridge is filled with packed microtubule fibers (green lines) that stretch all the way from the cell body to the center of the bridge. Microtubules coming from both cells overlap at the center of the bridge, forming a dense structure called the dark zone (dark gray). During abscission, the bridge is cut twice at narrow constriction sites, called abscission sites (red), located peripherally on both sides of the dark zone. The two asynchronous abscission events lead to the formation of a remnant that is released from the cells. Bottom panel: enlargement of the dark zone. The dark zone has a 1 μm diameter and ~600 nm width. Abscission sites are located 1 μm away from the center of the bridge, on each side of the dark zone. Taken from [6]

are ideal for applying high-resolution imaging techniques such as the super-resolution (SR) structured illumination microscopy (SIM) and cryo soft-X-ray tomography (cryo-SXT) [2].

In this section, we will provide a step-by-step protocol for applying SIM and cryo-SXT to study the spatial organization of the ESCRT machinery in cytokinetic abscission.

1.1 SIM Analysis of ESCRTs

SIM is a fluorescence imaging technique that can provide protein-specific information with a spatial resolution of ~100 nm in the *x* and *y* dimensions and ~350 nm in the *z* dimension—twice the resolution of conventional fluorescence microscopy methods [8, 9]. In SIM, series of images (15–25) that contain interference information are recorded using a grated pattern of light that is rotated and phase shifted between each image acquisition. Using the image series a high-resolution image is reconstructed. Combining SIM images captured at consecutive focal planes can create volumetric data sets (for more information on SIM, refer to [9]).

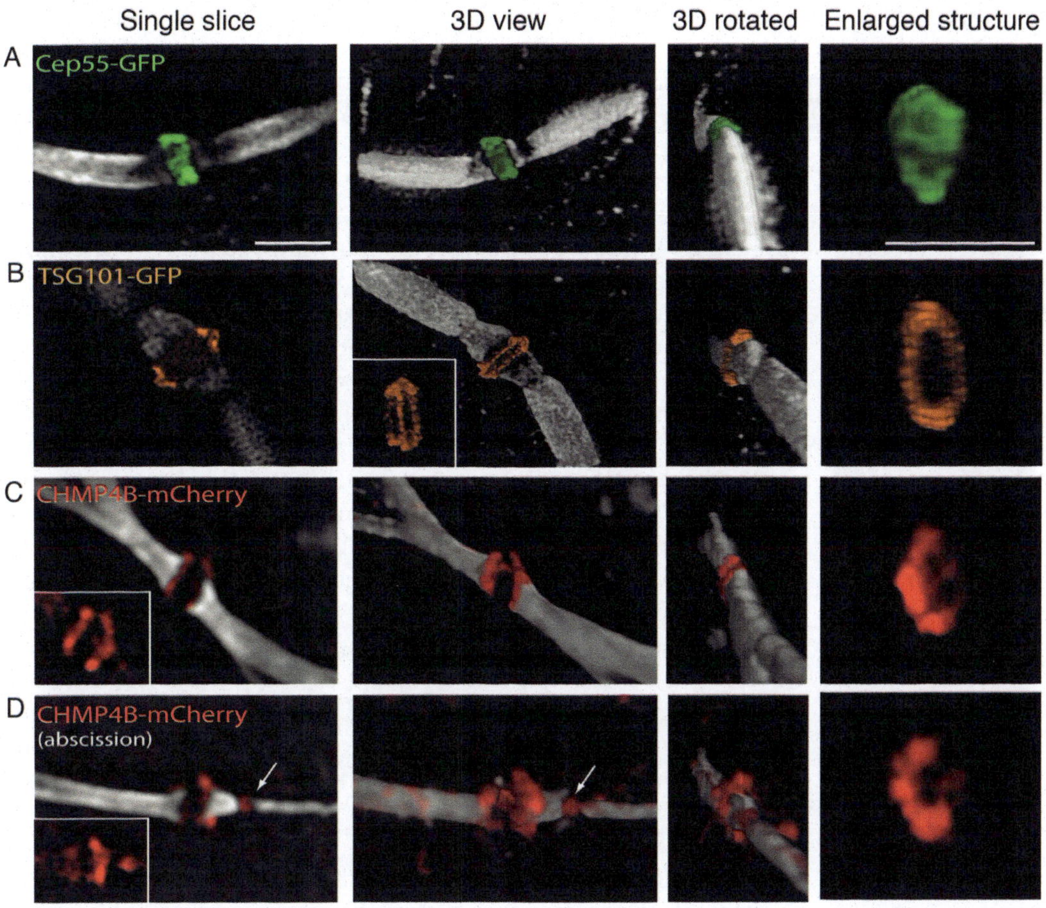

Fig. 2 Spatial organization of the ESCRT complex in the midbody as determined by SIM. MDCK cells expressing CEP55-GFP (**a**), TSG101-GFP (**b**), or CHMP4B-mCherry (**c** and **d**) were synchronized, fixed, stained with anti–α-tubulin antibodies and imaged by SIM. (**a–d**) Each panel shows (from left to right) a single slice, a 3D rendering, a 3D rendering rotated 90°, and a zoomed-in image of the structure. Microtubules are colored in white, CEP55-GFP in green, TSG101-GFP in orange, and CHMP4B-mCherry in red. (**a**) CEP55 forms a diffusely filled structure that is 1.4 ± 0.15 μm in diameter and 0.75 ± 0.07 μm in width. (**b**) TSG101 forms a tightly packed double-ring structure surrounding the microtubules at the center of the midbody dark zone (width = 0.82 ± 0.03 μm). The rings are 1.7 ± 0.07 μm in their outer diameter and are 0.23 ± 0.02 μm apart. Inset: A rotated image demonstrating the existence of two separate rings. (**c** and **d**) CHMP4B concentrates in two large diameter rings that are 0.43 ± 0.08 μm apart. The diameter of each ring is 1.25 ± 0.18 μm. In late stages of abscission (**d**) CHMP4B shows an additional pool that is located asymmetrically 1.2 μm away from the center of the dark zone (arrow in **d**). This pool colocalized with the site of microtubule constriction. Insets: CHMP4B signal alone. Scale bars: 2 μm. Scale bar in the top left panel applies to the first to third columns in **a–d**. Scale bar in the top right panel applies to right columns in **a–d**. Taken from [1]

Proteins can be fluorescently labeled for SIM, by fluorescent protein tags or through antibody staining. Using SIM, ESCRTs were shown to organize in large-diameter cortical rings at the rims of the dark zone and to redistribute to the abscission sites late in abscission (Fig. 2) [1, 4].

1.2 Cryo-SXT of the Intercellular Bridge

Cryo-SXT offers a complementary approach to light and electron-based microscopy techniques. Like the light microscope, it provides the opportunity to image whole cells intact, without sectioning, while the ultrastructural organization of cellular components can be resolved with ~30 nm resolution [10, 11]. SXT takes advantage of the X-ray spectral window between atomic absorption transitions of carbon and oxygen, the so-called "water window." At these energies, carbon absorbs the illuminating radiation much more strongly than oxygen, so that carbon-rich cellular ultrastructure appears dark on a relatively bright background [10]. Such microscopes typically operate at synchrotron facilities that provide a bright X-ray source at the suitable energy, although lab sources are beginning to appear. Additionally, cryo-SXT stations are equipped with an in-line fluorescence microscope port, which makes correlative imaging as precise and simple as possible [12]. SXT is similar to electron tomography, in that a series of projection images is recorded with the sample tilted to different angles around an axis. A 3D volume is reconstructed computationally. (For general information on the methodology, refer to [10, 12].) In spite of the reduced resolution (~30 nm in SXT and ~2 nm in EM), employing cryo-SXT to study abscission has some advantages over traditional EM. The intracellular bridges are typically too thick for cryo-TEM tomography, and therefore, they should first be isolated from the cell or sectioned, which can produce artifacts [4, 13]. By imaging intercellular bridges in their intact cellular environment, realistic ultrastructural description of the topology of abscission can be achieved. Using this correlative approach, large-diameter cortical filaments were resolved at the dark zone region in early stages of abscission, and helical filaments stretching between the dark zone and the abscission sites were resolved in late stages (Fig. 3) [14].

Based on the experience in our laboratory, we find SIM and cryo-SXT highly suitable for structural studies of the intercellular bridge. Additionally, the applicability of cryo-SXT to correlative studies with fluorescence imaging techniques opens the opportunity for performing correlative SR cryo-SXT studies, as recently demonstrated for the SR technique STORM [15].

2 Materials

2.1 Cell Culture, Transfection, and Fixative Reagents

All buffers are sterilized.

1. Cell culture incubator (37 °C + 5% CO_2).
2. HeLa cells growth medium: Dulbecco's modified Eagle's medium (DMEM) supplemented with 10% (v/v) fetal bovine serum (FBS), 2 mM L-glutamine, 10,000 U/mL penicillin, 10 mg/mL streptomycin, 4.5 g/L glucose.

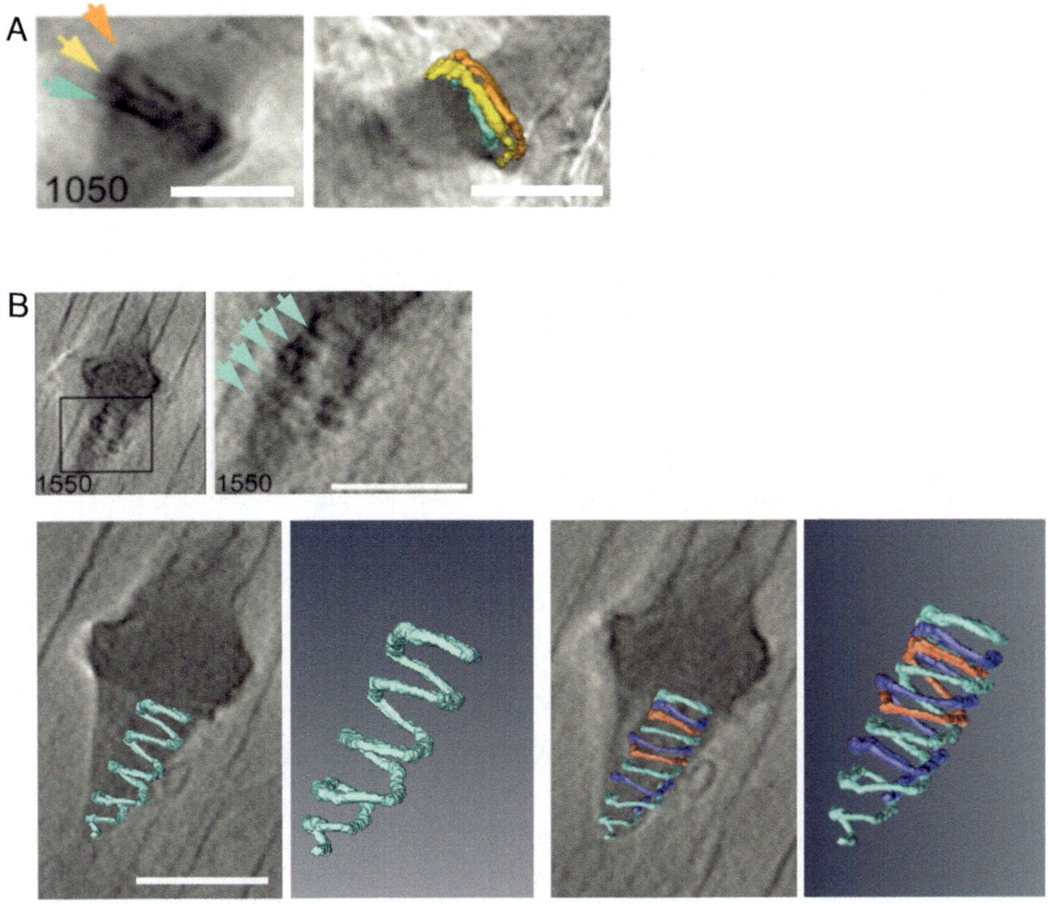

Fig. 3 Resolving high-ordered structures in the intercellular bridge using cryo-SXT. (**a**) Left panel: a *Z* section from a reconstructed tomographic 3D volume of an intercellular bridge. Arrows on tomogram correspond to the rendered high-ordered dark zone structures. Right panel: high-ordered structures in the dark zone overlaid on tomographic data from left panel. Scale bar: 1 μm. (**b**) Top panel: a *Z* section from a reconstructed tomographic 3D volume and intercellular bridge late in abscission. Right enlarged view of the area marked in black square. Arrows on tomogram correspond to the rendered high-ordered spiral structure (cyan). Bottom panel: from left to right: a single helical filament (cyan) rendered in the abscission site overlaid on tomographic data; a single rendered helical filament; all (three) helical filaments rendered in the abscission site overlaid on tomographic data; rendered helical filaments; Color represents individual filaments. Numbers on images indicate the *Z* section height in nm. Scale bar: 1 μm. Adapted from [14]

3. MDCK cells growth medium: Minimum Essential Medium (MEM) supplemented with 5% FBS, 2 mM L-glutamine, 10,000 U/mL penicillin-streptomycin, 10 mg/mL streptomycin.
4. Trypsin EDTA solution B: 0.25% (v/v) trypsin, 0.05% (v/v) EDTA.
5. Aphidicolin solution: MDCK cells growth medium supplemented with 2.5 mg/mL aphidicolin.

6. Thymidine synchronization solution: HeLa cells growth medium supplemented with 2 mM thymidine.
7. Lipofectamine 2000 (Life Technologies, transfection in MDCK cells).
8. jetPRIME® (Polyplus, transfection in HeLa cells).
9. Dulbecco's phosphate-buffered saline X10 buffer (PBS): dilute with DDW to PBS X1.
10. Fixation solution: 4% (v/v) PFA in PBS.
11. Storage solution: 0.1% (v/v) PFA in PBS.

2.2 Immunostaining Reagents for SIM

1. A list of commercially available ESCRT antibodies that were verified for immunostaining in abscission is provided in Table 1.
2. Anti-tubulin antibodies DM1A (Sigma, cat# T9026).
3. Secondary antibodies: Alexa-Fluor 405 (dilution 1:500), Alexa-Fluor 488 (dilution 1:1000), Alexa-Fluor 568 (dilution 1:1000), and Alexa-Fluor 647 (dilution 1:1000) anti-mouse or anti-rabbit (Invitrogen).
4. KHM buffer, pH 7.4: 110 mM CH_3CO_2K, 20 mM HEPES-NaOH, 2 mM MgCl.
5. Prefixation permeabilization solution: 0.01% (v/v) digitonin in KHM buffer.
6. Permeabilization solution: 0.5% (v/v) Triton-X 100 in PBS.
7. Blocking solution: 10% (v/v) FBS in PBS.
8. Mounting media: Fluoromount-G (Southern Biotech).

2.3 Dishes, Grids, and Coverslips

1. Coverslip number 1.5 high resolution (Marienfeld, cat# 0117530) [SIM].
2. Twenty-four-well dishes [SIM].

Table 1
Commercially available ESCRT antibodies verified for immunostaining of ESCRTs in abscission

Species	ESCRT	Catalog number	Company	HeLa	MDCK	Dilution	Digitonin 0.01%
Rabbit	IST1	510021-AP	PROTEIN TECH	+	+	1:50	No
Rabbit	CHMP4B	13683-1-AP	PROTEIN TECH	+	–	1:50	Yes
Rabbit	CHMP2A	10477-1-AP	PROTEIN TECH	+	–	1:50	Yes
Rabbit	CHMP1A	15761-1-AP	PROTEIN TECH	+	–	1:50	Yes
Rabbit	CHMP2B	ab33174	abcam	+	–	1:50	Yes
Rabbit	CHMP4A	H-52, SC 67229	Santa Cruz	+	–	1:50	Yes

+ suitable, – unsuitable

3. Microscope glass slide [SIM].
4. Tweezers [cryo-SXT].
5. Six-well dishes [cryo-SXT].
6. Eight-well dishes (ibidi, cat# 81806) [cryo-SXT].
7. QUANTIFOIL® R 2/2 or R1/4 on HZB-2 grids (Quantifoil Micro Tools GmbH) [cryo-SXT].

3 Methods

3.1 Cell Culture, Synchronization, and Fixation

We routinely use MDCK or HeLa cells to study ESCRTs in abscission. MDCK cells are relatively large, flat, and adhesive, which enable visualization of flat, planar intercellular bridges. HeLa cells still meet the imaging requirement and are sometimes more useful due to availability of antibodies and techniques for genome editing.

3.1.1 Cell Culture

Cells are grown in a 60 mm tissue culture dish with 5 mL growth medium (MEM for MDCK cells and DMEM for HeLa cells).

1. When cells reach 80%–100% confluency, remove the growth medium and add 1 mL trypsin.
2. Incubate the cells with trypsin solution until cells are detached.
3. Add 4 mL growth medium to neutralize trypsin solution.
4. Transfer 10% of the cells to a new dish and seed for further growth.

To enrich the percentage of intercellular bridges, cells are synchronized. MDCK cells are synchronized with aphidicolin, and HeLa cells are synchronized with a double thymidine block as described below (*see* Subheadings 3.1.2 and 3.1.3, respectively).

3.1.2 Synchronization of MDCK Cells with Aphidicolin

1. Plate cells on a glass bottom dish, depending on the use—see SIM and cryo-SXT Subheadings 3.2.1 and 3.3.1, respectively. Place in cell culture incubator.
2. The following day, replace the growth medium with aphidicolin solution.
3. Place cells for 16 h in cell culture incubator.
4. Remove aphidicolin by washing three times with warm (37 °C) sterile PBS.
5. Add fresh, warm growth medium, and place cells for 10–11 h in cell culture incubator.
6. Fix cells as described in Subheading 3.1.4.

3.1.3 Synchronization of HeLa Cells with a Double Thymidine Block

1. Plate cells on a glass bottom dish depending on the use—see SIM and cryo-SXT Subheadings 3.2.1 and 3.3.1, respectively. Place in cell culture incubator.
2. The following day, replace growth medium with thymidine synchronization solution.
3. Place cells for 18 h in cell culture incubator.
4. Remove thymidine by washing three times with warm (37 °C) sterile PBS.
5. Add warm growth medium and incubate for 9 h.
6. Subject cells to a second thymidine block as in **step 2** and place them in cell culture incubator for 15 h.
7. Remove thymidine by washing three times with warm sterile PBS.
8. Add fresh growth medium and place cells for 10–11 h in cell culture incubator (*see* **Note 1**).
9. Fix cells as described in Subheading 3.1.4.

3.1.4 Cell Fixation

1. Remove growth medium and wash once with PBS.
2. Fix cells: add fixation solution and incubate 15 min, at 25 °C.
3. Wash cells three times with PBS.
4. Cells can be stored at 4 °C up to a week in storage solution.

3.2 Imaging ESCRTs with SIM

3.2.1 Labeling by Overexpressing Fluorescently Tagged Proteins

Transient plasmid transfection can be used to label ESCRTs in cells. We transfect cells with expression plasmids containing an ESCRT component with mCherry conjugated to its N terminal (mCherry-ESCRT). Tubulin-GFP transfection or cells stably expressing tubulin-GFP are used for visualizing the intercellular bridge. Alternatively, tubulin can be stained post fixation (*see* Subheading 3.2.2) (*see* **Note 2**).

1. Plate cells at 10%–20% confluency on 1.5 high-resolution coverslips. Coverslips can be arranged in a 24-well dish. One coverslip in each well.
2. For MDCK cells: transfect with Lipofectamine 2000. For HeLa cells: transfect with jetPRIME. In both cases, follow manufacturer's guidelines for plasmid transfection. Overexpression of fluorescently tagged ESCRT components can result in cellular artifacts. To minimize that, use up to 400 ng of DNA for 24 wells in a single transfection and 500 ng total DNA for double transfection (250 ng each).
3. Optional: MDCK cells can be synchronized (*see* Subheading 3.1.2).
4. Twenty-four hours post transfection (or after synchronization), fix cells as described in Subheading 3.1.4. Make sure the

solutions entirely cover the coverslip (add at least 500 μl to each well in a 24-well dish). This applies to all the steps detailed below.

5. Mount the sample as described in Subheading 3.2.3. (**Notes 2–4**, refer to this section).

3.2.2 Labeling by Immunostaining

1. Plate cells at 10%–20% confluency on 1.5 high-resolution coverslips.

   Optional: Prefixation permeabilization: ESCRT proteins are mostly cytosolic; this can result in high background staining, which can interfere with visualizing the ESCRT pool at the intercellular bridge. Mild permeabilization of the cells prior to fixation can greatly reduce this cytosolic background. To do so, replace the growth medium with warm (37 °C) KHM buffer. Add prefixation permeabilization solution for 1 min. Wash with PBS three times and continue immediately to fixation.
2. Fix cells as described in Subheading 3.1.4. Cells should be at ~60% density at the time of fixation. It is difficult to find isolated bridges that are suitable for imaging at higher densities.
3. Permeabilize cells by incubating them with permeabilization solution for 10 min, at 25 °C.
4. Wash cells with PBS (three quick washes, three washes for 5 min each at 25 °C).
5. Block by incubating cells in blocking solution for 15 min at 25 °C. Continue to the next step without washing.
6. Primary antibody (Ab): dilute the primary Ab in blocking solution. Data of verified antibodies and recommended dilutions are provided in Table 1.
7. Incubate for 3 h at 25 °C or overnight at 4 °C. To detect the intercellular bridge, we routinely stain cells with anti-tubulin antibodies (DM1A, 1:1000, Sigma, T9026).
8. Wash as in **step 4**.
9. Secondary Ab: dilute the Ab in blocking solution. See recommended dilutions in Subheading 2.2.3.
10. Incubate for 1 h at 25 °C. Use Alexa-Fluor secondary fluorescent antibodies; make sure the combination of secondary antibodies matches your microscope setup to avoid bleed-through and cross-excitation.
11. Wash as in **step 4**.
12. Mount the sample as described in Subheading 3.2.3.

3.2.3 Mounting

1. Take the coverslip out of the PBS using tweezers; gently dry PBS leftovers using kimwipe paper.
2. Mount samples by applying a 5 μl mounting media drop onto a glass slide and place the coverslip on top of the drop, with cells facing the mounting media.
3. Let the mounting media dry and polymerize overnight.
4. Seal the coverslip with nail polish and let it dry for approximately 30 min.

3.2.4 Image Acquisition

1. Search for suitable cells with isolated planar bridges that show accumulation of ESCRT proteins at the midbody.
2. Image cells that express low-medium levels in which the cellular distribution of ESCRT proteins is mostly cytosolic. Avoid high-expressing cells (see example in Fig. 4). Place selected cells in the center of the field of view (*see* **Note 5**).
3. Use the following acquisition guideline: Signal intensity should be at least at half the saturation value. Saturation interferes with image reconstruction and should be avoided in the entire image, not just the intercellular bridge. In ELYRA PS.1 microscope (Carl Zeiss MicroImaging), we use gain: 30–50, exposure 100–200 ms, laser power 1%–10%, rotations 5. Five rotations provide better resolution but can cause more photobleaching. When signal is dim, use three rotations.

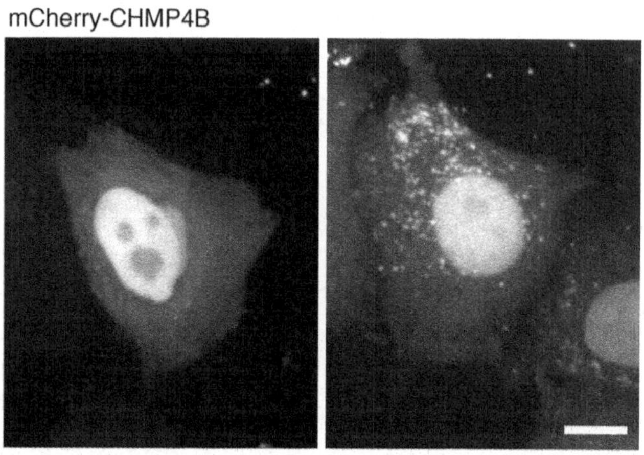

Fig. 4 Cellular distribution of mCherry-CHMP4B in low- and high-expressing cells. Representative HeLa cells transfected with mCherry-CHMP4B (white) were imaged using a spinning-disk confocal microscope. Cells expressing low levels of mCherry-CHMP4B (left) exhibit mostly cytosolic localization. Cells expressing high levels of mCherry-CHMP4B (right) show accumolation of the protein in multivesicular bodies (MVBs). Cells exhibiting the cellular distribution shown in left panel should be selected for imaging and analysis, while cells exhibiting the cellular distribution shown in right panel should be avoided. Scale bar: 10 μm

4. To capture the entire ESCRT structure at the intercellular bridge, Z sectioning is needed. To maximize axial resolution, use optimal step intervals (0.15 μm for X63 objective, numerical aperture, 1.4). In addition, take one to two extra Z slices above and beneath the structure for accurate reconstruction (*see* **Note 6**).
5. All images should be reconstructed post SIM acquisition. On the Elyra system, image series are reconstructed using ZEN software (Carl Zeiss MicroImaging). We use the following parameters for SIM reconstruction of ESCRTs:

   Noise filter (Wiener): ESCRT I and ESCRT II, -5.5; ESCRT III, -5.

   SR Frequency weighting: ESCRT I and ESCRT II, 1.7; ESCRT III, 1.

   Baseline shifted mode.

   Sectioning: 100, 95, 95.

   For other ESCRT components, it is recommended to use the automatic mode first and then fine-tune the parameters manually (*see* **Note 8**).

   (**Notes 5–8**, refer to this section).

3.2.5 Data Analysis

1. 3D rendering: we use Volocity 6 (PerkinElmer) for 3D rendering of the intercellular bridge. This software allows to rotate the bridge and to observe the protein organization in all dimensions. We find that looking through the bridge (Figs. 2a–d and 3d rotated) is very useful for distinguishing between a ring and a disk organization (Fig. 2a right panel, disk; b right panel, ring). Other programs such as Imaris (Bitplane) are equally suitable.
2. Measurements: All measurements should be performed on reconstructed data. We perform most geometrical measurements on a single slice from the central focal plane, using an intensity profile of a line drawn along the structure (Fig. 5a). To determine the localization of ESCRT proteins during abscission in relation to the intercellular bridge, the intensity profile of both the ESCRT component and the bridge microtubules can be plotted along the same line (Fig. 5a). To measure the thickness of a structure (Fig. 5, intensity plot), we calculate the distance between the lowest-intensity pixel located at the beginning of the structure and the lowest-intensity pixel located at the end of the structure, based on a line intensity profile along the bridge. Measurements of the net distance between different structural elements (Fig. 5b, intensity plot) are done by calculating the distance between the lowest-intensity pixel located at the end of one structure and the lowest-intensity pixel located at the beginning of the other structure, based on a line intensity profile along the bridge.

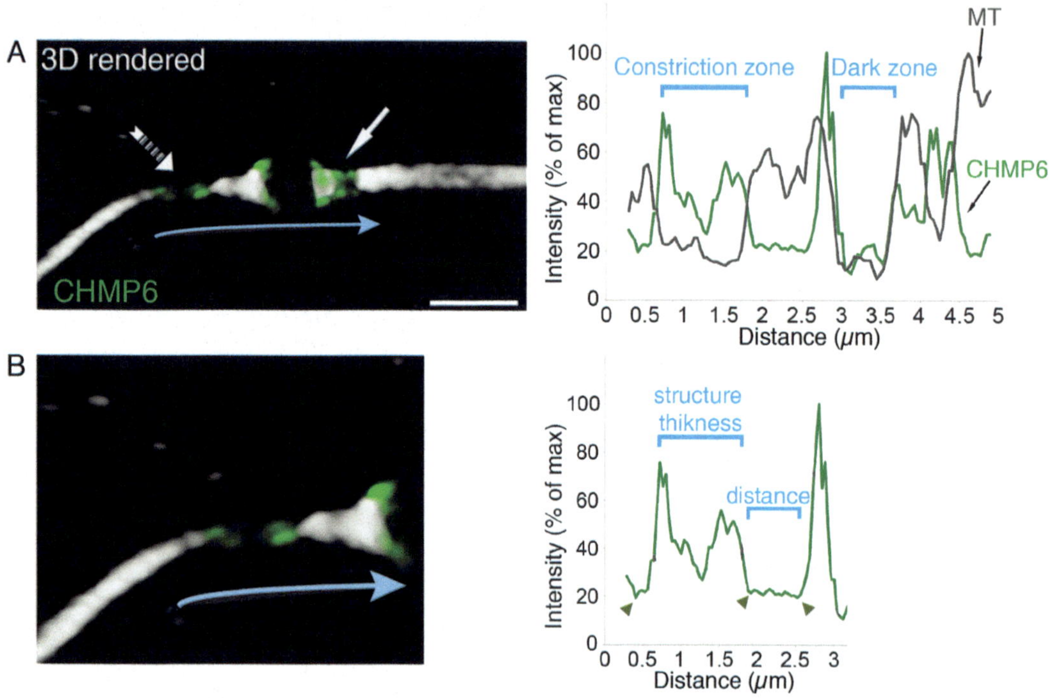

Fig. 5 Measuring ESCRT-III structures at the intercellular bridge. (**a**) Determining ESCRT-III localization along the intercellular bridge. MDCK cells expressing FLAG-CHMP6 were fixed, stained with anti–α-tubulin and anti-FLAG antibodies, and imaged by SIM. Shown is a 3D reconstruction of an overlay of the ESCRT III component CHMP6 (green) and tubulin (white). Plot on the right shows tubulin (gray) and CHMP6 (green) line intensity profiles along the intercellular bridge (indicated by a blue arrow in the left image). Distinct features of the intercellular bridge can be identified based on tubulin intensity profile and can be used for determining the position of the protein at high accuracy. Line intensity from left to right: first drop in tubulin intensity represents the abscission site (corresponds to dashed arrow on the left image). CHMP6 intensity peaks at the abscission site indicating CHMP6 localization at the abscission site. The second drop in tubulin intensity represents the dark zone. CHMP6 intensity peaks before and after the drop in tubulin intensity at the dark zone, indicating that CHMP6 is localized in two separated pools at the rims of the dark zone. Scale bar: 2 μm. (**b**) Measuring ESCRT -III structures' widths and distances. Shown in left is a zoomed-in image of one side of the intercellular bridge (corresponds to dashed arrow in **a**). Plot on the right shows CHMP6 line intensity profile in this region. Lowest intensity pixels (arrow head) are used to calculate the structure width and to determine the distance between two structures along the intercellular bridge. From left to right, intensity peak starts at 0.5 μm and ends at 1.75 μm, indicating that the width of the pool at the abscission site is 1.25 μm. Intensity drop starts at the end of the left structure—1.75 μm and signal begins to rise again at 2.6 μm, where the next structure starts. This indicates that the net distance between the pool in the abscission site and the pool in the dark zone is 0.85 μm. Adapted from [6]

3.3 Imaging with Cryo-SXT

Here we provide a specific cryo-SXT protocol, which was adapted to study the ultrastructure of the intercellular bridge using the cryo-SXT station in BESSY II, Helmholtz–Zentrum (Berlin, Germany). This protocol can be adapted to other cryo-SXT stations that are equipped with a cryo holder (cryo-SXT stations that use capillaries are less suitable for adherent cells).

3.3.1 Sample Preparation

To visualize the intercellular bridge and determine the stage in abscission, the intercellular bridge is labeled with fluorescently tagged tubulin. For this, MDCK cells stably expressing GFP-tubulin or HeLa cells stably expressing RFP-tubulin are used [14] (*see* **Note 9**).

Preparation of grids (*see* **Notes 10–11**):

1. Clean the grids before use. In a biological hood, dip grids in 100% ethanol and dry them on a glass slide for 20 min.
2. Optional: glow-discharge the grids to make the carbon film hydrophilic (*see* **Note 12**).
3. Place the grids in a six-well tissue culture dish, with the carbon side facing up.
4. Inspect the grids using a light microscope to ensure that the carbon foil is intact.
5. Incubate the grids, before seeding the cells, in growth medium for 1 h.
6. Plate cells on grids. When plating cells, consider that confluency must be correlated to the time of fixation, aim for ~60% confluency at the time of fixation (*see* **Note 13**).
7. On the next day, synchronize the cells as described in Subheadings 3.1.2 or 3.1.3.
   (**Notes 9–13**, refer to this section).

3.3.2 Fixation

1. Prewash the grids with warm (37 °C) PBS.
2. Fix grids with fixation solution for 15 min at 25 °C.
3. Wash cells three times with PBS. Grids can be stored in 4 °C up to 4 days in storage solution.

3.3.3 Grid Mapping

1. Place grids in eight-well dishes. For high precision and efficient mapping, place all grids at the same orientation. If using an inverted microscope, make sure the carbon is facing the bottom of the dish.
2. Image the grids using a fluorescence microscope. Systematically, acquire images of the entire grid; fluorescence images of the tubulin and brightfield images of the quantifoil grid. We routinely image the grids in a confocal spinning-disk microscope (Marianas; Intelligent Imaging, Denver, CO), using a 20× air objective (numerical aperture, 0.8).
3. Identify cells that are undergoing abscission by the tubulin fluorescence and register their location on a grid map (see example in Fig. 6a). Make sure all bridges are well documented to allow determining their stage in abscission. This will become important in the analysis step.

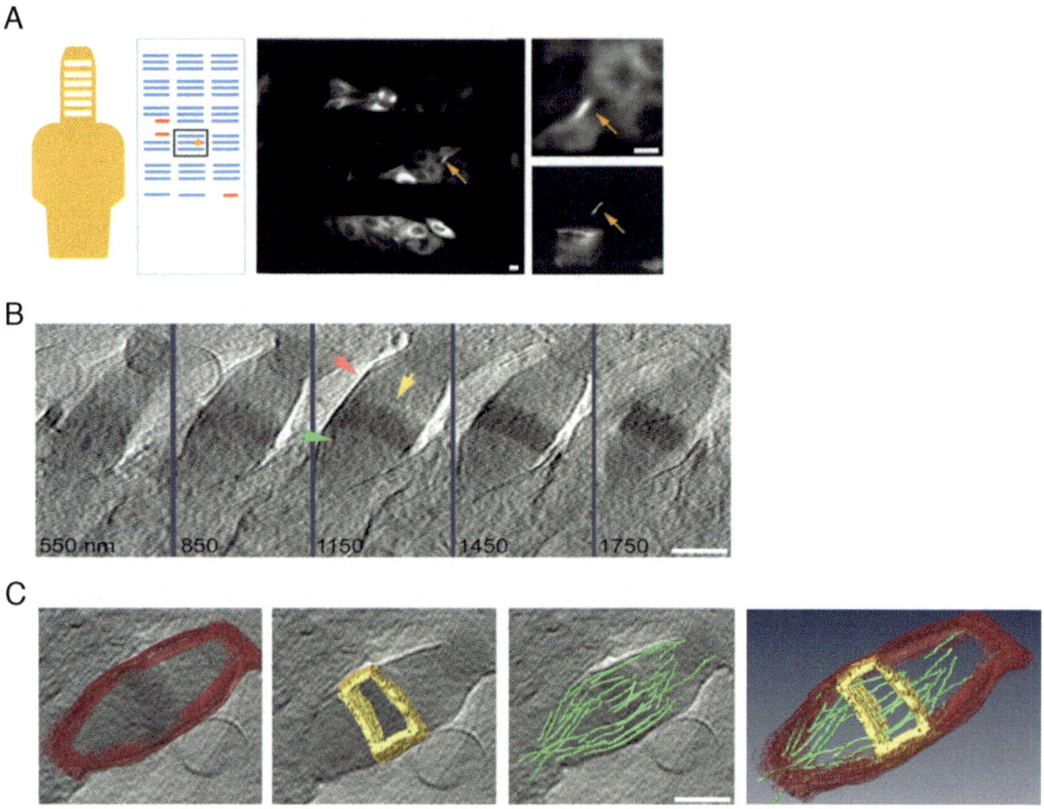

Fig. 6 Imaging cytokinetic abscission with cryo-SXT. (**a**) Shown from left to right are the tomography grid, the grid map, fluorescence spinning disk image of the area in map highlighted in a black square, a zoomed-in image of a selected bridge and a fluorescence image of the same bridge acquired at BESSY. Scale bar: 10 μm. (**b**) Cryo-SXT data set. An X-ray tilt series of a selected intercellular bridge was acquired and reconstructed. Shown are tomographic *Z* sections of the intercellular bridge shown in (**a**). Numbers at the bottom left corner represent *Z* section height in nm. Scale bar: 1 μm. (**c**) 3D rendering. Selected features of the intercellular bridge were rendered (red, intercellular bridge membrane; yellow, dark zone; green, microtubules). Shown are rendered structures overlaid on tomographic data and in right an image integrating the rendered structures. Scale bar: 1 μm. Adapted from [14]

4. Determine the microtubule diameter: draw a line intensity profile perpendicular to the intracellular bridge at the narrowest point along the bridge located about 1 μm peripherally to the center of the midbody [7, 16].

3.3.4 Vitrification

The technique is very sensitive to ice contamination. Avoid exposing the grids to the air and precool all instruments in liquid nitrogen before use.

1. Optional: Apply gold or polystyrene beads as fiducial markers (*see* **Note 14**).
2. Vitrify cells by plunging in liquid ethane. We use the Leica EM GP automated plunger, but other systems may be used as well.

Take care to blot opposite to the side on which the cells were seeded because face-on blotting may rupture the cells.

3. Immediately store grids in liquid nitrogen. HZB-2 grids do not fit standard transmission EM grid boxes. They may be stored in small PCR tubes, but make sure to puncture those to avoid explosion upon removal from liquid nitrogen (*see* **Note 15**).
4. Grids can be stored in liquid nitrogen for several months. (**Notes 14–15**, refer to this section).

3.3.5 Cryo-SXT acquisition

Grids should be shipped to synchrotron station in a cryogenic "dry shipper" and must be kept below −140 °C at all times (e.g., Taylor-Wharton model CX100B).

1. At the BESSY II station, the grids are mounted onto the cryo holder and inserted into the soft X-ray microscope.
2. Grid positioning is determined using the in-line light microscope.
3. Based on the stored grid positions, cells of interest are located. Stored positions are then verified based on tubulin fluorescence using the in-line epi-fluorescence microscope.

   It is recommended to capture a fluorescence image of the tubulin for your record.
4. Locate the intercellular bridge in SXT mode and focus on it. The intercellular bridge is then placed exactly at the center of the field for X-ray acquisition (*see* **Notes 16** and **17**).
5. Collect a single-axis tilt series over a range of up to ±65°, at 1° or 2° intervals.

   (**Notes 16–19**, refer to this section).

3.3.6 Reconstruction, 3D Rendering, and Data Integration

1. The tilt series of projection images should be aligned and reconstructed to a 3D volume (see example in Fig. 6b) [14]. Popular software packages for alignment include IMOD, Bsoft, and TomoJ (a plug-in for ImageJ) [17–19]. A preliminary alignment is performed on a binned version of the dataset and then refined using the full-resolution dataset. Volume reconstruction by weighted back-projection may be done either within the alignment software package or using the stand-alone program tomo3d [20] (*see* **Notes 20** and **21**).
2. To rotate the bridge and to observe structures in all dimensions, use 3dmod (part of the IMOD package), UCSF Chimera, or commercial Amira software (Thermo Fisher Scientific).
3. 3D volume surface rendering is performed using 3dmod or Amira. In most cases, we were unable to use automatic segmentation to render specific features of the intercellular bridge. Objects of interest should therefore be manually segmented

and then rendered by isosurface contours to show the minimum volume intact surface. Examples for rendered structures at the intercellular bridge are provided in Figs. 3 and 6b. Measurements of rendered data can be performed in ImageJ on masked intensities.

4. SXT data should eventually be correlated to their respective fluorescence images in order to determine the stage in abscission (*see* **Note 22**).
(**Notes 20–22**, refer to this section).

4 Notes

1. Transfected HeLa cells do not synchronize well with a double thymidine block.
2. To avoid cellular artifacts, ESCRT proteins should be tagged at the N terminus. Additionally, we found that mCherry conjugation provides better results than GFP. Recently, it has been shown that adding a long flexible linker (LAP) between the fluorescent tag and the ESCRT protein provides better results [3]. Expressing the protein at endogenous levels using the CRISPR/Cas9 technology can also be used to avoid cellular artifacts.
3. To verify that the fluorescently tagged ESCRTs do not interfere with abscission, it is recommended to perform live imaging and compare abscission timing and morphology to control cells (For more information, refer to [16]).
4. SIM imaging demands bright fluorescence. Therefore, when using fluorescently tagged proteins, it is recommended to image the cells shortly after fixation (not more than a few days post fixation). Additionally, exposing the slides to light should be avoided at all times. If signal intensity is not sufficient for SIM, the fluorescent tag can be replaced with a smaller tag such as FLAG, followed by anti-FLAG staining.
5. To avoid overexpression artifacts, we select low-expressing cells based on the overall organization of ESCRTs in these cells. Cells in which ESCRTs are distributed mostly in the cytosol are selected for imaging and data analysis, while cells in which ESCRTs are accumulated in discrete foci (most probably multivesicular bodies) are avoided (Fig. 4).
6. Number of *Z* stacks is a matter of a tradeoff between larger *Z* range to sample brightness and bleaching. When more than 25 slices are acquired, there is a risk of losing information due to bleaching. To set the *Z*-stack parameters, it is recomended to use the brightest channel; we typically use the microtubule staining at the intercellular bridge. Some of the ESCRTs

exhibit a ring (Fig. 2b, right panel) or disk (Fig. 2a, right panel) morphology that has a larger diameter than the microtubules in the bridge. In this case, it is recommended to use the ESCRT fluorescence when setting the *Z*-stack parameters. Alternatively, a few extra *Z*-steps above and below the microtubule staining should be taken.

7. When imaging ESCRT proteins at the intercellular bridge, the signal is often very dim and accumulates in a single condensed region. Such localization pattern can result in artifacts during reconstruction. This can be avoided by including other regions of the cells that include a fluorescence signal in the captured image.
8. SIM reconstruction is sensitive to saturation. Therefore, saturation should be avoided over the entire field of view. If the ESCRT staining at the bridge is too dim to allow capturing it while avoiding saturation in other regions, the imaging parameters should be optimized to allow capturing the ESCRT structure. The region of interest that does not contain saturated pixels should then be cropped and reconstructed separately.
9. Transient transfection can be also used to label tubulin. However, the percentage of labeled cells in each grid will be smaller. Additionally, transiently transfected HeLa cells fail to synchronize using a double thymidine block.
10. For cryo-SXT at BESSY II, we use QUANTIFOIL® R 2/2 or R1/4 on special format HZB-2 gold grids (Quantifoil Micro Tools GmbH) [12]. These fit the Gatan model 630 side-entry cryo-tomography holder. Standard 3 mm grids for electron microscopy may also be used. Copper grids must be avoided because of toxicity to the cells.
11. Handle the delicate grids very gently with thin tweezers, hold them gently at the edge, and avoid bending them. It is imperative to practice this in advance.
12. Based on our experience, glow-discharge does not influence the adherence of MDCK or HeLa cells and therefore can be avoided. This step should be considered when using other cell lines. Additionally, other cell types may adhere better if the grids are preincubated in a solution containing an adhesion protein such as fibronectin.
13. Cell density on the day of fixation is critical for the success of the experiment. The cells on the grid should be sub-confluent in order not to "shadow" the intercellular bridge with a thick layer of neighboring cells but should be dense enough for productive acquisition. Calibrating cell density is not always straightforward and requires some calibrations and adjustments. We recommend calibrating this step for each cell line and conditions used in the study.

14. Gold or polystyrene beads may be added as fiducial markers to assist in the reconstruction of the tomogram. These beads should be rather large (150–250 nm) for good visibility in the X-ray images. Adjusting their density is tricky without feedback from the cryo-imaging, but the target density is to have about ten such markers visible in a field of view (13 or 21 μm at BESSY II, depending on the X-ray objective in use). Where a cryo-light microscope is available, it is useful to calibrate fluorescent polystyrene beads density.
15. The grids are not marked. Therefore, make sure to place the grids in prelabeled tubes.
16. Similarly to EM, the dark zone appears dark in SXT because of the high concentration of proteins, which absorbs the soft X-rays.
17. It is sometimes difficult to detect the intercellular bridge in SXT mode. Identifying the dense dark zone region can help in this stage. Also, note that in most cases, the bridge is not in the same focal plan as the rest of the cell. Therefore, scan different focal plans when trying to locate the dark zone region.
18. In principle, it is not obligatory to visualize the bridge in SXT prior to acquisition of the tomogram. However, based on our experience, the ability to clearly visualize the dark zone in SXT mode provides a good indication for successful, high-resolution data acquisition. It is best to find bridges that are not obscured by cellular outgrowths.
19. It is not recommended to acquire a tilt series in regions that are covered with thick ice or that appear to be covered by ice contamination.
20. IMOD is the most familiar to electron microscopists. Bsoft is particularly useful for SXT when fiducial particles are visible throughout the tilt series. Often fiducial markers are not present, but image features may be used for alignment of the projections. In that case, we recommend TomoJ using the "critical minimum" detection mode to follow dark features in the image [21, 22].
21. In some cases, we found that expansion of the reconstruction to full size does not improve the level of detail. In such cases, the binned images are used for further analysis.
22. It is highly recommended to keep a record of all file names that are related to the same cell. This way, the fluorescence image acquired during mapping can be easily correlated to the fluorescence image acquired at the synchrotron and to the SXT 3D rendered data set.

Acknowledgments

We thank the all members of our laboratory and especially Dr. Dikla Nachmias for contributing technical tips. We acknowledge the Helmholtz-Zentrum Berlin–electron storage ring BESSY II for the allocation of synchrotron radiation beamtime at beamline U41-FSGM. This work is funded in part by the Instruct R&D pilot award for integrating biology, the Israeli Science Foundation (ISF) grant no. 455/13 (to N.E.) and 1285/14 (to M.E.), and the Marie Curie Integration Grant (CIG) (to N.E.).

References

1. Elia N, Sougrat R, Spurlin TA, Hurley JH, Lippincott-Schwartz J (2011) Dynamics of endosomal sorting complex required for transport (ESCRT) machinery during cytokinesis and its role in abscission. Proc Natl Acad Sci U S A 108(12):4846–4851. https://doi.org/10.1073/pnas.1102714108
2. Elia N, Ott C, Lippincott-Schwartz J (2013) Incisive imaging and computation for cellular mysteries: lessons from abscission. Cell 155 (6):1220–1231. https://doi.org/10.1016/j.cell.2013.11.011
3. Mierzwa BE, Chiaruttini N, Redondo-Morata-L, Moser von Filseck J, Konig J, Larios J, Poser I, Muller-Reichert T, Scheuring S, Roux A, Gerlich DW (2017) Dynamic subunit turnover in ESCRT-III assemblies is regulated by Vps4 to mediate membrane remodelling during cytokinesis. Nat Cell Biol 19 (7):787–798. https://doi.org/10.1038/ncb3559
4. Guizetti J, Schermelleh L, Mantler J, Maar S, Poser I, Leonhardt H, Muller-Reichert T, Gerlich DW (2011) Cortical constriction during abscission involves helices of ESCRT-III-dependent filaments. Science 331 (6024):1616–1620. https://doi.org/10.1126/science.1201847
5. Van Engelenburg SB, Shtengel G, Sengupta P, Waki K, Jarnik M, Ablan SD, Freed EO, Hess HF, Lippincott-Schwartz J (2014) Distribution of ESCRT machinery at HIV assembly sites reveals virus scaffolding of ESCRT subunits. Science 343(6171):653–656. https://doi.org/10.1126/science.1247786
6. Goliand I, Nachmias D, Gershony O, Elia N (2014) Inhibition of ESCRT-II-CHMP6 interactions impedes cytokinetic abscission and leads to cell death. Mol Biol Cell 25 (23):3740–3748. https://doi.org/10.1091/mbc.E14-08-1317
7. Elia N, Fabrikant G, Kozlov M, Lippincott-Schwartz J (2012) Computational model of cytokinetic abscission driven by ESCRT-III polymerization and remodeling. Biophys J 102:2309–2320
8. Heintzmann R, Cremer CG (1999) Laterally modulated excitation microscopy: improvement of resolution by using a diffraction grating. Proc SPIE 3568(1):185–196
9. Gustafsson MG (2000) Surpassing the lateral resolution limit by a factor of two using structured illumination microscopy. J Microsc 198(Pt 2):82–87
10. Schneider G, Guttmann P, Heim S, Rehbein S, Mueller F, Nagashima K, Heymann JB, Muller WG, McNally JG (2010) Three-dimensional cellular ultrastructure resolved by X-ray microscopy. Nat Methods 7(12):985–987. https://doi.org/10.1038/nmeth.1533
11. Do M, Isaacson SA, McDermott G, Le Gros MA, Larabell CA (2015) Imaging and characterizing cells using tomography. Arch Biochem Biophys 581:111–121. https://doi.org/10.1016/j.abb.2015.01.011
12. Hagen C, Guttmann P, Klupp B, Werner S, Rehbein S, Mettenleiter TC, Schneider G, Grunewald K (2012) Correlative VIS-fluorescence and soft X-ray cryo-microscopy/tomography of adherent cells. J Struct Biol 177(2):193–201. https://doi.org/10.1016/j.jsb.2011.12.012
13. Elad N, Abramovitch S, Sabanay H, Medalia O (2011) Microtubule organization in the final stages of cytokinesis as revealed by cryo-electron tomography. J Cell Sci 124 (Pt 2):207–215. https://doi.org/10.1242/jcs.073486
14. Sherman S, Kirchenbuechler D, Nachmias D, Tamir A, Werner S, Elbaum M, Elia N (2016) Resolving new ultrastructural features of cytokinetic abscission with soft-X-ray cryo-

tomography. Sci Rep 6:27629. https://doi.org/10.1038/srep27629

15. Varsano N, Dadosh T, Kapishnikov S, Pereiro E, Shimoni E, Jin X, Kruth HS, Leiserowitz L, Addadi L (2016) Development of correlative cryo-soft X-ray tomography and stochastic reconstruction microscopy. A study of cholesterol crystal early formation in cells. J Am Chem Soc 138(45):14931–14940. https://doi.org/10.1021/jacs.6b07584
16. Gershony O, Sherman S, Adar S, Segal I, Nachmias D, Goliand I, Elia N (2017) Measuring abscission spatiotemporal dynamics using quantitative high-resolution microscopy. Methods Cell Biol 137:205–224. https://doi.org/10.1016/bs.mcb.2016.03.032
17. Kremer JR, Mastronarde DN, McIntosh JR (1996) Computer visualization of three-dimensional image data using IMOD. J Struct Biol 116(1):71–76. https://doi.org/10.1006/jsbi.1996.0013
18. Heymann JB, Belnap DM (2007) Bsoft: image processing and molecular modeling for electron microscopy. J Struct Biol 157(1):3–18. https://doi.org/10.1016/j.jsb.2006.06.006
19. Schneider CA, Rasband WS, Eliceiri KW (2012) NIH Image to ImageJ: 25 years of image analysis. Nat Methods 9(7):671–675
20. Agulleiro JI, Fernandez JJ (2011) Fast tomographic reconstruction on multicore computers. Bioinformatics 27(4):582–583. https://doi.org/10.1093/bioinformatics/btq692
21. Messaoudii C, Boudier T, Sanchez Sorzano CO, Marco S (2007) TomoJ: tomography software for three-dimensional reconstruction in transmission electron microscopy. BMC Bioinformatics 8:288. https://doi.org/10.1186/1471-2105-8-288
22. Sorzano CO, Messaoudi C, Eibauer M, Bilbao-Castro JR, Hegerl R, Nickell S, Marco S, Carazo JM (2009) Marker-free image registration of electron tomography tilt-series. BMC Bioinformatics 10:124. https://doi.org/10.1186/1471-2105-10-124

Chapter 11

Three-Dimensional Surface Rendering of ESCRT Proteins Microscopy Data Using UCSF Chimera Software

Romain Le Bars, Michele W. Bianchi, and Christophe Lefebvre

Abstract

Visualization of subcellular localization of ESCRT proteins and their interactions with different cellular compartments are critical to understand their function. This approach requires the generation of an important amount of 3D fluorescence microscopy data that is not always easy to visualize and apprehend.

We describe a step-by-step protocol for 3D surface rendering of confocal microscopy acquisitions using the free software UCSF-Chimera, generating snapshots and animations to facilitate analysis and presentation of subcellular localization data.

Key words 3D fluorescence confocal microscopy, UCSF Chimera, ESCRT, Surface rendering, 3D animation

1 Introduction

The endosomal sorting complexes required for transport (ESCRT-0 to ESCRT-III) were initially identified, in eukaryotes, as key proteins involved in the sorting of ubiquitylated membrane proteins into intraluminal vesicles during the maturation of multivesicular bodies [1, 2]. Although the ESCRT machinery is well conserved in all eukaryotes, its complexity has arisen during the evolution of multicellular organisms. Notably, duplication of several genes in mammals [3] raises the possibility of additional cellular roles for ESCRT components. Indeed, beyond their canonical role in the formation of MVBs during endosomal maturation, ESCRT complexes in metazoans have also been shown to be involved in other fundamental cellular mechanisms: membrane remodeling during exosome biogenesis [4], enveloped virus budding [5, 6], cell cytokinesis membrane abscission [2], immunological synapse formation [7], plasma membrane repair [8], biogenesis of the nuclear membrane [9–11], and sarcoplasmic reticulum shaping

Emmanuel Culetto and Renaud Legouis (eds.), *The ESCRT Complexes: Methods and Protocols*, Methods in Molecular Biology, vol. 1998, https://doi.org/10.1007/978-1-4939-9492-2_11,

[12] (Fig. 2). Recently, Stenmark and colleagues have shown that ESCRT-III and VPS4 cooperate at sites of nuclear envelope and microtubules intersection during nuclear envelope reassembly at mitotic exit [11]. Direct observation, using confocal microscopy, of ESCRT subunits within the cell played a decisive role to reveal novel key cellular functions of ESCRTs. More recent technical developments, such as spinning disk microscopy, resonant scanners, or light sheet microscopy, allow resolutive imaging of entire cells within few seconds, enabling analysis of live samples and dynamic studies in 3D.

However, exploitation of obtained data is not always straightforward, and the user can be puzzled by the abundance of tools available for visualization of 3D acquisitions. In this chapter, we describe a step-by-step protocol to create 3D representations of confocal microscopy data (*z*-stacks), and the procedure to export them as a dynamic movie using the free software UCSF-Chimera (hereafter called Chimera) [13].

One way to represent microscopy data in 3D is with surfaces separating voxels considered as signal from those considered as background. Surface rendering computes isosurfaces approximating the positions of voxels, whose intensity is equal to a threshold value, and is well suited for the representation of small and dense objects. A second approach to the generation of 3D representations is volume rendering. In this case, every voxel above threshold is represented with two main properties: color and opacity. Since it allows the observation of inner structures by transparency, volume rendering is adapted to larger objects with complex organization.

The UCSF Chimera software is developed by the RBVI (Resource for Biocomputing, Visualization, and Informatics at the University of California San Francisco, USA) for the interactive visualization and analysis of molecular structures. However, among the multiple options offered, high-quality surface rendering and the possibility to record animations make it also a very practical tool for the easy representation of cellular confocal microscopy data.

Recent results indicate that ESCRT-II proteins are involved in the control and/or maintenance of the sarcoplasmic reticulum structure in *C. elegans* muscle cells [12]. Confocal microscopy analysis and correlative light and electron microscopy (CLEM) indicate that ESCRT-II proteins are localized nearby actin anchoring zones corresponding to the sarcoplasmic reticulum. Employing Chimera on original confocal microscopy data of VPS-22 and VPS-36 in *C. elegans* pharyngeal and body wall muscle cells, we were able to better understand the localization of these proteins by generating 3D surface reconstructions under different angles of view. VPS-22 can be visualized close to the actin attachment structure at the basal membranes of pharyngeal cells (Fig. 2a–d). In body wall muscle, VPS-36 can be observed not only along dense bodies but also very close to the muscle cell plasma membrane

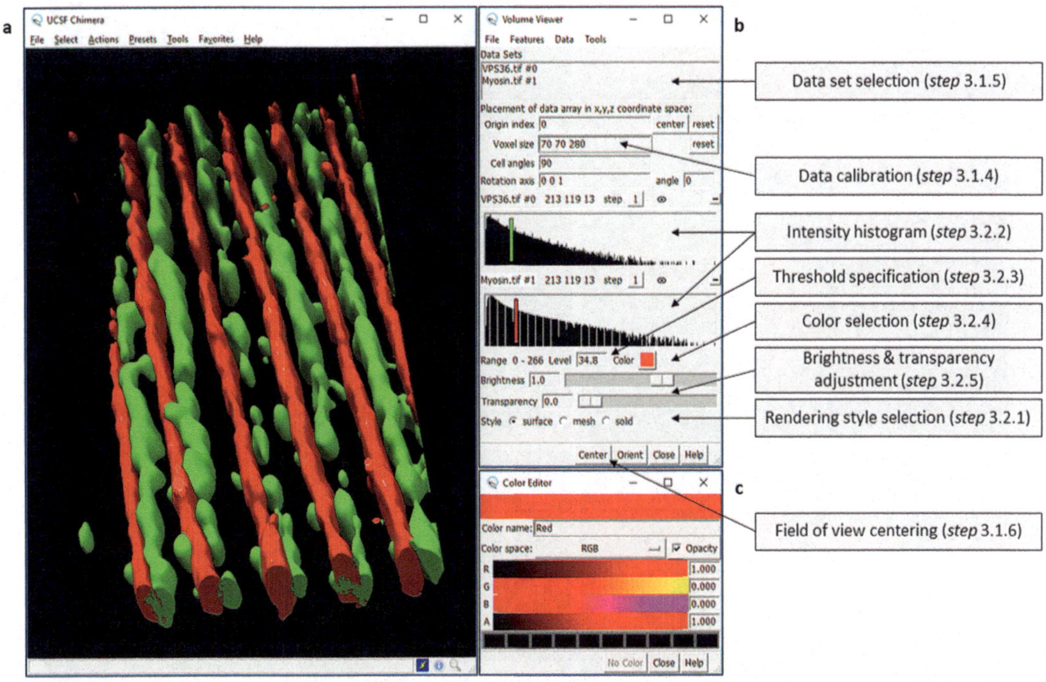

Fig. 1 Screen captures of the "UCSF Chimera" main window (**a**), the "Volume Viewer" (**b**) and the "Color Editor" (**c**) boxes. This snapshot presents the Chimera interface used for 3D rendering of confocal *z*-stack of ESCRT-II VPS-36 (green) and myosin (red) in *C. elegans* body wall muscle cells. This organization is consistent with the fact that in this tissue VPS-36 has been shown to be located in close proximity to the actin anchoring structures

(Figs.1 and 2e–h). This localization corresponds to the sarcoplasmic reticulum according to published data. The step-by-step protocol described in this paper will describe how to generate 3D representations and animations to investigate the spatial localization of proteins of interest.

2 Materials

2.1 Confocal Microscopy z-Stacks

3D rendering can be applied to image *z*-stacks acquired with any kind of optical system. However, in the field of cell biology, confocal microscopy, with its increased resolution and contrast, is often preferred.

Like always in image processing, but especially true for 3D reconstructions, the better the quality of original images, the easier their processing and analysis will be. Therefore, we strongly advice to ensure that acquisition steps are optimized before starting 3D reconstruction. Compared to 2D acquisitions, important additional aspects must be taken into account when acquiring *z*-stacks: an adequate *z*-step (*see* **Note 1**), sample movements made possible

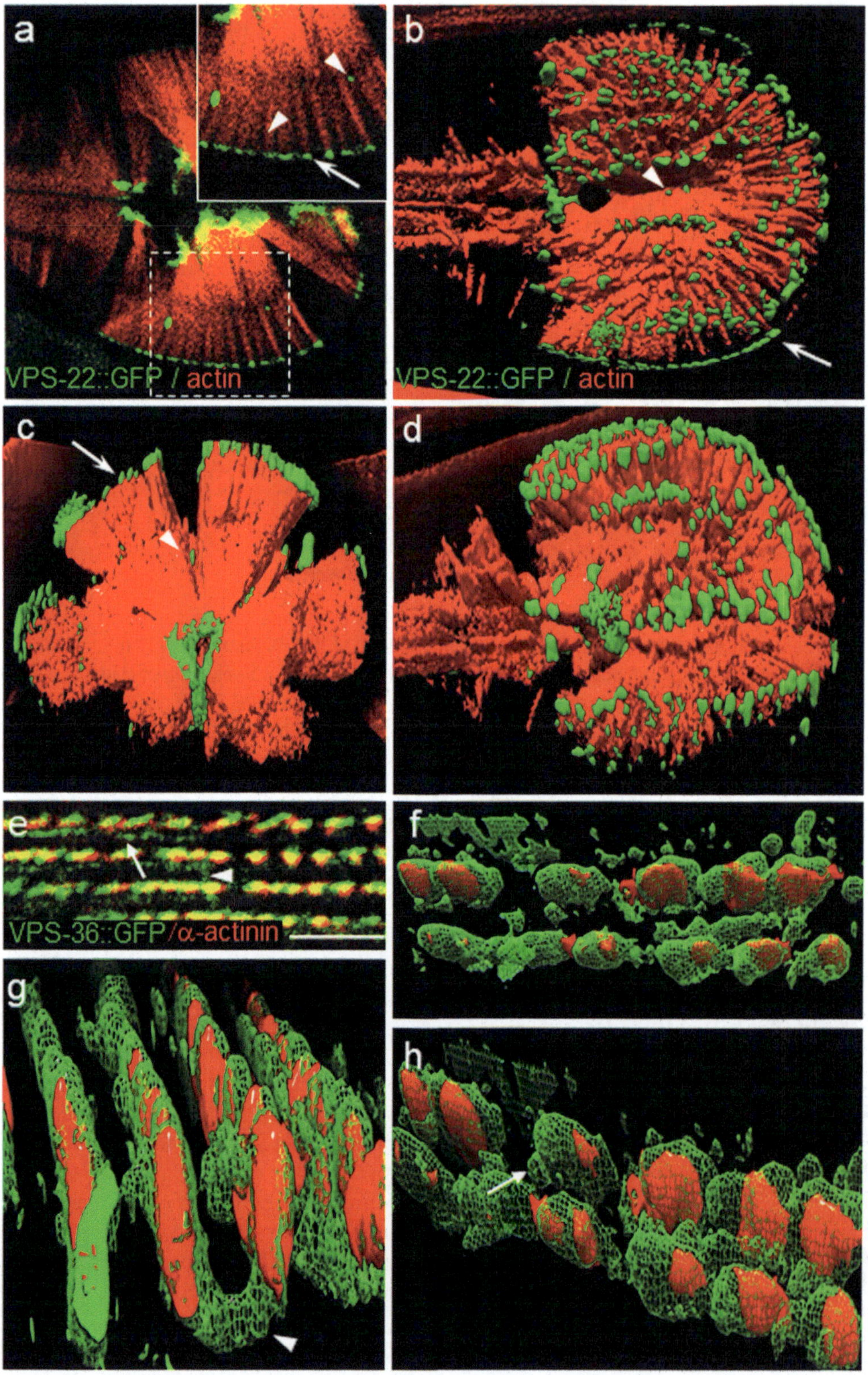

Fig. 2 3D localization of ESCRT-II to sarcomeric reticulum network in muscles of *C. elegans*. (**a**) Single confocal plan of the pharyngeal terminal bulb from worms expressing VPS-22::GFP (green) and immunolabelled for F actin filaments (red). The inset shows a higher magnification of the selected area

by the longer acquisition time required (*see* **Note 2**), mechanical or optical deformations along the *z* axis (*see* **Note 3**), and the loss of signal intensity along the *z* axis due to fluorochrome bleaching, or photon diffraction/refraction/absorption (*see* **Note 4**). On the bright side of things, *z*-stacks make it possible to employ user-friendly image deconvolution software to efficiently improve image quality by increasing resolution and signal-to-noise ratio (*see* **Note 5**).

2.2 UCSF Chimera Software

The last version of Chimera can be downloaded free of charge for academic, government, nonprofit, and personal use at https://www.cgl.ucsf.edu/chimera/index.html. Chimera versions are available for Windows, Mac OS, and Linux system and can be run on virtually any current desktop or laptop computer. To dwell further into the multiple possibilities of Chimera, do not hesitate to refer to the user guide (https://www.cgl.ucsf.edu/chimera/current/docs/UsersGuide).

3 Methods

3.1 Data Input and Calibration

1. Chimera can read stacks of grayscale images in any of the bit depths and image file types recognized by the *Python Imaging Library* (http://www.pythonware.com/products/pil/), such as .tif, .tiff, .png, etc. To reach the "Open File in Chimera" window, select: *File* > *Open*... from the menu. Navigate through the filepath of your stack by clicking (from left to right) on the different folders and subfolders, and press the "Open" button once you reach the image file. In the case of multiple color channels, it is recommended to prepare one image stack for each channel and to repeat the above procedure for each.
2. Once the image stack is loaded, Chimera automatically detects its format and opens the "Volume Viewer" window (Fig. 1).

Fig. 2 (continued) of the pharyngeal terminal bulb. Arrows point to the actin attachment structure localized to the basal membranes of pharyngeal cells, arrowheads point to the signal localized between the F actin bundles. (**b–d**) A Chimera 3D surface rendering corresponding to original images from panel (**a**) allows a more precise visualization of VPS-22 throughout the pharynx. (**e**) Single confocal plan from a *z*-stack of body wall muscles expressing VPS-36::GFP (green) and stained with dense body markers α-actinin (red). Arrow and arrowhead point to VPS-36::GFP signal, localized between dense body and just below the sarcolemma. (**f–h**) A Chimera 3D surface rendering of the same *z*-stack. The visualization angle chosen highlights the specific localization of VPS-36 between dense bodies in the proximity of the basal membrane (arrowhead) and around dense bodies (arrow). This localization corresponds to the sarcoplasmic reticulum network close to the plasma membrane facing hypodermis. Scale bar: 5 μm

This window manages almost all the tools needed for 3D rendering. Should you close it by mistake, it can be opened again following the main window menus: *Tools* > *Volume Data* > *Volume Viewer.*

3. Configure the "Volume Viewer" window by selecting the following options in the "Features" menu: "Brightness and Transparency," "Coordinates," "Data set list," "Display style," "Threshold and Color." Save this display for your next Chimera sessions by clicking: *Features* > *Save default panels.*
4. Calibrate the data by setting the proper voxel size (*see* **Note 1**). Type in the *x y z* values, separated by spaces, within the "Voxel size" box of the "Volume Viewer" window, and then press "Enter." *x* and *y* values are identical in the common case of square pixels, and the *z* value corresponds to the distance between two consecutive planes in your stack. For example, images with 65 nm wide square pixels acquired with a 200 nm *z*-step will be properly calibrated in micrometers by typing "0.065 0.065 0.2". Keep in mind the units employed, as this information will be required to configure the scale bar in a later step.
5. If you have opened multiple stacks (e.g., one for each color), switch to each channel listed in the "Data Sets" box of the "Volume Viewer" window to adjust individually voxel sizes, so that all loaded datasets are calibrated properly.
6. After this calibration step, objects will be rescaled, and sometimes disappear from the field of view. Should this happen, re-center the view by pressing the "Center" button at the bottom of the "Volume Viewer" window.

3.2 Setting Threshold Values and Colors

1. Select the rendering style for each data set in the "Volume Viewer window." It is possible to choose between "surface," "mesh," and "solid." Based on the threshold value(s) you will set, isosurfaces are computed and rendered in the selected style. It can be useful to apply a different rendering style to each data set/color channel.
2. Set the threshold value to generate a 3D representation of your data. This step is one of the most important and must be executed thoughtfully, as it shapes the final object. As a word of caution, note that these 3D representations should be considered a mean of illustrating properties, which have already been well established through careful analysis of many *z*-stacks and their orthogonal projections, rather than a tool for the discovery of new features.
3. To adjust the threshold value, slide the bar on the histogram (*see* **Note 6**) to move the threshold either toward the lowest values (left end of the histogram) for generous inclusion of

voxels in the objects or toward the highest intensities (right end of the histogram) to apply a more selective criterion. If you prefer to reproduce the same threshold multiple times, type the desired pixel intensity in the "Level" box under the histogram to specify a threshold value and validate by pressing "Enter" (*see* **Note 7**).

4. Select a display color for each dataset, by pressing the "Color" button of the "Volume Viewer" window to open the Color Editor. A color can then be selected either by selecting a color space (RGB, CMYK, ...) and adjusting the different sliders to obtain the desired combination of each component, or by directly typing a color name ("red", "green", "blue", ...).
5. Adjust brightness and transparency of the rendered surface for each channel. The "Brightness" slider allows to modulate color saturation, and the "Transparency" slider allows to display objects as completely opaque (transparency = 0) up to fully transparent (transparency = 1).
6. Remove unwanted small objects resulting from the thresholding step. Open the "Hide Dust" window (*Volume Viewer window > Tools > Hide Dust*). Set a cutoff value and press the "Hide" button. This will remove from the display all objects with a size, area, or volume (depending on the option you choose) smaller than the cutoff value. You can also remove or perform sections using the "Volume Eraser" and the "Side View" tools (*see* **Note 8**).
7. Finally, rendering of the 3D surface is adjusted by applying smoothing iterations and/or increasing the subdivision or your objects (*see* **Note 9**).

3.3 Generating a Snapshot

1. Inclusion of a scale bar: before exporting a 2D image of your objects, you may add a scale bar (*Chimera main window > Tools > Higher-Order Structure > Scale Bar*). Activate the "Show scale bar" option to display the scale bar (note that switching *off* and *on* this option helps to refresh the display after a modification). Set length and width of the bar (units will be those employed at the calibration step). In the "Label" field, type the proper unit preceded by a "#" sign to recall the length of the bar. Use the "Offset" boxes to move the label relatively to the bar. Turn on the "Preserve screen position" option and then set the position of the scale bar in the screen by typing its coordinates in the *X* and *Y* fields (from −1 to 1 with 0;0 as the center of the field of view).
2. Create an image: *Chimera main window > File > Save Image...*. Choose the output image file type and the image size (in pixels). Then in the image options, select "Chimera" rendering mode (best for most of representations), introducing

an oversampling as needed ("supersample" box) to enhance definition of the exported snapshot. Before saving, check, at the bottom left of the "Save Image" window, that the "effective maximum line width" has a value of at least 1 (this is a system-dependent limit on line drawing). If this is not the case, reduce the number of pixels of the desired snapshot, either by selecting lower "Supersample" values or by reducing image width and height. Save the image by pressing the "Save" button.

3. Save the session: to save all the work previously done, you can create a backup of your session as a ".py" file (*Chimera main window* > *File* > *Save Session As...*). Later on, you can restore a session (*Chimera main window* > *File* > *Restore Session...*) if you wish to modify your 3D rendering, extract new images, or create animations. Note that when restoring a session, the data files used to generate the rendering must be available again to Chimera. Any change of location of the image files or the .py session file may cause trouble, so we suggest that images files employed and the Chimera .py session file are kept in the same folder.

3.4 Record an Animation

1. Open the "Animation" window: *Chimera main window* > *Tools* > *Utilities* > *Animation* (Fig. 3).
2. Set up the display of your 3D objects (Orientation, zoom) to prepare the first scene of the animation, and add it to the

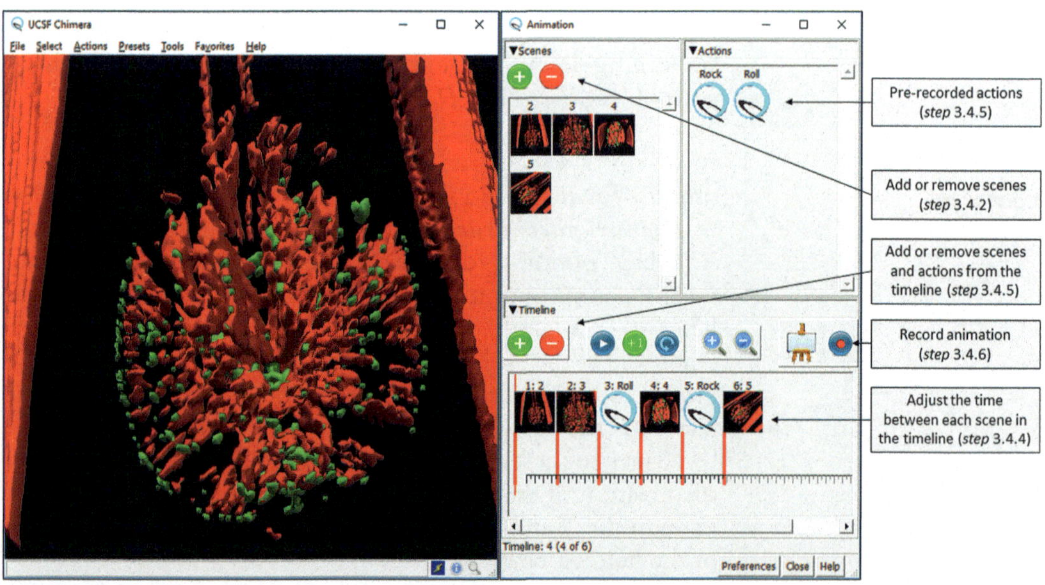

Fig. 3 Chimera's animation recording interface: presentation of the different tools used to generate and record an animation in Chimera. This snapshot represents 3D localization of ESCRT-II VPS-22 protein (green) and the actin filaments (red) in *C. elegans* pharyngeal muscle cells. This representation illustrates well the specific localization of VPS-22 near the sarcolemma at podosome-like sites of actin attachment

"Scenes" box by pressing the green "+" button in the "Animation" window.

3. Generate subsequent scenes changing, for example, the point of view (sections can be performed with the "Volume Eraser" tool) and add them to the "Scenes" box as described above.
4. Place scenes onto the "Timeline" box: select the scene (multiple selection is allowed) and press the green "+" button in the "Timeline" box of the "Animation" window. Chimera will automatically generate a transition to go from one scene to the next one. To set the speed of this transition, it is possible to adjust the distance between each scene by sliding them along the timeline or by specifying the duration of each transition (*right click on a timeline thumbnail* > *Properties* > *Duration*). Note that the movie is encoded at 25 frames per second.
5. You can also add predefined animation "Actions" such as "Rock" (rotation 60° left then right) or "Roll" (360° rotation) by dragging the corresponding icon from the "Action" box to the timeline after a scene.
6. Record your animation by pressing the record button (at the right of the timeline). In the "Record Animation" window, specify the output path and the file type of your animation. Choose "Chimera" as rendering mode and increase the "Supersample" value to obtain the desired rendering quality.

4 Notes

1. Axial sampling: As for the *x*-*y* pixel size, depth of the *x*-*y*-*z* voxel should ideally be 2.3 times smaller than the diffraction-limited *z*-resolution of the system, according to what is known by microscopists as the Niquist's criterion. However, larger sizes can still give acceptable results. In addition, bear in mind that, to estimate the actual depth of voxels, it may be necessary to consider some *z*-scaling due to refraction index mismatches in your system. In practice, voxel depth, which is particularly important when planning to apply image deconvolution, is often chosen as a compromise between the Niquist's criterion and the necessity to reduce the number of planes in the *z*-stack to control bleaching and/or movement in your sample.
2. Stack registration: When live imaging, some degree of whole-sample movement/vibration may be at times difficult to avoid. It may be possible in these cases to realign image planes to each other employing image registration algorithms, such as those implemented in the Fiji distribution of ImageJ [14] (navigate through Plugins/Registration/Rigid Registration). Fast movement of intracellular objects can be more problematic

and solutions, which will depend on the properties of the confocal system employed and on the number of different fluorochromes to be acquired, cannot be discussed here. We find that a spinning disk system, if necessary, equipped with a beam splitter device to acquire two colors simultaneously, can yield satisfactory results in most cases, especially after image deconvolution.

3. 3D imaging: To avoid mechanical deformation of thick samples, for example, whole organs or organisms, opportune spacers can be employed when mounting between a microscope slide and coverslip. For example, 100 μm-thick double-sided scotch works well for roots of *Arabidopsis* seedlings, stabilizing at the same time the assembly. For in vivo observation in *C. elegans*, worms or embryos are deposited within a drop of medium on an agarose pad and covered with a coverslip. Alternatively, worms can be deposited in a drop of medium containing beads. The beads play the role of spacer and allow the immobilization of the worms without having resort to drugs. We use beads of 0.25 μm for embryos and 0.10 μm for worms (Polybead Polystyrene 25 μm Microspheres and 0.1 μm Microspheres, respectively; Polysciences, Inc.). The major source of artifacts in 3D acquisitions is, however, the possible mismatch between the objective employed and refractive indexes of the mounting media and sample. This typically occurs when employing oil-immersion objectives instead of water-immersion objectives with live samples. When acquiring a *z*-stack in these conditions, the refraction of light at the interface of media with different RI leads, among other disadvantages, to geometrical distortion: a mismatch between the displacement along the *z*-axis of the sample imposed by the microscope motor and the actual displacement of the focal plane of the images acquired, causing, for example, a spherical object to appear as stretched along the *z* axis. This distortion can be corrected empirically, for example, by employing fluorescent beads, or theoretically [15].

4. Correction of signal attenuation along the *z*-axis: Loss of signal intensity due to the accumulation of optical aberrations along the *z*-axis can be controlled by imaging, when applicable, as close as possible to the cover slip, and by keeping RI mismatches to a minimum. Since some systematic loss of signal is nevertheless unavoidable in thicker *z*-stacks, this should be corrected post-acquisition employing image processing software such as NIH ImageJ [16]. Note that some confocal microscopy systems allow to automatically adjust laser intensity during acquisition of *z*-stack as an attempt to compensate for signal loss.

5. Image preprocessing: As 3D reconstructions are based on image thresholding to isolate regions corresponding to objects of interest, image preprocessing can greatly improve the final result. Several approaches can be implemented with NIH ImageJ (or similar software), such as background subtraction or spatial filtering. However, we advise, if possible, to apply image deconvolution, which can lead to remarkable image improvement via image reconstruction based on diffraction theory. We routinely employ Huygens software (Scientific Volume Imaging, https://svi.nl/HomePage). Once the principles are understood, we find that the choice of parameters for image deconvolution is less complex or arbitrary than the successive application of different image filters and that the procedure is more easily mastered by the novice user.
6. Intensity histogram: In the "Volume Viewer" window, a frequency distribution histogram of voxel intensity is shown for each data set. The *y*-axis represents (in Log scale) the number of voxels with intensities corresponding to the *x* values. The actual range of *x* values in the data is indicated below the histogram.
7. Setting the threshold: When opening a dataset, an initial threshold value is automatically determined by Chimera so that 1% of voxels are retained. The applied threshold value is indicated in the "Level" box. It is possible to adjust it either by sliding the threshold bar drawn on the histogram or by typing a value in the "Level" box. As already mentioned, careful choice of this threshold value is crucial. It is important to avoid biases between different channels by applying similar corrections and to keep an accurate record of all image preprocessing steps. Since Chimera employs a single threshold value across all *z* positions of the image stack, any intensity or background bias along the *z*-axis should also be corrected beforehand for optimal results.
8. Generate a section of an object: using the Volume Eraser tool ("Volume Viewer" window > Tools > Volume Eraser), a sphere can be moved over a part of your object and its radius adjusted with the "Radius" slider. By pressing the "Erase" button, all the voxels within the sphere will be removed from the representation. Note that a copy of the dataset is generated by this operation, preserving the original data. The display state of each dataset can be modified by clicking on the eye symbol, at the right of each data set name. Using the "Side View" (Chimera Main Window > Tools > Viewing Controls > Side View), you can position two vertical yellow lines, representing the front and back clipping planes, to remove all the pixels not comprised between these two markers. This tool is very useful to generate sections of the whole dataset.

9. Surface and Mesh options: To improve the display of your 3D data representation, additional options can be activated in the "Surface and Mesh options" feature of the volume viewer ("Volume Viewer" window > Features > Surface and Mesh options). The "Surface smoothing" option modifies the coordinates of points on the surface (surface or mesh) according to "iterations" and "factor" values. "Subdivide surfaces" allows to increase the definition of the surface or mesh, which are represented by triangles, by generating "subtriangles" (each side of the surface's triangles is divided the indicated number of times; e.g., 1 subdivision will increase the number of triangles by a factor 4. Beware of the impact on file size and visualization by the computer). "Smooth mesh lines" activates anti-aliasing, improving display of mesh lines. "Square mesh" activates display of only a subset of the triangular mesh lines to generate a lighter structure. Thickness of the mesh lines can be adjusted with the "Mesh line thickness" value (in pixels).

Acknowledgments

All figures were captured with the UCSF Chimera package. Chimera is developed by the Resource for Biocomputing, Visualization, and Informatics at the University of California, San Francisco (supported by NIGMS P41-GM103311).

The present work has benefited from the light microscopy facility of Imagerie-Gif, member of IBiSA and supported by "France-BioImaging" (ANR-10-INBS-04-01).

References

1. Hanson PI, Cashikar A (2012) Multivesicular body morphogenesis. Annu Rev Cell Dev Biol 28:337–362. https://doi.org/10.1146/annurev-cellbio-092910-154152
2. McCullough J, Colf LA, Sundquist WI (2013) Membrane fission reactions of the mammalian ESCRT pathway. Annu Rev Biochem 82:663–692. https://doi.org/10.1146/annurev-biochem-072909-101058
3. Stefani F, Zhang L, Taylor S et al (2011) UBAP1 is a component of an endosome-specific ESCRT-I complex that is essential for MVB sorting. Curr Biol 21:1245–1250. https://doi.org/10.1016/j.cub.2011.06.028
4. Colombo M, Moita C, van Niel G et al (2013) Analysis of ESCRT functions in exosome biogenesis, composition and secretion highlights the heterogeneity of extracellular vesicles. J Cell Sci 126:5553–5565. https://doi.org/10.1242/jcs.128868
5. Garrus JE, von Schwedler UK, Pornillos OW et al (2001) Tsg101 and the vacuolar protein sorting pathway are essential for HIV-1 budding. Cell 107:55–65. https://doi.org/10.1016/S0092-8674(01)00506-2
6. Carlton JG, Martin-Serrano J (2007) Parallels between cytokinesis and retroviral budding: a role for the ESCRT machinery. Science 316:1908–1912. https://doi.org/10.1126/science.1143422
7. Choudhuri K, Llodrá J, Roth EW et al (2014) Polarized release of T-cell-receptor-enriched microvesicles at the immunological synapse. Nature 507:118–123. https://doi.org/10.1038/nature12951
8. Jimenez AJ, Maiuri P, Lafaurie-Janvore J et al (2014) ESCRT machinery is required for plasma membrane repair. Science 343:1247136. https://doi.org/10.1126/science.1247136

9. Webster BM, Colombi P, Jäger J, Lusk CP (2014) Surveillance of nuclear pore complex assembly by ESCRT-III/Vps4. Cell 159:388–401. https://doi.org/10.1016/j.cell.2014.09.012
10. Olmos Y, Hodgson L, Mantell J et al (2015) ESCRT-III controls nuclear envelope reformation. Nature 522:236–239. https://doi.org/10.1038/nature14503
11. Vietri M, Schink KO, Campsteijn C et al (2015) Spastin and ESCRT-III coordinate mitotic spindle disassembly and nuclear envelope sealing. Nature 522:231–235. https://doi.org/10.1038/nature14408
12. Lefebvre C, Largeau C, Michelet X et al (2016) The ESCRT-II proteins are involved in shaping the sarcoplasmic reticulum in *C. elegans*. J Cell Sci 129:1490–1499. https://doi.org/10.1242/jcs.178467
13. Pettersen EF, Goddard TD, Huang CC et al (2004) UCSF Chimera—a visualization system for exploratory research and analysis. J Comput Chem 25:1605–1612. https://doi.org/10.1002/jcc.20084
14. Schindelin J, Arganda-Carreras I, Frise E et al (2012) Fiji: an open-source platform for biological-image analysis. Nat Methods 9:676–682. https://doi.org/10.1038/nmeth.2019
15. Besseling T, Jose J, Blaaderen AV (2015) Methods to calibrate and scale axial distances in confocal microscopy as a function of refractive index. J Microsc 257:142–150. https://doi.org/10.1111/jmi.12194
16. Schneider CA, Rasband WS, Eliceiri KW (2012) NIH image to ImageJ: 25 years of image analysis. Nat Methods 9:671–675

Chapter 12

Transient Expression of ESCRT Components in Arabidopsis Root Cell Suspension Culture-Derived Protoplasts

Marie-Kristin Nagel, Karin Vogel, and Erika Isono

Abstract

Localization studies are important to understand the function of diverse proteins. The endosomal trafficking pathway is very complex, and a lot of proteins function in this pathway, primarily the endosomal sorting complexes required for transport (ESCRTs). Some of the ESCRT-related proteins or mutant variants cannot be stably expressed in planta due to the toxicity of their expression. Therefore, a transient expression system is necessary to study their function. Transient expression in protoplasts from *Arabidopsis* root cell-derived culture serves as a fast and reliable method for the expression and cell biological and biochemical analyses of otherwise toxic constructs.

Key words Transient expression, Protoplast, ESCRT, Microscopy, *Arabidopsis* cell culture

1 Introduction

The endosomal membrane trafficking of cargo proteins to the vacuole is coordinated by the ESCRT complexes [1]. In plants, three ESCRT complexes, namely, ESCRT-I, -II, and -III, are conserved [2, 3], whereas plants lack homologs of the ESCRT-0 complex that recognizes and binds ubiquitinated cargo proteins to guide them to the subsequent ESCRT machinery [1, 3]. The function of ESCRT-0 seems to be controlled by multiple ubiquitin adaptor proteins, TARGET OF MYB (TOM)-like proteins (TOLs), Fab1, YOTB, Vac1 and EEA (FYVE) domain-containing protein 1 (FYVE1)/FYVE DOMAIN PROTEIN REQUIRED FOR ENDOSOMAL SORTING 1 (FREE1), and Src homology-3 (SH3) domain-containing protein 2 (SH3P2) [4–6]. ESCRT-I and -II also contain ubiquitin-binding subunits that recruit ubiquitinated cargo proteins at the endosomal membranes to lead them to the multivesicular bodies (MVBs) [1]. ESCRT-III sorts the cargo proteins into the intraluminal vesicles (ILVs) of the MVBs [1, 3], for which the disassembly of ESCRT-III is also important [7, 8]. ESCRT-III disassembly depends on the function of the

Emmanuel Culetto and Renaud Legouis (eds.), *The ESCRT Complexes: Methods and Protocols*, Methods in Molecular Biology, vol. 1998, https://doi.org/10.1007/978-1-4939-9492-2_12,

AAA-ATPase SUPPRESSOR OF K+ TRANSPORT GROWTH DEFECT 1 (SKD1)/VACUOLAR PROTEIN SORTING 4 (VPS4) that is recruited to the membrane of MVBs [1, 7, 9, 10].

Proper function of the ESCRTs is essential for endosomal protein degradation as well as other intracellular trafficking pathways in plants. Mutations in the ESCRT pathway lead to severe developmental phenotypes in plants [5, 9–11]. For example, knockout of ESCRT-I and ESCRT-III and related components has been shown to cause embryo lethality [5, 9]. The overexpression of an inactive version of *SKD1*, *SKD1(E232Q)*, causes enlargement of MVBs and impaired internalization of ILVs and cannot be constitutively expressed without having lethal effects [7]. ESCRT-III associated proteins accumulate on SKD1(E232Q)-induced structures [9], which is specific to late endosome-localized proteins that do not include clathrin and early endosomal markers [12].

The toxicity of the dominant-negative SKD1(E232Q) construct as well as the nonfunctionality of fluorescent fusion protein of ESCRT-III subunits [7, 9] make it difficult to use them for cell biological studies in planta. A transient expression system of these factors in leaf- or cell culture-derived protoplasts, Tobacco BY-2 cells, or tobacco leaves is therefore a good alternative for the analysis of ESCRT components in plant cells. *Arabidopsis* root cell culture-derived protoplasts can be transformed without *Agrobacterium tumefaciens*, are easy to handle, and, as they do not contain chloroplasts, are suitable for multicolored cell biological analysis. Expression of the fusion of the yellow fluorescent protein YFP-VPS2.1 in *Arabidopsis* root cell suspension culture-derived protoplasts is shown in Fig. 1. *YFP-VPS2.1* in co-expression with

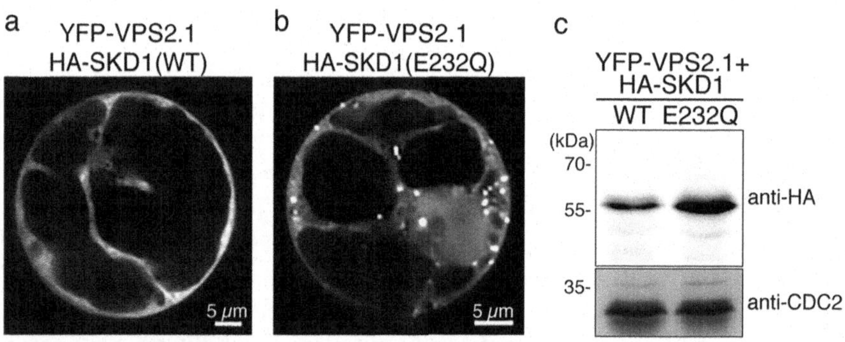

Fig. 1 Expression of fluorescent fusion proteins in *Arabidopsis* root cell suspension culture-derived protoplasts. (**a**) A confocal micrograph of a protoplast transformed with plasmids expressing *YFP-VPS2.1* and HA-tagged wild-type SKD1 (*HA-SKD1(WT)*). (**b**) A confocal micrograph of a protoplast transformed with plasmids expressing *YFP-VPS2.1* and HA-tagged inactive version SKD1 (*HA-SKD1(E232Q)*). Note that YFP-VPS2.1 relocalizes to aggregates in the cytosol upon co-expression with SKD1(E232Q). Scale bar: 5 μm. (**c**) Immunoblot analysis of total extracts from transformed protoplasts. Expression of HA-SKD1(WT) and HA-SKD1(E232Q) was verified by immunoblotting with an anti-HA antibody on total protein extract of transformed protoplasts. An anti-CDC2 antibody was used to verify loading

HA-SKD1(WT) was localized in the cytosol (Fig. 1a), whereas the co-expression of the inactive version *HA-SKD1(E232Q)* caused aggregation of YFP-VPS2.1 in SKD1(E232Q)-induced structures (Fig. 1b). The expression of *HA-SKD1(WT)* and *HA-SKD1 (E232Q)* can be verified by immunoblotting (Fig. 1c).

In this protocol, we describe the cultivation and protoplastation of *Arabidopsis thaliana* root cell suspension culture and the polyethylene glycol (PEG)-mediated transformation of protoplasts and the analysis of *HA-SKD1* expression by immunoblotting. Such strategy has been employed to analyze the expression of the fluorescent fusion proteins of the ESCRT pathway in *Arabidopsis* root cell suspension culture-derived protoplasts and to investigate the effect of co-expression of the inactive version HA-SKD1(E232Q) that causes the formation of the SKD1(E232Q)-induced structures.

2 Materials

2.1 Plant Material and Medium

1. Cell culture of *Arabidopsis thaliana* root cells, incubated at 21 °C in a rotary shaker at 120 rpm (*see* **Note 1**).
2. *A. thaliana* root cell culture medium: 4.33 g/l Murashige and Skoog (MS) medium with Gamborg B5 vitamins (e.g., Murashige and Skoog medium including B5 vitamins, [Duchefa Biochemie]), 30 g/l sucrose, 1 mg/l 2,4-dichlorophenoxyacetic acid, 0.34 g/l KH_2PO_4, adjust to pH 5.8 with KOH (*see* **Note 2**). Divide into 50 ml medium in 300 ml Erlenmeyer flasks and close the flasks with a cellulose plug and aluminum foil on top of the cellulose plug. Autoclave the flasks filled with 50 ml medium at 120 °C for 15 min.

2.2 Plasmid DNA Preparation

1. Plasmid containing your gene of interest with a promoter for expression transformed in *E. coli* cells, for example, DH5α or TOP10 (*see* **Note 3**).
2. LB medium: 10 g/l tryptone, 5 g/l yeast extract, and 10 g/l NaCl, adjust to pH 7 with NaOH. For plates, prepare solid medium supplemented with 15 g/l agar. Autoclave the medium and add antibiotics for vector resistance after cooling down of the medium. Store the medium at 4 °C.
3. Plasmid DNA purification kit (*see* **Note 4**).
4. Photometer to control the concentration and purity.

2.3 Protoplastation

Prepare all stock solutions and filtrate stock solutions through a 0.2 μm syringe filter (*see* **Note 5**).

1. Enzyme solution without enzymes: 0.4 M mannitol (20 ml/ 50 ml of 1 M mannitol), 5 mM EGTA (2.5 ml/50 ml of 100 mM EGTA, pH 8.2), fill up to 50 ml with distilled water.

2. Enzyme solution with enzymes: 0.4 M mannitol (4 ml/10 ml of 1 M mannitol), 5 mM EGTA (0.5 ml/10 ml of 100 mM EGTA, pH 8.2), supplement with 1% cellulase (100 mg) and 0.25% Macerozyme (25 mg). Heat the solution for 10 min at 50 °C. Filtrate the solution through a 0.2 μm syringe filter cap.
3. Solution A: 0.4 M mannitol (20 ml/50 ml of 1 M mannitol), 70 mM $CaCl_2$ (3.5 ml/50 ml of 1 M $CaCl_2$), 5 mM MES (500 μl/50 ml of 0.5 M MES, pH 5.7).
4. PEG-calcium transfection solution: 40% polyethylene glycol (PEG) 6000 (4 g/10 ml), 0.4 M mannitol (4 ml/10 ml of 1 M mannitol), 0.1 M $Ca(NO_3)_2$ (1 ml/10 ml of 1 M $Ca(NO_3)_2$) (*see* **Note 6**).
5. MMG solution: 15 mM $MgCl_2$ (150 μl/10 ml of 1 M $MgCl_2$), 5 mM MES (100 μl/10 ml of 0.5 M MES, pH 5.7), 0.4 M mannitol (4 ml/10 ml), fill up to 10 ml with distilled water.
6. Dilution solution: 0.4 M mannitol (20 ml/50 ml of 1 M mannitol), 125 mM $CaCl_2$ (6.25 ml/50 ml of 1 M $CaCl_2$), 5 mM glucose (250 μl/50 ml of 1 M glucose), 5 mM KCl (250 μl/50 ml of 1 M KCl), 1.5 mM MES (150 μl/50 ml of 0.5 M MES, pH 5.7), fill up to 50 ml with distilled water.
7. MS-Mannitol solution: 4.6 g/l MS including Gamborg B5 medium with vitamins (e.g., Murashige and Skoog medium including B5 vitamins, [Duchefa Biochemie]), (prepare 46 mg/10 ml) with 0.4 M mannitol (4 ml/10 ml of 1 M mannitol). Filtrate solution through a 0.2 μm syringe filter.
8. Nylon mesh filters with 60–100 μm pore size.

2.4 Microscope Equipment

1. For controlling the protoplastation and counting the cells, use a standard stereomicroscope with a 10× objective.
2. For analyzing of the protoplasts transformed with fluorescent fusions, use a fluorescence microscope (e.g., BX61 [Olympus]) equipped with 10× and 60× objectives and the required fluorescence filters for GFP, YFP, or RFP. Alternatively, use a confocal laser scanning microscope (e.g., FV-1000/IX81 [Olympus]) equipped with 10× and 60× objectives and with suitable lasers.

2.5 Sample Preparation for Western Blotting

1. Phosphate-buffered saline (PBS): 137 mM NaCl, 2.7 mM KCl, 10 mM Na_2HPO_4, 2 mM KH_2PO_4, pH 7.4.
2. 5× Laemmli buffer [13]: 250 mM Tris–HCl pH 6.8, 10% (w/v) sodium dodecyl sulfate (SDS), 50% (w/v) glycerol, 0.05% (w/v) bromophenol blue (BPB), 5% (v/v) β-mercaptoethanol. Store at room temperature.

2.6 SDS-Polyacrylamide Gel Electrophoresis

1. Separation gel buffer: 1.5 M Tris–HCl pH 8.8 and 0.4% (w/v) SDS. Store at room temperature.
2. Stacking gel buffer: 0.5 M Tris–HCl pH 6.8 and 0.4% (w/v) SDS. Store at room temperature.
3. 30% acrylamide-bisacrylamide solution (37.5:1), 10% (w/v) ammonium persulfate (APS), and *N,N,N,N′*-tetramethyl ethylenediamine (TEMED). Store at 4 °C.
4. 100% Isopropanol for overlaying the separation gel.
5. 10× SDS running buffer: 250 mM Tris, 1.92 M glycine, 0.4% (w/v) SDS. Dilute 10× stock in deionized water to prepare 1× running buffer.
6. A prestained molecular mass marker.
7. Minigel and electrophoresis system.

2.7 Western Blot

1. Semidry transfer buffer: 25 mM Tris pH 8.3, 192 mM glycine, 20% (v/v) methanol, 0.04% (w/v) SDS. Store at room temperature.
2. Semidry transfer apparatus.
3. Four filter papers (1.2 mm thick) cut in the size of 0.5 cm larger than that of the protein gel.
4. Polyvinylidene difluoride (PVDF) membrane with pore size of 0.45 μm.
5. 100% methanol (analytical grade).
6. 10× Tris-buffered saline (TBS): 0.5 M Tris–HCl pH 7.5, 1.5 M NaCl, 10 mM $MgCl_2$.
7. Tris-buffered saline with Tween-20 (TBST): use 10× TBS stock to prepare 1× working solution. Add Tween-20 to 0.05% (v/v). Store at room temperature.
8. Prepare 100 ml blocking buffer with TBST, supplemented either with 5% (w/v) skimmed milk powder or 5% (w/v) bovine serum albumin (BSA).
9. Anti-HA antibody (e.g., anti-HA, monoclonal, clone 3F10 from ROCHE). Used for detection of HA-tagged proteins.
10. Anti-CDC2 antibody (e.g., anti-CDC2 p34 (17): sc-54, monoclonal, from Santa Cruz). Used for detection of endogenous CDC2 as loading control.
11. Anti-rat HRP-conjugated antibody.
12. Anti-mouse HRP-conjugated antibody.
13. Enhanced chemiluminescent (ECL) reagents (e.g., SuperSignal West Femto Chemiluminescent Substrate or with ECL Western Blotting Substrate [Thermo Fisher Scientific]).
14. Development: imaging device (e.g., ImageQuant™ LAS 4000 [GE Healthcare]).

3 Methods

3.1 Plant Culture

1. Culture cells at 21 °C (light or dark) on a rotary shaker at 100–120 rpm (*see* **Note** 7).
2. Subcultivate the culture every 7 days (*see* **Note 8**).
3. For protoplastation, use suspension cell culture 4–5 days after subculturing.

3.2 Plasmid Preparation

1. Culture *E. coli* cells harboring the plasmid of interest in 50 to 100 ml of liquid LB medium supplemented with the necessary antibiotic(s). Incubate the culture overnight at 37 °C on a rotary shaker.
2. Harvest the cells by centrifuging at 4000 × *g* for 10 min.
3. Purify the plasmid DNA with a plasmid DNA purification kit (e.g., PureLink HiPure Plasmid Midiprep Kit [Thermo Fisher Scientific]) following the manufacturer's instruction (*see* **Note 9**). Dilute the plasmid DNA to 1–2 μg/μl, and store the plasmid DNA at −20 °C until use.

3.3 Protoplastation

1. Collect 10 ml of the cell culture into a 15-ml tube under sterile conditions and spin down cells in a centrifuge at 300 × *g* for 2 min. This should yield c.a. 2 ml cell volume.
2. Discard supernatant and wash the cell pellet in 10 ml enzyme solution **without** enzymes to remove the culture medium.
3. Spin down cells in a centrifuge at 300 × *g* for 2 min.
4. Discard supernatant and resuspend the cell pellet in 10 ml enzyme solution with enzymes.
5. Incubate the tube with cells for digesting at room temperature (20–25 °C) in the dark or wrap an aluminum foil around the tube for 2–3 h on a shaker at 50–55 rpm. Invert the tube from time to time (*see* **Note 10**).
6. Control digest efficiency of the cells every hour. For this end, pipette 20 μl of the cells on a glass slide and assess the cells under a light microscope. Most of the cells should be digested and appear as roundish protoplasts after 2 h (*see* **Note 11**).

3.4 Plasmid Transformation

1. Filtrate protoplasts through a 60–100 μm nylon mesh and collect the filtrate in a 15-ml tube. Undigested cells remain in the filter mesh.
2. Spin down cells at 100 × *g* for 2 min in a 15-ml tube. The protoplasts will be collected at the bottom of the tubes.
3. Remove supernatant by pipetting. Wash the protoplasts carefully in 10 ml of Solution A. Do not pipette the solution directly on top of protoplasts but rather to the tube wall.

4. Centrifuge at 100 × *g* for 2 min, and remove the supernatant.
5. Repeat the washing step with 10 ml Solution A.
6. Resuspend the cells in 1 ml Solution A and transfer the protoplasts carefully with a cut tip to two 2-ml tubes.
7. Centrifuge protoplasts at 100 × *g* for 2 min in the 2-ml tubes, and remove the supernatant.
8. Add slowly 1 ml of MMG solution and mix by tapping the tube with your fingers.
9. Incubate the protoplasts on ice for 30 min.
10. During the incubation, count the protoplasts under a microscope using a Fuchs-Rosenthal counting chamber. Fill both fields with 20 μl protoplast solution, and count the cells in four big squares in the upper chamber and in four big squares in the lower chamber. Calculate the number of cells as follows (*see* **Note 12**):

$$\frac{\text{Number of cells per square (average)}}{0.0002} = \text{Number of protoplasts per 1 ml}$$

11. Centrifuge at 100 × *g* for 2 min.
12. Remove the supernatant, and add the MMG solution to a final cell density of approximately 106 cells per ml. Pipette the MMG solution carefully to the wall of the tube and mix by tapping the tube with your fingers.
13. Transfer 100 μl of the protoplast solution in a new 2-ml tube and add 20–40 μg purified plasmid DNA. In case of double or triple transformation, add 20 μg of each construct.
14. Add 400 μl PEG solution to the protoplast/DNA mixture. Pipette PEG solution drop by drop to the tube wall and mix the solution after every drop by tapping and occasionally inverting the tube. Make sure that the PEG is completely mixed. The solution must be homogeneous and free of streaks.
15. Incubate cells in PEG solution on ice for 20–30 min. Because of toxicity of PEG, protoplasts should not be incubated longer.
16. Add 1 ml of dilution solution to the incubated cells, drop by drop, and mix the solution after each drop by tapping and inverting the tube. Make sure that the solution is completely mixed after each drop. The solution must be homogeneous and free of streaks.
17. Centrifuge at 100 × *g* for 4 min. Make sure that the protoplasts pellet at the bottom. If not, repeat the centrifugation step (*see* **Note 13**).
18. Remove the supernatant as much as possible.

19. Wash the protoplasts gently in 2 ml of dilution solution, and centrifuge at 100 × *g* for 2 min. Remove the supernatant.
20. Repeat the washing step with 2 ml of dilution solution once.
21. Resuspend the cells gently in 200–500 μl of storage solution by tapping the tube.
22. For the expression of the transformed constructs, incubate the protoplasts at 21 °C in the dark for 12–20 h in a sterile 24-well cell culture plate with gentle shaking (50–55 rpm).
23. Analyze expression of the transformed constructs (*see* **Note 14**).

3.5 Sample Preparation for Western Blot

1. Dilute the protoplasts with 1 ml PBS and spin down 5 min at 400 × *g*.
2. Remove supernatant and resuspend the pellet in 100 μl of 1× Laemmli buffer.
3. Boil the sample for 5 min at 95 ° C.
4. Centrifuge for 10 min at 16,000 × *g* and transfer the supernatant to a new 1.5-ml tube. Keep samples at room temperature for SDS-Polyacrylamide Gel Electrophoresis (PAGE) (*see* **Note 15**).

3.6 SDS-Polyacrylamide Gel Electrophoresis (PAGE)

1. Use a minigel system with 1-mm thick gels for the SDS-PAGE. Clean the plates with ethanol before assembling.
2. Prepare a 10% separation gel by mixing in a 50-ml tube: 3.4 ml water, 1 ml separation gel buffer, 2.6 ml 30% acrylamide-bisacrylamide solution, 50 μl 10% APS, and 6 μl TEMED. Pour the gel immediately after mixing the components and leave 2 cm space for the stacking gel. Overlay the separation gel with 100% isopropanol. Leave the gel at room temperature for 15–30 min to polymerize.
3. Discard the isopropanol overlay. Remove rest of the isopropanol solution with a paper towel.
4. Prepare the stacking gel by mixing 2.4 ml water, 1 ml stacking gel buffer, 0.6 ml 30% acrylamide-bisacrylamide solution, 50 μl 10% APS, and 6 μl TEMED. Mix the components and pour the gel. Insert the comb immediately avoiding the formation of air bubbles. Leave the gel for 15 min at room temperature to polymerize. At this point, the gels can be stored for several days at 4 °C wrapped in wet paper towels and a plastic bag.
5. Remove the sample comb carefully and assemble the gel apparatus. Fill the inner and outer chambers with 1× running buffer.

6. Load a prestained molecular mass marker, for example, PageRuler™ Plus (Thermo Fisher Scientific). Load 10 μl of total protein samples (*see* **Note 16**).
7. Run the gel at 25 mA (constant current) until the dye reaches the bottom of the gel.

3.7 Western Blot

1. Incubate the PVDF membrane for 30 s in 100% methanol before immersing it in semidry transfer buffer.
2. Soak the filter papers in the semidry transfer buffer and assemble a stack in the following order (from bottom to top): two filter papers, PVDF membrane, gel, and two filter papers. Eliminate air bubbles by rolling a glass tube over the transfer package after adding each filter paper.
3. Transfer the proteins from the gel to the PVDF membrane for 1 h at a constant current of 2 mA per 1 cm^2 membrane.
4. After the transfer, cut the top left corner of the membrane to find the right orientation of the membrane in the following steps.
5. Place the membrane into the blocking buffer and incubate for 15 min at room temperature on a shaker. Alternatively, the membrane can be incubated overnight at 4 °C on a shaker (*see* **Note 17**).
6. Prepare a proper dilution of the primary antibody in blocking buffer.
7. Incubate the membrane in the primary antibody solution at room temperature, shaking for at least 1 h or overnight at 4 °C.
8. Remove the solution with the primary antibody. Wash the membrane for 15 min with 1× TBST buffer. Repeat the washing step three times, using fresh 1× TBST buffer each time.
9. Prepare a proper dilution of the secondary antibody in 1× TBST.
10. Incubate the membrane in secondary antibody solution at room temperature, shaking for at least 30 min or overnight at 4 °C.
11. Wash the membrane with 1× TBST as in **step** 7.
12. Take the membrane from the washing solution and remove excess liquid. Incubate the membrane with the ECL solution for 3 min (e.g., SuperSignal West Femto Chemiluminescent Substrate [Thermo Fisher Scientific]). Remove excess ECL solution and expose the membrane in a CCD camera system. Optimal exposure time varies between experiments (*see* **Note 18**).

4 Notes

1. For the cell suspension culture, use a wide swing shaker with an orbit of around 25 mm, for example, Celltron or Ecotron (Infors HT). Root cell-derived cultures can be cultivated either in light with 110–120 μmol m^{-2} s^{-1} light intensity or without light.
2. Prepare a stock solution of 2,4-dichlorophenoxyacetic acid with a concentration of 100 mg/l, solved in 10% ethanol. For a concentration of 1 mg/l in the medium, use 10 ml/l of this stock solution. Prepare a stock solution of KH_2PO_4 with a concentration of 100 g/l. For a concentration of 0.34 g/l in the medium, use 3.4 ml/l of this stock solution.
3. The plasmid DNA to be used for transformation should not be larger than 10 kbp. Both bacterial and binary vectors can be used, whereas large binary vectors are often disadvantageous for transformation because of their comparatively large size. Common promoters such as 35S and UBQ10 can be used to drive expression. Native promoters can also be used as long as all factors required for the expression exist in root cell suspension culture.
4. Plasmid DNA can be purified also by other methods if the purity and concentration of the purified DNA are sufficient for the transformation.
5. Use only chemicals with high purity. All stock solutions can be prepared the day before. The solutions used during the transformation (Enzyme solution, Solution A, MMG solution, PEG-calcium transfection solution, Dilution solution, and MS-Mannitol solution) must be prepared freshly on the day of transformation. Heat the 1 M mannitol stock solution to 40–50 °C to solve the mannitol.
6. PEG may take some time to resolve. Vortex the solution vigorously and rotate for 1 h at RT overhead. Filtrate solution through a 0.2 μm syringe filter cap to remove residual PEG particles.
7. The light conditions do not seem to be relevant. The temperature should be stable at 21 °C, and the shaker should have a wide swing.
8. Sterilize all materials that are necessary for the inoculation process appropriately. Open the cell culture flasks exclusively under a sterile hood. For the subculture, best use a sterile one-way plastic pipette. Transfer 5 ml of 1-week-old cell suspension culture into 50 ml of fresh medium under sterile conditions.

9. For the transformation of protoplasts, use highly pure plasmid DNA with a minimum concentration of 1 μg/μl. Control the concentration and purity of the plasmid DNA with photometric measurement.
10. Cells can be incubated in the 15-ml tube or, for better protoplastation efficiency, incubated in a small petri dish.
11. Preparations of glass slides for protoplasts: To avoid squeezing of protoplasts, place several layers of adhesive tape on the glass slide. Cut a window of 10 mm × 10 mm out of the tape and pipette 20 μl of the protoplast solution into the window.
12. Fuchs-Rosenthal counting chamber: A small square has a length of 250 μm, and usually, an *Arabidopsis* root cell-derived protoplast also has a size between 20 and 40 μm. The big square is 1 mm^2. Depth of the chamber is 0.2 mm; accordingly, the volume of the big square is 0.2 mm^3. If 1 mm^3 = 1 μl, then accordingly 0.2 mm^3 = 0.2 μl.
13. If you see no pellet after the first centrifugation step, turn the tube 180° around before centrifuging again to pellet the cells efficiently.
14. The transformation efficiency depends on the size and purity of the plasmid but is usually between 10% and 25% when a plasmid DNA of around 8 kbp is used. When smaller plasmids are used, the transformation efficiency may be higher.
15. For long time storage, store the samples in 1× Laemmli buffer at −20 °C.
16. Alternatively, if it is necessary to measure the protein concentration of your sample, sonicate the protoplast for 5 s in PBS or TBS and measure the protein concentration by the Bradford method or a method of your choice and add 5× Laemmli buffer before boiling for 5 min.
17. Make sure that the PVDF membrane is always immersed in solution, since otherwise the membrane surface will dry.
18. Alternatively, the immunoblot can be developed by a detection method of your choice.

Acknowledgments

This work conducted in the authors' laboratory is supported by grants from the German Science Foundation (DFG) to E.I. (IS 221/4-1). The authors thank Prof. Takashi Ueda (NIBB, Japan) and Prof. Tomohiro Uemura (The University of Tokyo, Japan) for advices to optimize the protocol.

References

1. Henne WM, Buchkovich NJ, Emr SD (2011) The ESCRT pathway. Dev Cell 21(1):77–91. https://doi.org/10.1016/j.devcel.2011.05.015
2. Paez Valencia J, Goodman K, Otegui MS (2016) Endocytosis and endosomal trafficking in plants. Annu Rev Plant Biol 67:309–335. https://doi.org/10.1146/annurev-arplant-043015-112242
3. Winter V, Hauser MT (2006) Exploring the ESCRTing machinery in eukaryotes. Trends Plant Sci 11(3):115–123. https://doi.org/10.1016/j.tplants.2006.01.008
4. Korbei B, Moulinier-Anzola J, De-Araujo L, Lucyshyn D, Retzer K, Khan MA, Luschnig C (2013) Arabidopsis TOL proteins act as gatekeepers for vacuolar sorting of PIN2 plasma membrane protein. Curr Biol 23 (24):2500–2505. https://doi.org/10.1016/j.cub.2013.10.036
5. Nagel MK, Kalinowska K, Vogel K, Reynolds GD, Wu Z, Anzenberger F, Ichikawa M, Tsutsumi C, Sato MH, Kuster B, Bednarek SY, Isono E (2017) Arabidopsis SH3P2 is an ubiquitin-binding protein that functions together with ESCRT-I and the deubiquitylating enzyme AMSH3. Proc Natl Acad Sci U S A 114(34):E7197–E7204. https://doi.org/10.1073/pnas.1710866114
6. Gao C, Luo M, Zhao Q, Yang R, Cui Y, Zeng Y, Xia J, Jiang L (2014) A unique plant ESCRT component, FREE1, regulates multivesicular body protein sorting and plant growth. Curr Biol 24(21):2556–2563. https://doi.org/10.1016/j.cub.2014.09.014
7. Haas TJ, Sliwinski MK, Martinez DE, Preuss M, Ebine K, Ueda T, Nielsen E, Odorizzi G, Otegui MS (2007) The Arabidopsis AAA ATPase SKD1 is involved in multivesicular endosome function and interacts with its positive regulator LYST-INTERACTING PROTEIN5. Plant Cell 19(4):1295–1312. https://doi.org/10.1105/tpc.106.049346
8. Babst M, Sato TK, Banta LM, Emr SD (1997) Endosomal transport function in yeast requires a novel AAA-type ATPase, Vps4p. EMBO J 16 (8):1820–1831. https://doi.org/10.1093/emboj/16.8.1820
9. Katsiarimpa A, Anzenberger F, Schlager N, Neubert S, Hauser MT, Schwechheimer C, Isono E (2011) The Arabidopsis deubiquitinating enzyme AMSH3 interacts with ESCRT-III subunits and regulates their localization. Plant Cell 23(8):3026–3040. https://doi.org/10.1105/tpc.111.087254
10. Spitzer C, Reyes FC, Buono R, Sliwinski MK, Haas TJ, Otegui MS (2009) The ESCRT-related CHMP1A and B proteins mediate multivesicular body sorting of auxin carriers in Arabidopsis and are required for plant development. Plant Cell 21(3):749–766. https://doi.org/10.1105/tpc.108.064865
11. Spitzer C, Schellmann S, Sabovljevic A, Shahriari M, Keshavaiah C, Bechtold N, Herzog M, Muller S, Hanisch FG, Hulskamp M (2006) The Arabidopsis elch mutant reveals functions of an ESCRT component in cytokinesis. Development 133(23):4679–4689. https://doi.org/10.1242/dev.02654
12. Kolb C, Nagel MK, Kalinowska K, Hagmann J, Ichikawa M, Anzenberger F, Alkofer A, Sato MH, Braun P, Isono E (2015) FYVE1 is essential for vacuole biogenesis and intracellular trafficking in Arabidopsis. Plant Physiol 167 (4):1361–1373. https://doi.org/10.1104/pp.114.253377
13. Laemmli UK (1970) Cleavage of structural proteins during the assembly of the head of bacteriophage T4. Nature 227 (5259):680–685

Chapter 13

Crystallization and Biophysical Approaches for Studying the Interactions Between the Vps4-MIT Domain and ESCRT-III Proteins

Takayuki Obita, Rieko Kojima, and Mineyuki Mizuguchi

Abstract

The AAA ATPase Vps4 disassembles the ESCRT complex from the endosomal membrane. Vps4 contains an N-terminal MIT (microtubule interacting and transport) domain and a C-terminal catalytic domain. The MIT domain binds to MIMs (MIT-interacting motifs), which exist at the C-terminus of ESCRT-III proteins, with a dissociation constant in the micromolar range. Five MIMs have been identified by structural and biophysical methods to date, and the recognition motifs have been refined. Among biophysical approaches used to analyze protein interactions, surface plasmon resonance (SPR) analysis is often suitable for weak interactions, and fluorescence-binding assay has an advantage in terms of sensitivity. We have introduced protein modification tags into the N-terminus of proteins with bacterial expression vectors for biotinylation and FlAsH (fluorescein arsenical hairpin binder) fluorescent labeling. Here, we describe how to purify the MIT domain of Vps4 and the MIMs of ESCRT-III proteins and how to conduct crystallography, SPR, and fluorescence-binding assays.

Key words Crystallization, SPR, Fluorescence-binding assay, MIT domain, Vps4, MIM, FlAsH-EDT_2

1 Introduction

The endosomal sorting complex required for transport (ESCRT) proteins, which include ESCRT-0, -I, -II, -III, and Vps4, are conserved from yeast to humans [1, 2]. The ESCRT proteins are required for a number of biological processes, including multivesicular body biogenesis [3], budding of enveloped viruses [4], and cytokinesis [5]. Proteins homologous to Vps4 and ESCRT-III are also present in Archaea of the genus *Sulfolobus* and are involved in cell division [6, 7]. Vps4 contains an N-terminal MIT (microtubule interacting and transport) domain and a C-terminal catalytic domain. The MIT domain consists of a three-helix bundle and binds short sequences known as MIMs (MIT-interacting motifs). ESCRT-III proteins, which consist of an N-terminal core domain

Emmanuel Culetto and Renaud Legouis (eds.), *The ESCRT Complexes: Methods and Protocols*, Methods in Molecular Biology, vol. 1998, https://doi.org/10.1007/978-1-4939-9492-2_13,

and a C-terminal MIM, are thought to assemble into oligomeric filaments and are likely to catalyze the vesicle scission reaction [8]. Twelve, seven, and four subunits of ESCRT-III proteins have been identified in humans, yeast, and archaea, respectively. Previous studies suggest that Vps4 maintains the function of ESCRT-III via the MIT–MIM interaction by promoting its ATPase-catalyzed disassembly from the membrane [9]. The first characterizations of MIT–MIM complex structures revealed that the MIM adopts a short α-helical conformation and binds into a groove formed between the second and third helixes of the MIT domain (MIM1 mode) [10, 11]. A second round of MIT–MIM structural analysis revealed that the MIM adopts an extended structure and binds into a groove formed between the first and third helixes of the MIT domain (MIM2 mode) [7, 12]. So far, five MIM modes, which are classified by the shape of the MIMs and the binding interface of the MIT domain, have been identified [7, 10–16].

MIMs, which consist of around 20–50 residues, are normally unstructured in the absence of the core region and are often expressed with various affinity tags using bacterial expression systems. We previously introduced a lipoyl domain (LD) as a solubilizing tag in the interaction studies using biophysical methods. We showed that the LD, which exists as a monomer (molecular weight: 10 kDa), had no effect on the binding experiments, which included surface plasmon resonance (SPR), fluorescence binding assay, and isothermal titration calorimetry (ITC) experiments [17, 18]. For the biophysical studies of protein–protein interactions, protein labeling is often a key step in designing an experiment. In the event that the proteins need to be modified, this will be easier if the expressed proteins contain an enzymatic recognition sequence. We have therefore introduced bacterial expression vectors for a protein fused with various tags: a biotinylation tag for SPR, a fluorescence labeling tag for fluorescence binding assay.

Here we report our technological strategies used to investigate interactions between the MIT domain and MIMs using crystallography, SPR, and fluorescence-binding assay.

2 Materials

Expressions of ESCRT proteins have been carried out in bacteria using the pOP expression vectors developed by Dr. Roger Williams' group at the MRC-LMB in Cambridge (UK). pOP vectors include pOPTH (6xHis tag), pOPTG (GST tag with tobacco etch virus (TEV) protease cleavage), pOPHLT (6xHis-LD tag with TEV protease cleavage), pOPHBL (6xHis-biotinylation-LD tag), and pOPHFL (6xHis-FlAsH-LD tag). The biotinylation tag contains the amino acid sequence of LHHILDAQKMVWNHRH, within which the side chain of a lysine residue is specifically biotinylated by

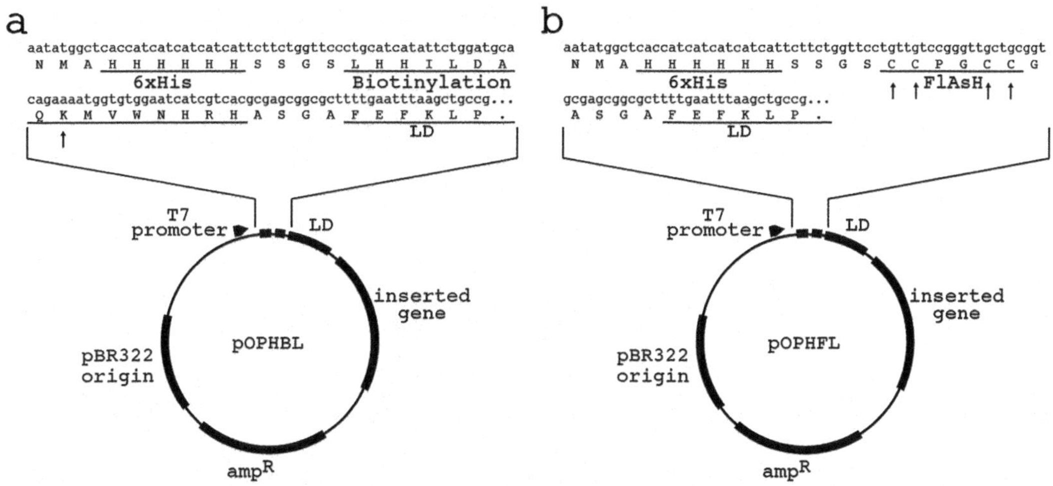

Fig. 1 pOPHBL and pOPHFL bacterial expression vectors. (**a**) A map of the pOPHBL plasmid. DNA sequences and their amino acid coding sequences are shown. The lysine residue that is biotinylated by the BirA enzyme is indicated by an arrow. The following are underlined: *6xHis* hexa-histidine, *biotinylation* biotinylation motifs for BirA, *LD* the lipoyl domain. (**b**) A map of the pOPHFL plasmid. DNA sequences and their amino acid coding sequences are shown. The tetra-cysteine residues that are covalently linked to FlAsH fluorescein are indicated by arrows. The following are underlined: *6xHis* hexa-histidine, *FlAsH* FlAsH fluorescein binding motifs, *LD* the lipoyl domain

the biotin protein ligase BirA (Avidity) (Fig. 1a). The fluorescein arsenical hairpin binder (FlAsH) fluorescein tag contains the amino acid sequence of CCPGCC, within which the tetracysteine motif is labeled with FlAsH-ETD2 fluorescein (Fig. 1b).

2.1 Protein Expression and Purification

1. Ampicillin 100 mg/ml (stock solution): Dissolve 5 g of ampicillin in 50 ml of pure water. Filter with a 0.22 μm pore size membrane, and store at −20 °C (*see* **Note 1**).
2. C41(DE3) or C43(DE3) bacterial strains (Lucigen).
3. LB medium: Dissolve 10 g of tryptone, 5 g of yeast extract, and 10 g of NaCl in 1000 ml of deionized water. Autoclave and store at room temperature.
4. LB agar plate: Prepare LB medium as mentioned above, but add 17.5 g of agar before autoclaving. After autoclaving, cool to circa 60 °C, add ampicillin to a final concentration of 100 μg/ml, and pour into petri dishes. Store at 4 °C.
5. Bacterial expression vectors (pOP vectors): pOPTH, 6xHis-tag; pOPTG, GST-tag; pOPHLT, 6xHis-LD-tag; pOPHBL, 6xHis-biotinylation-LD-tag; pOPHFL, 6xHis-FlAsH-LD-tag.
6. 1 M IPTG: Dissolve 11.9 g of isopropyl-beta-D-1-thiogalactopyranoside (IPTG) in 50 ml of pure water. Filter with a 0.22 μm pore size membrane, and store at −20 °C.

7. 0.5 M Tris(2-carboxyethyl)phosphine hydrochloride solution pH 7.0 (TCEP) (Sigma-Aldrich).
8. 1 M Tris–HCl pH 8.0 at 4 °C: Dissolve 121.4 g of Tris-base in circa 800 ml of pure water in a glass beaker. Titrate to pH 7.41 at a lab temperature of 25 °C using 1 M HCl. Bring to 1000 ml with pure water, and filter with a 0.22 μm pore size membrane (*see* **Note 2**).
9. 1 M Tris–HCl pH 7.4 at 25 °C or pH 8.5 at 25 °C: Dissolve 121.4 g of Tris-base in circa 800 ml of pure water in a glass beaker. Titrate to pH 7.4 or 8.5 at the lab temperature of 25 °C using 1 M HCl. Bring to 1000 ml with pure water, and filter with a 0.22 μm pore size membrane.
10. 1 M HEPES pH 7.4: Dissolve 238.3 g of HEPES in circa 800 ml of pure water. Titrate to pH 7.4 at a lab temperature of 25 °C using 1 M NaOH. Bring to 1000 ml with pure water, and filter with a 0.22 μm pore size membrane.
11. 5 M NaCl: Dissolve 292.2 g of NaCl in 1000 ml of pure water. Filter with a 0.22 μm pore size membrane.
12. 1 M Imidazole: Dissolve 68.1 g of imidazole in circa 700 ml of pure water. Titrate to pH 8.0 using 1 M HCl. Bring to 1000 ml with pure water, and filter with a 0.22 μm pore size membrane.
13. HisTrap buffer A: 20 mM Tris–HCl pH 8.0 at 4 °C, 100 mM NaCl, 20 mM imidazole pH 8.0.
14. HisTrap buffer B: 20 mM Tris–HCl pH 8.0 at 4 °C, 100 mM NaCl, 500 mM imidazole pH 8.0.
15. Ion exchange buffer Q-A: 20 mM Tris–HCl pH 8.5 at 25 °C.
16. Ion exchange buffer Q-B: 20 mM Tris–HCl pH 8.5 at 25 °C, 1 M NaCl.
17. Gel filtration buffer Xtal: 20 mM Tris–HCl pH 7.5 at 25 °C, 100 mM NaCl, 1 mM DTT.
18. Gel filtration buffer Bio: 20 mM Tris–HCl pH 8.0 at 4 °C, 20 mM NaCl.
19. Gel filtration buffer SPR: 10 mM HEPES pH 7.4, 150 mM NaCl, 3 mM ethylenediaminetetraacetic acid (EDTA).
20. Gel filtration buffer Fluo: 20 mM Tris–HCl pH 8.0 at 4 °C, 100 mM NaCl, 1 mM Tris(2-carboxyethyl)phosphine (TCEP).
21. Dialysis buffer Fluo: 20 mM Tris–HCl pH 8.0 at 4 °C, 100 mM NaCl, 1 mM DTT.
22. Tobacco etch virus (TEV) protease 10 units/μl (Thermo Fisher Scientific): Dissolve 10,000 units of TEV protease in 1 ml of pure water.
23. 5 ml HisTrap HP (GE Healthcare Life Sciences).

24. 5 ml HiTrap Q HP (GE Healthcare Life Sciences).
25. HiLoad 16/60 Superdex 75 Gel Filtration Column (GE Healthcare Life Sciences).
26. Superdex 75 10/300 Gel Filtration Column (GE Healthcare Life Sciences).
27. ÄKTA Purification System (GE Healthcare Life Sciences).
28. Amicon® Ultra-15 Concentrators (Millipore).
29. Crystal Screen I/II (Hampton Research).
30. JCSG-*plus*™ (Molecular Dimensions).
31. JBScreen Classic (Jena Bioscience).
32. Cryschem plate (Hampton Research).
33. MRC 2-well crystallization plate (Swissci).

2.2 Surface Plasmon Resonance

1. BirA-500: BirA biotin-protein ligase standard reaction kit (Avidity).
2. Biotinylated Bovine Serum Albumin (BSA) 10 mg/ml (Sigma-Aldrich): Dissolve 10 mg of Biotinylated Bovine Serum Albumin in 1 ml of pure water.
3. Streptavidin from Streptomyces avidinii 1 mg/ml (Sigma-Aldrich): Dissolve 1 mg of streptavidin in 1 ml of pure water.
4. Amine Coupling Kit (GE Healthcare Life Sciences). Contents: 0.4 M 1-ethyl-3-(3-dimethylaminopropyl)carbodiimide hydrochloride (EDC): Dissolve 750 mg of EDC in 10 ml of pure water, and store at −20 °C; 0.1 M N-hydroxysuccinimide (NHS): Dissolve 115 mg of NHS in 10 ml of pure water, and store at −20 °C; 1 M Ethanolamine hydrochloride-NaOH pH 8.5.
5. 1× HBS-EP buffer (GE Healthcare Life Sciences): 0.01 M HEPES pH 7.4, 0.15 M NaCl, 3 mM EDTA, 0.005% (v/v) Surfactant P20.
6. 10 mM sodium acetate buffer pH 5.0 (GE Healthcare Life Sciences).
7. 10 mM glycine-HCl buffer pH 2.0 (GE Healthcare Life Sciences).
8. Sensor Chip CM5/Series S Sensor Chip CM5 (GE Healthcare Life Sciences).
9. Biacore 2000/Biacore T200 (GE Healthcare Life Sciences).

2.3 Fluorescence Binding Assay

FlAsH-EDT$_2$ is nonfluorescent, but becomes fluorescent on binding with the tetracysteine motif (Cys-Cys-Xaa-Xaa-Cys-Cys, where Xaa denotes any amino acid) [19]. Previous studies have used FlAsH fluorescein for protein localization [20], protein oligomeric state analysis [21], and FRET assay [22]. We have investigated protein–protein interactions using a FlAsH fluorescein-labeled protein that includes a genetically encoded tetracysteine motif [10, 23].

1. TC-FlAsH™ II In-Cell Tetracysteine Tag Detection Kit (Thermo Fisher Scientific).
2. Slide-A-Lyzer™ dialysis cassette (Thermo Fisher Scientific).
3. Hellma® fluorescence cuvette 109.004F (Hellma Analytics).
4. NanoDrop™ 1000 (Thermo Fisher Scientific).
5. LS55 fluorescence spectrometer (PerkinElmer).

3 Methods

For crystallization, affinity tags should be removed, except for a short His-tag (pOPTH; MAHHHHHHM). MIMs of ESCRT-III are often unstructured in the absence of the core region. It is recommended that MIMs be expressed with affinity tags and cleaved with a protease during the purification steps. For SPR measurements, we normally use biotinylated proteins as the ligands to be immobilized. The ligand is expressed with a His-biotinylation-LD tag (pOPHBL; MAHHHHHHSSGSLHHIL-DAQKMVWNHRH-"LD"-M) and is biotinylated using the enzyme BirA. A His-LD tag (pOPHLT) can be useful for the purification and quantification of MIMs of ESCRT-III and can be removed with TEV protease if needed (*see* **Note 3**).

3.1 Protein Expression

1. Transform a pOP expression vector into C41(DE3)-competent cells, and spread these cells onto an LB agar plate with ampicillin (100 μg/ml). Incubate the plate at 37 °C for 14 h (*see* **Note 4**).
2. Isolate a single colony and incubate into 5 ml of LB medium with ampicillin (100 μg/ml). Incubate in a shaker (180 rpm) at 37 °C until the OD600 is 0.5–1.0 (*see* **Note 5**).
3. Transfer the whole volume of the starter culture into 1000 ml of LB medium with ampicillin. Incubate for 2–4 h until the OD600 reaches 0.5–1.0. Induce expression by adding IPTG at a final concentration of 0.3 mM. Incubate for 3 h and harvest the cell pellet by centrifugation (15 min at 7000 × *g*). Remove the supernatant and freeze the pellet in a 50 ml conical centrifuge tube at −20 °C.

3.2 Purification and Crystallization

1. Resuspend the cell pellet in buffer, typically 50 ml of HisTrap buffer A for the purification of 1–2 l of cells.
2. Sonicate the cell pellet on ice with a probe sonicator (Instrument, Ultrasonic disruptors UD-211 (TOMY); Probe,

Standard tip TP-012 (recommended a sample volume of 10–500 ml), typically for 4 min, with a 10 s burst of sonication followed by 10 s of rest.

3. Centrifuge the lysate at 35,000 × *g* for 30 min at 4 °C.
4. Remove the supernatant from the ultracentrifuge vial and filter through a 0.45 μm pore size filter. Load the filtered supernatant onto a HisTrap HP column equilibrated with HisTrap buffer A.
5. After washing with 100 ml of HisTrap buffer A, elute with 150 ml of HisTrap buffer B using a linear gradient. Run an SDS-PAGE of fractions to determine the location and purity of proteins.
6. In order to obtain MIT-MIM complex crystals, it is recommended that the His-LD-tag be removed. Add TEV protease in a ratio of 1:100 of TEV to His-LD-tag-fused MIM (w/w) and incubate for 14 h at 4 °C before size-exclusion chromatography. The His-tag derived from pOPTH vector does not inhibit the crystal formation.
7. If the protein purity is low, apply samples to an ion exchange column (e.g., a HiTrap Q HP equilibrated in ion exchange buffer Q-A). Elute with 150 ml of ion exchange buffer Q-B using a linear gradient. Run an SDS-PAGE of fractions to determine the location and purity of proteins.
8. Concentrate a sample to 2 ml using an Amicon Ultra15 concentrator and load onto a HiLoad 16/60 Superdex 75 gel filtration column equilibrated in gel filtration buffer Xtal.
9. Run an SDS-PAGE of fractions and concentrate samples to nearly 10 mg/ml using an Amicon Ultra15 concentrator.
10. Mix the MIT domain and MIM peptide in a ratio of 1:1–1:2. An excess amount of MIM might help with the complex crystal formation because the MIT–MIM interactions might be weak.
11. Set up crystallization plates for initial screening. We normally test several sparse matrix screening kits (e.g., Crystal Screen I/II (Hampton Research), JCSG-*plus*™ (Molecular Dimensions), or JBScreen Classic (Jena Bioscience)) (*see* **Note 6**).
12. Optimize the precipitant concentrations, pH (in steps of 0.1 pH units), additive reagent concentrations, and protein concentrations.

3.3 Purification and Biotinylation for Surface Plasmon Resonance

1. His-biotinylation-LD-tagged protein (pOPHBL) is used for the ligands of SPR experiments and the purification steps are the same as those in Subheading 3.2 except that Bio is used as the gel filtration buffer.

2. Use the BirA-500 biotinylation kit. Mix as follows: 320 μl of 6xHis-biotinylation-LD-tagged protein (nearly 50 μM), 40 μl of Biomix A, 40 μl of Biomix B, and 1 μl of BirA enzyme (*see* **Notes** 7 and **8**).
3. Incubate the reaction mixture for 3 h at 30 °C. Load onto a Superdex 75 10/300 gel filtration column equilibrated with the gel filtration buffer SPR.
4. Run an SDS-PAGE of fractions from gel filtration column. Confirm that the protein has been biotinylated using MALDI-TOF mass spectroscopy. Concentrate samples to nearly 1 mg/ml using an Amicon Ultra4 concentrator.

3.4 Purification and Fluorescence Labeling

1. His-FlAsH-LD-tagged protein (pOPHFL) is used for the fluorescence-binding assay, and the purification steps are the same as in Subheading 3.2 except that Fluo is used as the gel filtration buffer.
2. Use the TC-FlAsH™ II kit. Mix as follows: 100 μl of 6xHis-FlAsH-LD-tagged protein (20 μM), 99 μl of the gel filtration buffer Fluo, and 1 μl of FlAsH-EDT_2 fluorescein (2 mM).
3. Incubate the reaction mixture for 2 h at 4 °C in the dark (*see* **Note 9**).
4. Dialyze against 3.5 l of the dialysis buffer Fluo using a Slide-A-Lyzer™ dialysis cassette for 14 h (o/n) at 4 °C in the dark.
5. Measure at 508 nm assuming a molar extinction coefficient for FlAsH fluorescein of 41,000 M^{-1} cm^{-1} for labeled protein [24] using a NanoDrop™ 1000. The protein concentration will be around 10 μM, if the fluorescence labeling and dialysis are successful.

3.5 Surface Plasmon Resonance Experiment

We performed SPR experiments with Biacore 2000/T200 instruments (GE Healthcare Life Sciences). Amine coupling to a CM5 (a carboxymethylated dextran matrix) sensor chip (GE Healthcare Life Sciences) is the first choice for coupling of a new molecule. The biotinylated ligand was immobilized on a CM5 sensor chip, onto which streptavidin was covalently coupled. As a negative control, either poly-biotinylated BSA or mono-biotinylated LD can be bound on a different surface.

1. "Prime" the instrument to flush the flow system with 1× HBS-EP buffer. The analysis temperature and sample compartment temperature are both 25 °C.
2. Inject a mixture of 35 μl of 0.1 M NHS and 35 μl of 0.4 M EDC to activate the surface of the CM5 sensor chip (Fig. 2).
3. Inject a mixture of 99 μl of 10 mM sodium acetate buffer pH 5.0 and 1 μl of 1 mg/ml streptavidin to couple to the surface matrix (*see* **Note 10**).

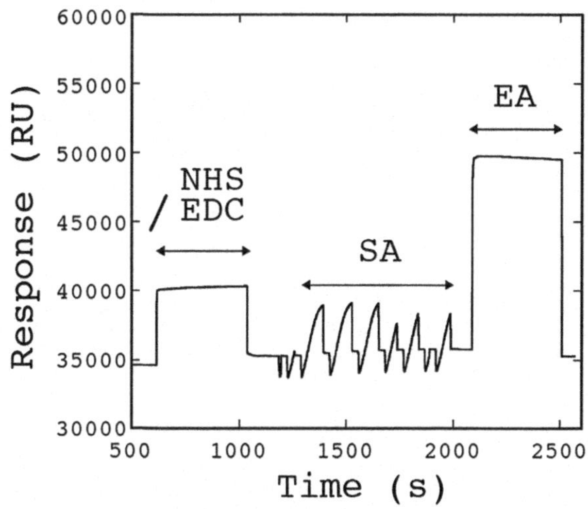

Fig. 2 A typical sensorgram of the immobilization of a ligand to a CM5 sensor ship. NHS/EDC shows the injection of NHS/EDC to activate the surface. SA shows the injections of streptavidin (ligand) to couple to the surface matrix. The small amounts of ligand injections were repeated in order to prevent the excess amount of ligand binding. EA shows the injection of ethanolamine to deactivate unreacted NHS-ester

4. Inject 70 μl of 1 M ethanolamine hydrochloride-NaOH pH 8.5 to deactivate unreacted NHS-ester.
5. Inject 100 μl of 10 mM glycine-HCl buffer pH 2.0.
6. Inject a mixture of 100 μl of 1 mg/ml biotinylated BSA and 100 μl of 1× HBS-EP buffer into the flow cell 1 until the resonance unit (RU) value reaches 500–1000 RU.
7. Inject a mixture of 100 μl of a biotinylated protein (1 mg/ml) and 100 μl of 1× HBS-EP buffer into the flow cell 2 until the resonance unit (RU) value reaches 500–1000 RU.
8. Inject analytes over the sensor surfaces (Fig. 3a).
9. The equilibrium dissociation constants (K_D) were determined by the nonlinear least squares method assuming a 1:1 Langmuir binding model (Fig. 3b).

3.6 Fluorescence-Binding Assay

1. Add the gel filtration buffer Fluo and FlAsH-labeled protein (to a final concentration of 50 nM) to the cuvette with a magnetic stirring bar, and place the cuvette in the sample holder at 25 °C (*see* **Note 11**).
2. Set the excitation and emission wavelengths to 490 and 530 nm, respectively. Set the slit size of excitation and emission to 15 and 20 nm, respectively. Set the PMT voltage to 950 V. Measure the G-factor.

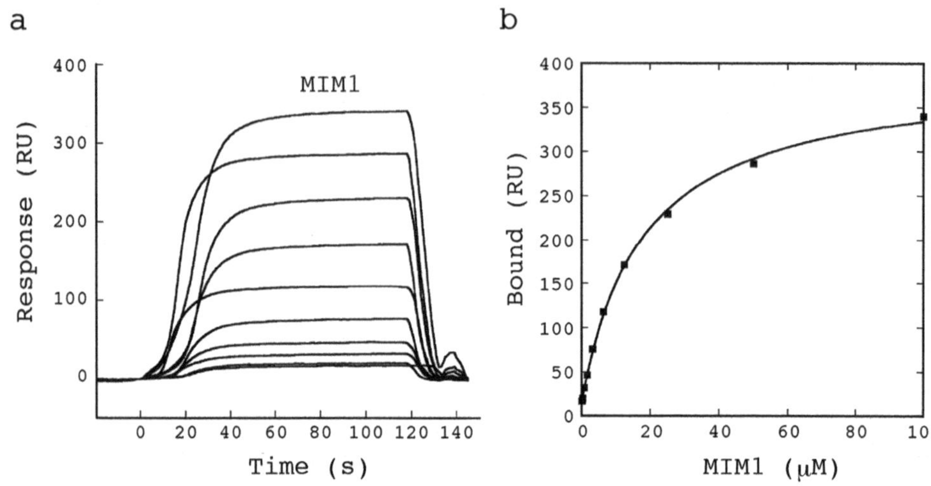

Fig. 3 SPR analysis of binding of MIM1 to immobilized Vps4-MIT domain. (**a**) Sensorgrams obtained when MIM1 at various concentrations were injected to the sensor chip. (**b**) Plots of the equilibrium binding responses of MIM1 to the Vps4-MIT domain

3. Titrate the sample into the cuvette, and measure the polarization (P) and anisotropy (R). Wait for 180 s after each titration for the solution to come to equilibrium after adding an aliquot of protein.
4. Determine the K_D values by the nonlinear least squares method assuming a 1:1 Langmuir binding model.

4 Notes

1. All buffers were prepared with pure water and analytic grade reagents. All buffers used in the purification of ESCRT proteins were chilled to 4 °C prior to use.
2. Solution pH was adjusted at room temperature (25 °C) to a given pH value that will give the right value at the temperature at which the solution is used.
3. First, we express a protein in the C41(DE3) strain from pOP vectors with different tags: pOPTH (6xHis tag), pOPTG (GST tag with TEV protease cleavage), pOPTM (MBP tag), pOPHLT (6xHis-LD tag with TEV protease cleavage), pOPHL (6xHis-LD tag), pOPHBL (6xHis-biotinylation-LD tag), pOPHB (6xHis-biotinylation tag), and pOPHFL (6xHis-fluoresnce-LD tag).
4. Second, we express a protein from a standard pOPTH vector in several different *E. coli* host strains: C41(DE3) (Lucigen), C41 (DE3) pLysS (Lucigen), C43(DE3) (Lucigen), C43(DE3) pLysS (Lucigen), BL21(DE3) (Agilent), BL21(DE3)

CodonPlus RIL (Agilent), and Rosetta(DE3) (Novagen). If the protein is toxic to the cell, the C41/C43 strains or strains containing a pLysS vector might promote the protein expression. If the protein contains a high number of rare codons, it is worthwhile to express it in strains that co-express the tRNAs for these rare codons.

5. It is important to pick a single colony from a freshly streaked plate. We do not recommend starting culture directly from a glycerol stock. In addition, in order to avoid full growth, culture the starter culture at 28 °C or lower for overnight, rather than at 37 °C. Inoculation of the starter culture into the main culture at OD600 of ~0.6 is recommended. Induction of protein expression is done by IPTG (final concentration of 0.2–0.5 mM). The cultures are incubated from 3 to 14 h depending on the induction temperature (e.g., at 37 °C for 3 h, at 30 °C for 5 h, or at 20 °C for 14 h). Lower temperature may help for soluble expression.
6. For crystallization, we use 24-well or 96-well sitting drop plates such as Cryschem plates (Hampton Research) or MRC 2-well crystallization plates (Swissci). Hanging drop plates are also effective.
7. BirA enzyme is commercially available from Avidity, but we have successfully used recombinant BirA (EC: 6.3.4.15) cloned from bacteria. We have confirmed that recombinant BirA can be used for biotinylation under normal reaction conditions: pH 6–8, 20–37 °C, 0–200 mM NaCl, 0–5 mM DTT.
8. We characterize the biotinylation using MALDI-TOF mass spectroscopy. The molecular weight of a biotinylated protein is increased by 226 Da compared to that of the unmodified one.
9. For a final concentration of FlAsH fluorescent labeling, we use a 10 μM FlAsH-EDT_2. Under our labeling condition, the stoichiometry of the FlAsH:tetracystein motif is 1:1. A slightly excess amount of FlAsH fluorescein may help complete the labeling, and unreacted fluorescein can be removed by dialysis.
10. For the amine coupling on sensor chip CM5, we immobilize 2000–4000 RU of streptavidin on the activated sensor surface. After deactivation and washing on the sensor surface, we inject 500–1000 RU of biotinylated proteins over the streptavidin surface.
11. Our working concentration for the fluorescence binding assay is 50 nM in the cuvette. It is important to wait for a few minutes after each titration because the binding reactions may require a minute to reach equilibrium.

Acknowledgments

We thank Dr. Olga Perisic (MRC-LMB, Cambridge, UK) for providing the pOP vectors. If you have an inquiry about the pOP vectors, please ask (Email: op1@mrc-lmb.cam.ac.uk (O.P.); obita@pha.u-toyama.ac.jp (T.O.)). This work was supported by a JSPS KAKENHI grant (number 16K07265).

References

1. Bowers K, Lottridge J, Helliwell SB, Goldthwaite LM, Luzio JP, Stevens TH (2004) Protein-protein interactions of ESCRT complexes in the yeast *Saccharomyces cerevisiae*. Traffic 5(3):194–210. https://doi.org/10.1111/j.1600-0854.2004.00169.x; pii: TRA169
2. Martin-Serrano J, Yarovoy A, Perez-Caballero D, Bieniasz PD (2003) Divergent retroviral late-budding domains recruit vacuolar protein sorting factors by using alternative adaptor proteins. Proc Natl Acad Sci U S A 100 (21):12414–12419. https://doi.org/10.1073/pnas.2133846100; pii: 2133846100
3. Katzmann DJ, Odorizzi G, Emr SD (2002) Receptor downregulation and multivesicular-body sorting. Nat Rev Mol Cell Biol 3 (12):893–905. https://doi.org/10.1038/nrm973; pii: nrm973
4. Martin-Serrano J, Zang T, Bieniasz PD (2001) HIV-1 and Ebola virus encode small peptide motifs that recruit Tsg101 to sites of particle assembly to facilitate egress. Nat Med 7 (12):1313–1319. https://doi.org/10.1038/nm1201-1313. pii: nm1201-1313
5. Carlton JG, Martin-Serrano J (2007) Parallels between cytokinesis and retroviral budding: a role for the ESCRT machinery. Science 316 (5833):1908–1912. https://doi.org/10.1126/science.1143422; pii: 1143422
6. Lindas AC, Karlsson EA, Lindgren MT, Ettema TJ, Bernander R (2008) A unique cell division machinery in the Archaea. Proc Natl Acad Sci U S A 105(48):18942–18946. https://doi.org/10.1073/pnas.0809467105; pii: 0809467105
7. Samson RY, Obita T, Freund SM, Williams RL, Bell SD (2008) A role for the ESCRT system in cell division in archaea. Science 322 (5908):1710–1713. https://doi.org/10.1126/science.1165322; pii: 1165322
8. Wollert T, Wunder C, Lippincott-Schwartz J, Hurley JH (2009) Membrane scission by the ESCRT-III complex. Nature 458 (7235):172–177. https://doi.org/10.1038/nature07836; pii: nature07836
9. Babst M, Wendland B, Estepa EJ, Emr SD (1998) The Vps4p AAA ATPase regulates membrane association of a Vps protein complex required for normal endosome function. EMBO J 17(11):2982–2993. https://doi.org/10.1093/emboj/17.11.2982
10. Obita T, Saksena S, Ghazi-Tabatabai S, Gill DJ, Perisic O, Emr SD, Williams RL (2007) Structural basis for selective recognition of ESCRT-III by the AAA ATPase Vps4. Nature 449 (7163):735–739. https://doi.org/10.1038/nature06171; pii: nature06171
11. Stuchell-Brereton MD, Skalicky JJ, Kieffer C, Karren MA, Ghaffarian S, Sundquist WI (2007) ESCRT-III recognition by VPS4 ATPases. Nature 449(7163):740–744. https://doi.org/10.1038/nature06172; pii: nature06172
12. Kieffer C, Skalicky JJ, Morita E, De Domenico I, Ward DM, Kaplan J, Sundquist WI (2008) Two distinct modes of ESCRT-III recognition are required for VPS4 functions in lysosomal protein targeting and HIV-1 budding. Dev Cell 15(1):62–73. https://doi.org/10.1016/j.devcel.2008.05.014; pii: S1534-5807(08)00239-6
13. Yang D, Rismanchi N, Renvoise B, Lippincott-Schwartz J, Blackstone C, Hurley JH (2008) Structural basis for midbody targeting of spastin by the ESCRT-III protein CHMP1B. Nat Struct Mol Biol 15(12):1278–1286. https://doi.org/10.1038/nsmb.1512; pii: nsmb.1512
14. Solomons J, Sabin C, Poudevigne E, Usami Y, Hulsik DL, Macheboeuf P, Hartlieb B, Gottlinger H, Weissenhorn W (2011) Structural basis for ESCRT-III CHMP3 recruitment of AMSH. Structure 19(8):1149–1159. https://doi.org/10.1016/j.str.2011.05.011; pii: S0969-2126(11)00205-X
15. Yang Z, Vild C, Ju J, Zhang X, Liu J, Shen J, Zhao B, Lan W, Gong F, Liu M, Cao C, Xu Z (2012) Structural basis of molecular recognition between ESCRT-III-like protein Vps60

and AAA-ATPase regulator Vta1 in the multivesicular body pathway. J Biol Chem 287 (52):43899–43908. https://doi.org/10.1074/jbc.M112.390724; pii: M112.390724

16. Skalicky JJ, Arii J, Wenzel DM, Stubblefield WM, Katsuyama A, Uter NT, Bajorek M, Myszka DG, Sundquist WI (2012) Interactions of the human LIP5 regulatory protein with endosomal sorting complexes required for transport. J Biol Chem 287 (52):43910–43926. https://doi.org/10.1074/jbc.M112.417899; pii: M112.417899
17. Kojima R, Obita T, Onoue K, Mizuguchi M (2016) Structural fine-tuning of MIT-interacting motif 2 (MIM2) and allosteric regulation of ESCRT-III by Vps4 in yeast. J Mol Biol 428(11):2392–2404. https://doi.org/10.1016/j.jmb.2016.04.007; pii: S0022-2836(16)30051-1
18. Miwa K, Kojima R, Obita T, Ohkuma Y, Tamura Y, Mizuguchi M (2016) Crystal structure of human general transcription factor TFIIE at atomic resolution. J Mol Biol 428 (21):4258–4266. https://doi.org/10.1016/j.jmb.2016.09.008; pii: S0022-2836(16)30372-2
19. Griffin BA, Adams SR, Jones J, Tsien RY (2000) Fluorescent labeling of recombinant proteins in living cells with FlAsH. Methods Enzymol 327:565–578. pii: S0076-6879(00)27302-3
20. Ignatova Z, Gierasch LM (2004) Monitoring protein stability and aggregation in vivo by real-time fluorescent labeling. Proc Natl Acad Sci U S A 101(2):523–528. https://doi.org/10.1073/pnas.0304533101; pii: 0304533101
21. Webber TM, Allen AC, Ma WK, Molloy RG, Kettelkamp CN, Dow CA, Gage MJ (2009) Conformational detection of p53's oligomeric state by FlAsH Fluorescence. Biochem Biophys Res Commun 384(1):66–70. https://doi.org/10.1016/j.bbrc.2009.04.073; pii: S0006-291X(09)00769-4
22. Granier S, Kim S, Shafer AM, Ratnala VR, Fung JJ, Zare RN, Kobilka B (2007) Structure and conformational changes in the C-terminal domain of the beta2-adrenoceptor: insights from fluorescence resonance energy transfer studies. J Biol Chem 282(18):13895–13905. https://doi.org/10.1074/jbc.M611904200; pii: M611904200
23. Mizuguchi M, Obita T, Serita T, Kojima R, Nabeshima Y, Okazawa H (2014) Mutations in the PQBP1 gene prevent its interaction with the spliceosomal protein U5-15 kD. Nat Commun 5:3822. https://doi.org/10.1038/ncomms4822; pii: ncomms4822
24. Krishnan B, Gierasch LM (2008) Cross-strand split tetra-Cys motifs as structure sensors in a beta-sheet protein. Chem Biol 15 (10):1104–1115. https://doi.org/10.1016/j.chembiol.2008.09.006; pii: S1074-5521(08)00363-3

Chapter 14

Biochemical Approaches to Studying *Caenorhabditis elegans* ESCRT Functions In Vitro

Samuel Block, Amber L. Schuh, and Anjon Audhya

Abstract

Our fundamental understanding of the roles played by the endosomal sorting complex required for transport (ESCRT) machinery in cells comes from interdisciplinary approaches that combine numerous in vivo and in vitro techniques. Here, we focus on methods used to biochemically characterize *Caenorhabditis elegans* ESCRT components in vitro, including the production and characterization of recombinant ESCRT complexes and their use in membrane interaction studies. Key methodologies used include gel filtration chromatography, glycerol density gradient analysis, multi-angle light scattering, liposome co-flotation, and single-liposome fluorescence microscopy. Collectively, these studies have enabled us to define subunit stoichiometry of soluble *C. elegans* ESCRT complexes and to demonstrate that the late-acting ESCRT-III complex facilitates membrane bending and remodeling, at least in part by virtue of its ability to sense the curvature of lipid bilayers.

Key words ESCRT, *C. elegans*, Biochemistry, Multi-angle light scattering, Liposome co-flotation

1 Introduction

Although *Caenorhabditis elegans* is generally known for its utility in genetic analysis of pathway function, we and others have found it to be equally amenable to biochemical approaches, including in vitro reconstitution-based studies [1–5]. In particular, components of the *C. elegans* ESCRT machinery can be purified recombinantly from *Escherichia coli* with limited proteolysis and, in some cases, superior stability as compared to analogous proteins from other species [6]. The early-acting ESCRT machinery is composed of three multi-subunit, soluble complexes (ESCRT-0, ESCRT-I, and ESCRT-II), which only assemble properly in vitro when their subunits are co-expressed. The advent of polycistronic expression systems facilitated the production of these complexes in large quantities [7]. In contrast, the late-acting ESCRT-III complex forms polymers, mostly on lipid bilayers in vivo [8, 9]. Individual subunits can be expressed and purified using *E. coli*, albeit at

Emmanuel Culetto and Renaud Legouis (eds.), *The ESCRT Complexes: Methods and Protocols*, Methods in Molecular Biology, vol. 1998, https://doi.org/10.1007/978-1-4939-9492-2_14, © Springer Science+Business Media, LLC, part of Springer Nature 2019

differing levels of solubility, and they can subsequently be added together in the presence of membranes to generate higher-order oligomers. Among the *C. elegans* ESCRT-III subunits, our findings indicate that Did2/CHMP1, Vps24/CHMP3, and Vps20/CHMP6 are soluble at millimolar levels, while Vps2/CHMP2, Vps32/CHMP4, and Vps60/CHMP5 precipitate at elevated concentrations (above 10 μM) when purified in the presence of physiological levels of salt. The final component of the core ESCRT machinery is Vps4, an ATPase that functions in concert with ESCRT-III to promote membrane scission reactions in cells [10]. Upon binding ATP, *C. elegans* Vps4 subunits assemble into hexameric ring complexes in vitro. However, these complexes exhibit minimal intrinsic ATPase activity, in contrast to their yeast counterpart, although *C. elegans* Vps4 can be stimulated by the presence of ESCRT-III subunits, similar to mammalian orthologues [3]. Here, we will describe approaches to purifying components of the *C. elegans* ESCRT machinery. In addition, we will highlight the impact of hydrodynamic studies and imaging approaches to our understanding of ESCRT assembly and function on membrane surfaces.

For early-acting ESCRT complexes, we co-express individual subunits from the polycistronic expression plasmid pST39. In all three cases, only a single subunit of each complex is polyhistidine tagged to enable isolation from bacterial extracts on nickel-nitrilotriacetic acid (Ni-NTA) resin. For ESCRT-III subunits and Vps4, the expression occurs using the plasmid pGEX6P-1, which encodes a glutathione S-transferase (GST) tag followed by a PreScission protease cleavage site upstream of each ESCRT component. Bacterially expressed proteins are initially recovered on glutathione agarose and then cleaved using the protease to enable release of the untagged, soluble protein. In most cases, proteins are further purified using size-exclusion chromatography.

Following purification, ESCRT proteins can be used in numerous experimental approaches, including hydrodynamic analyses and interaction studies with membranes. In particular, we have developed approaches to analyze subunit stoichiometry in the early-acting ESCRT machinery, both in solution and on lipid bilayers [2, 4]. Influence of specific membrane lipids and bilayer curvature on ESCRT protein association can also be tested rigorously, using chemically defined liposomes and parallel fluorescence-based measurements [1–5]. In this chapter, we will describe some of the key strategies used to understand how ESCRT complexes assemble in solution and on membranes.

2 Materials

2.1 Recombinant Protein Purification and Hydrodynamic Analysis

1. Terrific Broth-modified powder autoclaved in H_2O to manufacturer's specifications (500 mL in 2.8 L Erlenmeyer flasks).
2. Lysogeny broth (LB) autoclaved in H_2O (100 mL in 250 mL flasks).
3. 100 mg/mL ampicillin in H_2O (1000× stock).
4. Incubator shaker for bacterial growth (temperature range from 18 °C to 37 °C).
5. 400 mM isopropyl β-D-1thiogalactopyranoside (IPTG) in H_2O.
6. Large-capacity centrifuge (Avanti J-26 XPI with JLA-8.1000 6 L rotor and 1 L polypropylene bottles).
7. GST lysis buffer: 1× PBS, 10 mM EGTA, 10 mM EDTA, 0.1% Tween-20, and 250 mM NaCl.
8. GST wash buffer:1× PBS, 250 mM NaCl, 0.1% Tween-20, and 1 mM dithiothreitol (DTT).
9. 6xHis lysis buffer: 50 mM $NaHPO_3$ pH 8.0, 300 mM NaCl, 10 mM imidazole, and 0.15% Tween-20.
10. 6xHis wash buffer: 50 mM $NaHPO_3$ pH 8.0, 300 mM NaCl, 10 mM imidazole, 0.1% Tween-20, and 2-mercaptoethanol.
11. Vortexer (standard for falcon tubes).
12. 50 mL polypropylene centrifuge tubes.
13. Sonicator (Branson Sonifier with macrotip).
14. 1.0 M phenylmethane sulfonyl fluoride (PMSF) in ethanol.
15. 1.0 M benzamidine in ethanol.
16. 20 mg/mL lysozyme; egg white in 1× PBS.
17. 2-Mercaptoethanol.
18. Floor model centrifuge (Sorvall RC-5B centrifuge with SS-34 rotor).
19. Glutathione agarose.
20. Ni-NTA agarose.
21. Tube rotator.
22. Clinical centrifuge (Thermo IEC31R with swinging bucket rotor).
23. Aspirator.
24. 5 mL syringes (to assemble gravity flow columns).
25. 6xHis elution buffer: 50 mM HEPES (4-(2-hydroxyethyl)-1-piperazineethanesulfonic acid) pH 7.6, 100 mM NaCl, 300 mM imidazole, and 1 mM DTT.

26. GST elution buffer: 50 mM Tris pH 8.2, 75 mM KCl, 1 mM DTT, and 10 mM glutathione.
27. Cleavage buffer: 50 mM HEPES pH 7.6, 150 mM NaCl, 1 mM EDTA, 1 mM DTT, and 0.01% Tween-20.
28. PreScission Protease.
29. Tabletop ultracentrifuge (Beckman TL100 with TLA-100.3 rotor and polycarbonate centrifuge tubes).
30. Fast protein liquid chromatography (FPLC) buffers: 50 mM HEPES pH 7.6, 100–500 mM NaCl, and 1 mM DTT.
31. Disposable 3 mL syringes and needles.
32. FPLC with size-exclusion chromatography columns (AKTA purifier with Superose 6 and/or S200 columns from GE).
33. Sodium dodecyl sulfate–polyacrylamide gel electrophoresis (SDS-PAGE) supplies and Coomassie Blue.
34. Glycerol solutions: 50 mM Hepes pH 7.6, 100 mM NaCl, 1 mM DTT, and 10% or 30% glycerol.
35. Gradient Master and accessories (to pour glycerol gradients).
36. Gradient centrifuge tubes (7010, Seton).
37. Wyatt miniDAWN TREOS multi-angle light-scattering device.
38. Liquid nitrogen.

2.2 Protein and Membrane Interaction Studies

1. Avanti mini extruder.
2. Polycarbonate filters.
3. Purified lipids: phosphatidylcholine (POPC), phosphatidylethanolamine (POPE), phosphatidylserine (POPS), phosphatidylinositol 3-phosphate (PI3P), biotinyl-PE, and rhodamine-labeled PE.
4. Tweezers.
5. 200 proof ethanol.
6. Deionized H_2O.
7. N_2 gas.
8. Lyophilizer.
9. Co-flotation buffer: 50 mM HEPES pH 7.6, 100 mM NaCl, and 1 mM DTT.
10. Vortexer.
11. 100% accudenz (wt/vol) in deionized H_2O).
12. Accudenz solutions: 30%, 35%, and 80% accudenz in co-flotation buffer.
13. Tabletop ultracentrifuge (Beckman TL100 with TLA-100 rotor and polycarbonate tubes).
14. Boron-dipyrromethene (BODIPY)-FL-labeled maleimide.

15. Glutathione.
16. Spectrofluorometer (PTI QM40).
17. Swept-field confocal microscope (Nikon Ti-E equipped with a Roper CoolSNAP HQ2 CCD camera and a 60×, 1.4 numerical aperture Plan Apo oil objective lens).
18. Nikon Elements Software.
19. Glass coverslips.
20. Polyethylene glycol (PEG) and biotin-PEG (40:1 ratio) solutions.
21. 1.0 M Avidin in wash buffer.
22. Wash buffer: 50 mM HEPES pH 7.6, 100 mM NaCl.
23. MetaMorph software.
24. MicroCal ITC200 calorimeter.

3 Methods

3.1 Recombinant Protein Purification and Characterization

3.1.1 ESCRT-0, ESCRT-I, and ESCRT-II Complex Expression and Purification

1. Transfect pST39-based expression plasmid into BL21 (DE3) competent cells. For ESCRT-0, the Hrs subunit harbors an amino-terminal polyhistidine tag. For ESCRT-I, the Tsg101 subunit harbors a carboxyl-terminal polyhistidine tag. For ESCRT-II, the Vps25 subunit harbors a carboxyl-terminal polyhistidine tag.
2. Set up an overnight culture (100 mL LB, 1× ampicillin in a 250 mL flask) with a single BL21 (DE3) transfected colony. Grow cultures at 25 °C (shaking at 240 RPM [revolutions per minute]) for 12–16 h.
3. Inoculate an Erlenmeyer flask containing 500 mL of sterile Terrific Broth media (containing 1× ampicillin) using 10 mL of the overnight LB culture. Place the flask in an incubator shaker and grow at 25 °C (shaking at 240 RPM) to an OD_{600} of approximately 2.0. Turn down the incubator shaker to 18 °C and add 300 μL of 400 mM IPTG to the culture (final concentration: 0.24 mM). Let the culture shake for approximately 12–14 h.
4. Centrifuge the cells in a large-capacity centrifuge (4785 RCF [relative centrifugal force], 4 °C) for 15 min (*see* **Note 1**). Pour off the majority of the media and transfer the pelleted bacteria to a 50 mL polypropylene centrifuge tube (by resuspending the pellet in the remaining media). Centrifuge again, aspirate the media, and resuspend the pellet in 6xHis lysis buffer by vortexing. Freeze the pellet at −80 °C (*see* **Note 2**).
5. Thaw the pellet at room temperature. Once thawed, transfer the lysate to a 50-mL glass beaker on ice and add 1 mL of 1.0 M

benzamidine, 1 mL of 1.0 M PMSF, and 2 mL of 1 mg/mL lysozyme (*see* **Note 3**). Swirl gently on ice.

6. Sonicate the cells by using a 400-watt ultrasonic homogenizer (70% amplitude, 0.3 s pulse on, 0.7 s pulse off, macrotip) for a total of 60 s in an ice bath. Avoid creating bubbles or froth. Add 40 μL of 2-mercaptoethanol and swirl gently on ice (*see* **Note 4**).
7. Centrifuge the lysate in a floor model centrifuge at 27,000 RCF, 4 °C, for 1 h. Carefully collect the supernatant into a 50 mL polypropylene centrifuge tube containing 1 mL Ni-NTA resin. Place the tube on a rotator and rotate for 1 h at 4 °C (*see* **Note 5**).
8. Centrifuge the resin in a clinical centrifuge (1000 RCF, 4 °C) for 2 min. Aspirate to the top of the resin and then resuspend the resin with cold 6xHis lysis buffer through gentle inversion (*see* **Note 6**). Repeat two times.
9. Transfer the resin to a 5 mL syringe column (with a porous frit at the bottom to prevent resin from leaking out) and wash with 200 mL of cold 6xHis wash buffer by gravity flow at 4 °C (add PMSF to the wash buffer immediately prior to use). Wash with 50 mL of cold 6xHis wash buffer without Tween-20 by gravity flow at 4 °C (add PMSF to the wash buffer immediately prior to use). Avoid allowing the column to run dry.
10. Cap the bottom of the syringe and add 1 mL 6xHis elution buffer. Plug the top of the syringe and rotate for 5 min at 4 °C. Remove the top plug first, and collect the eluate containing the protein of interest by gravity flow. Repeat twice.
11. Concentrate the sample to 1 mL total volume using a Vivaspin centrifugal concentrator and gel filter using a Superose 6 or S200 (GE) size-exclusion column equilibrated in 6xHis elution buffer without imidazole. Elute in 1 mL fractions.
12. Analyze fractions using SDS-PAGE followed by Coomassie blue staining. Combine fractions containing intact ESCRT complexes and snap freeze aliquots in liquid nitrogen and store at −80 °C. In general, only one or two of the fractions recovered from the gel filtration column will contain intact ESCRT complexes.

3.1.2 Vps4 and ESCRT-III Subunit Expression and Purification

1. Transfect pGEX6P-1-based expression plasmid into BL21 (DE3) competent cells. For all ESCRT-III subunits and Vps4, an amino-terminal GST tag is present to enhance solubility and enable purification.
2. Set up an overnight culture (100 mL LB, 1× ampicillin in a 250 mL flask) with a single BL21 (DE3) transfected colony. Grow cultures at 25 °C (shaking at 240 RPM) for 12–16 h.

3. Inoculate an Erlenmeyer flask containing 500 mL of sterile Terrific Broth media (containing 1× ampicillin) using 10 mL of the overnight LB culture. Place the flask in an incubator shaker and grow at 25 °C (shaking at 240 RPM) to an OD_{600} of approximately 2.0. Turn down the incubator shaker to 18 °C and add 300 μL of 400 mM IPTG to the culture (final concentration: 0.24 mM). Let the culture shake for approximately 12–14 h.
4. Centrifuge the cells in a large-capacity centrifuge (4785 RCF, 4 °C) for 15 min. Pour off the majority of the media and transfer the pelleted bacteria to a 50 mL polypropylene centrifuge tube (by resuspending in the remaining media). Centrifuge again, aspirate the media, and resuspend the pellet in GST lysis buffer by vortexing. Freeze the pellet at −80 °C.
5. Thaw the pellet at room temperature. Once thawed, transfer the lysate to a 50 mL glass beaker on ice and add 1 mL of 1.0 M benzamidine, 1 mL of 1.0 M PMSF, and 2 mL of 1 mg/mL lysozyme. Swirl gently on ice. For Vps4, additionally add 1 mM ADP to the extract (*see* **Note** 7).
6. Sonicate the cells by using a 400-watt ultrasonic homogenizer (70% amplitude, 0.3 s pulse on, 0.7 s pulse off, macrotip) for a total of 60 s in an ice bath. Add 40 μL of 2-mercaptoethanol and swirl gently on ice.
7. Centrifuge the lysate in a floor model centrifuge at 27,000 RCF, 4 °C, for 1 h. Carefully collect the supernatant into a 50 mL polypropylene centrifuge tube with 1 mL glutathione agarose that has been equilibrated in GST lysis buffer. Place the tube on a rotator and rotate slowly for 1 h at 4 °C.
8. Centrifuge the resin in a clinical centrifuge (1000 RCF, 4 °C) for 2 min. Aspirate to the top of the resin and then resuspend the resin with cold GST lysis buffer through gentle inversion. Repeat two times.
9. Transfer the resin to a 5 mL syringe column (with a porous frit at the bottom to prevent resin from leaking out) and wash with 200 mL of cold GST wash buffer by gravity flow at 4 °C (add PMSF to the wash buffer immediately prior to use). Avoid allowing the column to run dry.
10. Wash with 20 mL of cleavage buffer by gravity flow at 4 °C. Avoid allowing the column to run dry.
11. Cap the bottom of the column and transfer resin using an equal volume of cleavage buffer to a screw-top polypropylene tube. For Vps4, supplement the cleavage buffer with 10 μM ADP. Add PreScission protease and rotate tube overnight at 4 °C (Fig. 1).

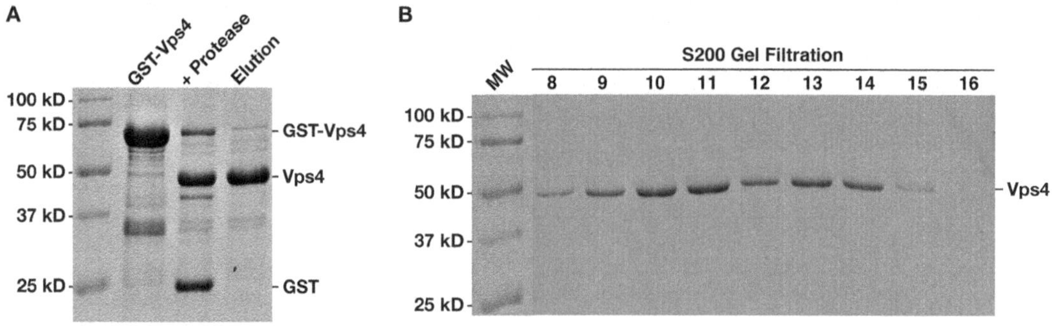

Fig. 1 Purification of *C. elegans* Vps4. (**a**) Following the expression from pGEX6P-1, GST-Vps4 is purified initially on glutathione agarose beads (lane 2), cleaved using PreScission protease (lane 3, a sample of boiled resin), and the eluate is resolved by SDS-PAGE (lane 4). Molecular weight marker is also shown (lane 1). (**b**) Recombinant Vps4 cleaved from glutathione resin following purification as a GST fusion protein was applied onto an S200 gel filtration column and elutes as two distinct peaks. Only peak fractions 13–15 harbor monomeric protein, which is capable of homotypic assembly into hexameric protein complexes when ATP is supplemented into the buffer

12. Centrifuge at 27,000 RCF and transfer the supernatant, which contains the cleaved protein, to a fresh tube.
13. Resuspend resin with 200 μL cleavage buffer, centrifuge, and add the supernatant to the previously collected, eluted protein.
14. Centrifuge the eluted protein in a tabletop ultracentrifuge (100,000 RCF, 4 °C) for 15 min. Collect the supernatant, being careful not to disturb any pellet that may have formed during centrifugation (Fig. 1).
15. Gel filter the sample using a Superose 6 or S200 (GE) size-exclusion column equilibrated in GST elution buffer without glutathione. Elute in 1 mL fractions. For Vps4, supplement the buffer with 10 μM ADP.
16. Analyze fractions using SDS-PAGE followed by Coomassie blue staining. Combine fractions containing intact ESCRT subunits and snap freeze aliquots in liquid nitrogen and store at −80 °C. In general, only one or two of the fractions will contain the intact ESCRT components. In the case of Vps4, oligomers that form during initial purification from bacterial extracts are aggregates and will not respond to the addition of ATP. Only retain the final fraction of Vps4 that elutes from the gel filtration column, as it contains mostly monomers, which are capable of assembling into hexameric ring complexes (Fig. 1).

3.2 Hydrodynamic Analysis of ESCRT Components

1. Use a Gradient Master system (Biocomp Instruments) to pour 10–30% glycerol gradients.
2. Mark the center of the gradient centrifuge tubes and fill to approximately 2 mm above this point with 10% glycerol.

3. Using a 10 mL syringe, layer the 30% glycerol solution underneath the 10% solution, until the interface between solutions reaches the center of the centrifuge tube. Be careful not to introduce any air bubbles into the tube.
4. Cap and rotate the tubes in the Gradient Master to generate linear gradients, following the manufacturers' instructions.
5. Cool the gradients for 1 h at 4 °C.
6. Carefully remove the caps and load 100 μL of purified ESCRT protein on top of one gradient. In parallel, load standards (with known sedimentation values such as thyroglobulin, apoferritin, BSA, etc.) onto other gradients that you will centrifuge at the same time.
7. Balance tubes by weight and centrifuge for 8 h at 50,000 RPM (260,000 RCF) in an SW60Ti rotor (swinging bucket).
8. Collect fractions (200 μL each) by hand from the top of the gradient, being careful not to disturb the gradient during this process.
9. Analyze fractions using SDS-PAGE followed by Coomassie blue staining.
10. Create a standard curve by plotting the peak elution of known standards against their sedimentation values. At least three standards must be run in parallel with your ESCRT protein, although additional standards will generate a more precise standard curve.
11. Plot the peak elution of the ESCRT component on the standard curve to identify its sedimentation value.
12. In combination with its Stokes radius, derived from size-exclusion chromatography studies, the native molecular weight of the ESCRT component can be estimated (within approximately 10%).
13. To calculate the native molecular weight, the following equation can be used: $M = 6\pi\eta Nas/(1 - \upsilon\rho)$, where M is the native molecular weight, η is the viscosity of the medium, N is Avogadro's number, a is the Stokes radius, s is the sedimentation value, υ is the partial specific volume, and ρ is the density of the medium [11].
14. In cases where ESCRT proteins can be purified to high concentrations (greater than 50 μM), molecular mass can also be determined with higher accuracy using a combination of size-exclusion chromatography and multi-angle light scattering. Use a Wyatt miniDAWN TREOS three-angle light-scattering device, coupled to a high-resolution size-exclusion column, for this purpose.
15. Apply 500 μL of purified ESCRT protein onto a Wyatt WTC-030S5 gel filtration column.

16. Use a flow rate of no more than 0.5 mL/min to maintain resolution.
17. Use Astra software to determine molecular mass, based on UV detection and in-line light scattering [12].

3.3 Interaction Studies with Membranes

3.3.1 Production of Liposomes

1. Using a glass micropipette, transfer stock lipid solutions (in 100% chloroform) into a glass test tube. Dry lipids using a steady stream of N_2 gas and then lyophilize overnight. The composition of liposomes can vary. In general, use a combination of POPC, POPS, and POPE to mimic biological membranes. However, the addition of cholesterol, phosphoinositides, or other lipids can be performed. Incorporate 0.5% of rhodamine-labeled PE into liposomes if they are used in co-flotation assays or other membrane interaction studies that leverage fluorescence microscopy.
2. Add 250 μL of co-flotation buffer and resuspend by gently vortexing at 5 min intervals for 30 min.
3. Set up the mini extruder as described by the manufacturer, making sure to soak the polycarbonate filter in deionized H_2O. Draw up 500 μL of deionized H_2O into one of the syringes and pass through the filter ten times. Discard the water and repeat with co-float buffer. The size of the filter used will determine liposome size. Hydrated lipid solutions will initially form large, multilamellar vesicles. After the initial pass through a membrane, the particle size distribution will tend toward bimodal. After sufficient passes through the membrane, a unimodal, normal distribution can be obtained. However, liposomes will not likely be homogenous in size using this method, and the use of dynamic light scattering to measure the size distribution of liposomes generated is highly recommended.
4. Draw up the resuspended lipids into one of the syringes and extrude them by passing them through the filter 19 times, making sure to collect the liposomes in the opposite syringe. Store the liposomes in a 1.5 mL polypropylene tube at 4 °C (*see* **Note 8**).

3.3.2 Co-flotation Assay

1. Thaw-purified ESCRT component on ice. While waiting, make up solutions of accudenz density media (30%, 35%, and 80%) in co-flotation buffer. Once the proteins are thawed, mix them with liposomes to a final volume of 60 μL and place the tube on ice for 30 min (*see* **Note 9**).
2. Add 60 μL of the 80% accudenz solution to the protein/liposome mixture, making sure to pipette up and down until the solution is homogenous. Add 100 μL of the solution to the bottom of a polycarbonate centrifuge tube (TLA 100).

3. Carefully layer 50 μL of the 35% accudenz solution on top. A clear phase boundary should develop. Repeat with the 30% accudenz solution. Then add 20 μL of co-float buffer to the top of the centrifuge tube (*see* **Note 10**).
4. Centrifuge in a tabletop ultracentrifuge for 1 h at 541,000 RCF. Collect the liposomes with associated proteins at the top of the tube (*see* **Note 11**).
5. Normalize the recovery of liposomes based on fluorescence intensity using a spectrofluorometer and analyze the samples using SDS-PAGE followed by Coomassie blue staining (*see* **Note 12**).
6. By varying the size of liposomes used, curvature sensing of ESCRT proteins can be measured (Fig. 2). By varying lipid composition, the dependence of ESCRT association with membranes on specific lipid headgroups can be determined.

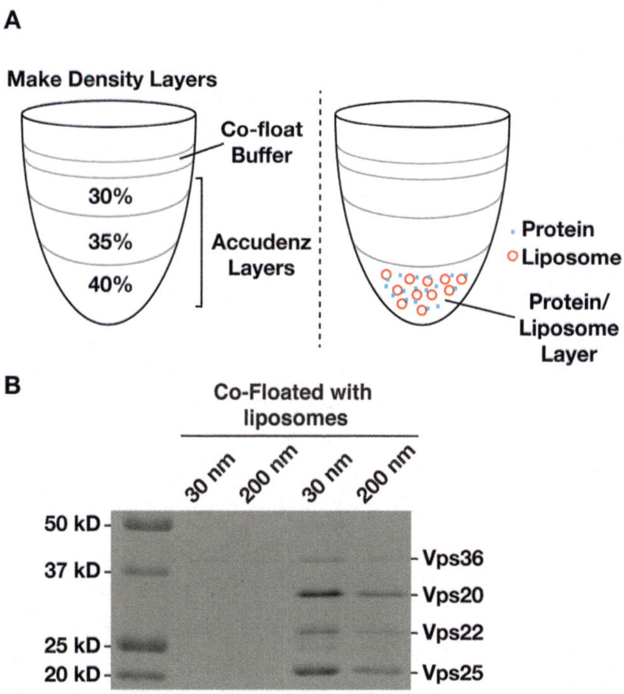

Fig. 2 Liposome co-flotation analysis to measure curvature sensing in the ESCRT machinery. (**a**) A cartoon illustrating the steps involved in conducting a co-flotation study. First, varying concentrations of accudenz are layered in a centrifuge tube, with liposomes and ESCRT proteins mixed in the 40% layer at the bottom. During centrifugation, the liposomes and associated ESCRT proteins float to the top of the gradient, while unbound ESCRT proteins remain at the bottom of the centrifuge tube. (**b**) Image of SDS-PAGE analysis followed by Coomassie blue staining of co-flotation assay. The ESCRT-II complex composed of Vps36, Vps22, and Vps25 co-floats modestly with liposomes (55% PC, 30% PE, and 15% PS), irrespective of their size (lanes 2 and 3). However, when the ESCRT-III subunit Vps20 is added, the set of proteins co-float with liposomes more robustly, with a preference for more highly curved liposomes (lanes 4 and 5). A molecular weight marker is also included (lane 1)

3.3.3 Fluorescent Labeling and Imaging of ESCRT Proteins

1. Several *C. elegans* ESCRT subunits harbor endogenous cysteine residues that are amenable to direct labeling with fluorescent dyes. Use dye maleimides, including BODIPY-FL-maleimide, to label recombinant ESCRT proteins.
2. Incubate a 20-fold molar excess of BODIPY-FL-maleimide with the ESCRT protein of interest. Rotate the mixture overnight.
3. Quench the reaction by adding excess glutathione (fivefold molar excess), and remove the excess unbound dye using size-exclusion chromatography.
4. Measure dye-labeling stoichiometry based on the final concentration of recovered protein and its absorbance.
5. To visualize ESCRT protein association with membranes, fluorescently labeled liposomes can be used in combination with fluorescently labeled ESCRT subunits.
6. To generate immobilized-, fluorescently labeled liposomes, incorporate both rhodamine-PE and biotin-PE (0.5%) into lipid mixtures prior to extrusion. These liposomes can then be immobilized on avidin-coated glass coverslips.
7. Clean glass coverslips with H_2O and coat them with a solution of PEG (5000 Da) and biotin-PEG (40:1 ratio).
8. Incubate the PEG-coated coverslips with 1.0 μM avidin for 10 min.
9. Wash the coverslips 2–3 times with wash buffer to remove excess avidin.
10. Mix 100 μL of the biotinylated, rhodamine-labeled liposomes with the BODIPY-FL-labeled ESCRT protein (250 nM) and incubate the mixture on the avidin-coated coverslips for 30 min.
11. Aspirate unbound liposomes and protein. Add 3 μL of wash buffer and invert onto a depression slide for imaging.
12. Use a 60×, 1.4 numerical aperture Plan Apo oil objective lens for imaging.

4 Notes

1. RCF can be converted to RPM using the following equation: $\text{RCF} = r(1.118 \times 10^{-5})$ RPM, where r is the rotational radius in cm.
2. Pellets can be stored at −80 °C for up to 3 months.
3. Benzamidine and PMSF are protease inhibitors that are added to prevent unwanted proteolysis of overexpressed protein.

Lysozyme facilitates cell rupture by degrading the bacterial cell wall.

4. To avoid creating bubbles or froth, the tip should be submerged approximately ½ in. 2-mercaptoethanol is added after sonication to reduce oxidized protein.
5. It is important to avoid collecting insoluble material in the supernatant. Contamination will compromise the subsequent wash steps.
6. After the first centrifugation step, the pelleted resin may be difficult to resuspend through inversion. If this occurs, do not shake the tube. Use a pipette tip to gently break up the resin.
7. ADP is required to stabilize monomeric Vps4 and prevent aggregation.
8. After extrusion, it is important to clean the extruder materials to avoid buildup of salts and lipids. Soak the extruder apparatus in 100% ethanol, and wash out the syringes with 100% ethanol, followed by deionized H_2O.
9. Accudenz is a nontoxic, inert, nonionic, tri-iodinated derivative of benzoic acid with three aliphatic hydrophilic side chains. It has low osmolality and low viscosity, making it useful for the fractionation of many proteins.
10. To avoid mixing the layers, touch the pipette tip to the side of the tube just below the meniscus of the previous layer and pipette slowly.
11. The liposomes/proteins should form a distinct layer at the top of the tube, but they may occasionally form an aggregate just below the surface. If this occurs, try and collect as much of the aggregate as possible.
12. If the liposome-associated proteins are not visible after Coomassie blue staining, more sensitive dyes can be used. SYPRO Ruby is recommended for staining less than 50 ng of protein.

Acknowledgments

This work was supported in part by NIH grant GM088151 (to A.A).

References

1. Fyfe I, Schuh AL, Edwardson JM, Audhya A (2011) Association of the endosomal sorting complex ESCRT-II with the Vps20 subunit of ESCRT-III generates a curvature-sensitive complex capable of nucleating ESCRT-III filaments. J Biol Chem 286:34262–34270
2. Mayers JR, Fyfe I, Schuh AL, Chapman ER, Edwardson JM, Audhya A (2011) ESCRT-0-assembles as a heterotetrameric complex on membranes and binds multiple ubiquitinylated cargoes simultaneously. J Biol Chem 286:9636–9645

3. Shen QT, Schuh AL, Zheng Y, Quinney K, Wang L, Hanna M, Mitchell JC, Otegui MS, Ahlquist P, Cui Q, Audhya A (2014) Structural analysis and modeling reveals new mechanisms governing ESCRT-III spiral filament assembly. J Cell Biol 206:763–777
4. Takahashi H, Mayers JR, Wang L, Edwardson JM, Audhya A (2015) Hrs and STAM function synergistically to bind ubiquitin-modified cargoes in vitro. Biophys J 108:76–84
5. Schuh AL, Hanna M, Quinney K, Wang L, Sarkeshik A, Yates JR 3rd, Audhya A (2015) The VPS-20 subunit of the endosomal sorting complex ESCRT-III exhibits an open conformation in the absence of upstream activation. Biochem J 466:625–637
6. Kern DM, Kim T, Rigney M, Hattersley N, Desai A, Cheeseman IM (2015) The outer kinetochore protein KNL-1 contains a defined oligomerization domain in nematodes. Mol Biol Cell 26:229–237
7. Tan S, Kern RC, Selleck W (2005) The pST44 polycistronic expression system for producing protein complexes in *Escherichia coli*. Protein Expr Purif 40:385–395
8. Babst M, Katzmann DJ, Estepa-Sabal EJ et al (2002) ESCRT-III: an endosome-associated heterooligomeric protein complex required for MVB sorting. Dev Cell 3:271–282
9. Hanson PI, Roth R, Lin Y, Heuser JE (2008) Plasma membrane deformation by circular arrays of ESCRT-III protein filaments. J Cell Biol 180:389–402
10. Babst M, Wendland B, Estepa EJ, Emr SD (1998) The Vps4p AAA ATPase regulates membrane association of a Vps protein complex required for normal endosome function. EMBO J 17:2982–2993
11. Siegel LM, Monty KJ (1966) Determination of molecular weights and frictional ratios of proteins in impure systems by use of gel filtration and density gradient centrifugation. Application to crude preparations of sulfite and hydroxylamine reductases. Biochim Biophys Acta 112:346–362
12. Wyatt PJ (1993) Light-scattering and the absolute characterization of macromolecules. Anal Chim Acta 272:1–40

Chapter 15

Purification of Recombinant ESCRT-III Proteins and Their Use in Atomic Force Microscopy and In Vitro Binding and Phosphorylation Assays

Luisa Capalbo, Ioanna Mela, Maria Alba Abad, A. Arockia Jeyaprakash, J. Michael Edwardson, and Pier Paolo D'Avino

Abstract

The endosomal sorting complex required for transport (ESCRT)-III proteins are known to assemble into filaments that mediate membrane remodeling and fission in various biological processes, including the formation of endosomal multivesicular bodies, viral budding, cytokinesis, plasma membrane repair, nuclear pore quality control, nuclear envelope reformation, and neuron pruning. The study of the regulation and function of ESCRT-III proteins is therefore crucial to understand these events and requires a combination of in vivo and in vitro experimental techniques. Here we describe two protocols for the purification of human and *Drosophila* ESCRT-III proteins from bacteria and their use in in vitro phosphorylation assays and atomic force microscopy experiments on membrane lipid bilayers. These protocols can also be applied for the purification of other proteins that are insoluble when expressed in bacteria.

Key words CHMP4C, Kinase assay, Protein purification, Atomic force microscopy, Protein binding

1 Introduction

The endosomal sorting complex required for transport (ESCRT) proteins form evolutionary conserved complexes that mediate a range of topologically similar membrane remodeling events, from archaea to humans. These include the formation of endosomal multivesicular bodies, viral budding, cytokinesis, the repair of both nuclear and plasma membranes, nuclear pore quality control, nuclear envelope reformation, and neuron pruning [1–5]. In all these events, the ESCRT machinery, composed of three sub-complexes (ESCRT-I, -II, and -III), promotes the formation and scission of membrane necks from the cytoplasm. The ESCRT-III complex is the core machinery that mediates the final membrane deformation and fission events, and Snf7/Vacuolar protein sorting-associated protein 32 (Vps32) components (known as Charged multivesicular

Emmanuel Culetto and Renaud Legouis (eds.), *The ESCRT Complexes: Methods and Protocols*, Methods in Molecular Biology, vol. 1998, https://doi.org/10.1007/978-1-4939-9492-2_15,

body protein (CHMP4) proteins in humans) have been shown to form spiral filaments that appear to remodel and constrict the membrane during the final step of cytokinesis [6–8]. The activity and recruitment of Snf7/Vps32 proteins at different cellular locations is finely regulated by post-translational modifications, which control the ability of these proteins to polymerize and to interact with other proteins. For example, the chromosomal passenger complex (CPC), composed of Aurora B kinase, Borealin, Survivin, and Inner centromere protein (INCENP), has been suggested to regulate both the membrane remodeling activity and the association with different partners of the ESCRT-III component CHMP4C in late cytokinesis [9]. These regulatory mechanisms serve to dictate the proper timing of the final abscission of the two daughter cells [4, 10–12].

The dissection of the functions and regulation of ESCRT-III proteins requires a combination of in vivo functional analysis and in vitro biochemical and structural studies. Unfortunately, despite their small size, recombinant ESCRT-III proteins tend to be highly insoluble and to form aggregates, which complicates their use in in vitro assays. Here we describe two protocols that we have successfully employed to purify recombinant human and *Drosophila* ESCRT-III proteins from *Escherichia coli*. These proteins can then be used for many in vitro applications, including atomic force microscopy (AFM), protein binding and phosphorylation assays. Our protocols can also be applied to the purification of other proteins known to be insoluble when expressed in bacteria.

2 Materials

2.1 Protein Expression in Escherichia coli

2.1.1 Solutions

1. Luria broth (LB) liquid media for 1 l: 10 g tryptone, 10 g NaCl, and 5 g yeast extract.
2. LB plates: add 15 g of agar to 1 l LB.
3. Ampicillin (AMP): 100 mg/ml ampicillin stock (1000× stock) filter sterilize with a 0.22 m filter.
4. Bacteria culture medium: Luria broth (LB) with appropriate antibiotics. Dilute AMP antibiotic into your LB medium in the ratio of 1:1000.
5. Isopropyl β-D-1-thiogalactopyranoside (IPTG): stock solution concentration 100 mM IPTG, usually we use 1 mM final concentration in the bacterial culture medium.
6. 20% (w/v) L(+)arabinose: add 2 g of L(+)arabinose to water to a final 10 ml volume.

2.1.2 Tools

1. Shaking incubator for bacteria.

2.2 GST Purification

2.2.1 Solutions

1. Glutathione-Sepharose 4B 17056–01 10 ml (GE Healthcare).
2. Lysis buffer: 20 mM Hepes pH 7.5, 500 mM NaCl, 1 mM EDTA, 1 mM Dithiothreitol (DTT), EDTA-free complete protease inhibitor cocktail (Roche), and 2% (v/v) sarkosyl detergent (*N*-lauroylsarcosine sodium salt solution).
3. Lysis buffer with Triton: 4% (v/v) Triton X-100 in lysis buffer.
4. Wash buffer—1 M NaCl: 20 mM Hepes, 1 M NaCl, 1 mM EDTA, 1 mM DTT, EDTA-free complete protease inhibitor cocktail, and 0.1% (v/v) Triton X-100.
5. Wash Buffer—100 mM NaCl: 20 mM Hepes, 100 mM NaCl, 1 mM EDTA, 1 mM DTT, EDTA-free complete protease inhibitor cocktail, and 0.1% (v/v) Triton X-100.
6. Laemmli sample buffer: 60 mM Tris–HCl pH 6.8, 2% (w/v) sodium dodecyl sulfate (SDS), 10% (v/v) glycerol, 5% (v/v) 2-mercaptoethanol, and 0.01% (w/v) bromophenol blue.
7. Phosphate-buffered saline (PBS) 10× (for 1 l): 2 g KCl, 2.4 g KH_2PO_4, 80 g NaCl, and 11.45 g Na_2HPO_4.
8. PBS 1×: dilute 1/10 the PBS 10× stock solution.
9. Coomassie staining solution: 0.1% Coomassie Brilliant Blue R-250, 50% methanol, and 10% glacial acetic acid (*see* **Note 1**).
10. 10% Glycerol.

2.2.2 Tools

1. Sonicator (Soniprep 150, MSE, Lower Sydenham, UK).
2. Ear defenders.
3. Pre-chilled 1.5 ml microfuge tubes and 15 ml conical tubes.
4. Benchtop centrifuge Heraeus Biofuge Pico75003235 24 Place Microlitre centrifuge, 17,000 × *g* (RCF).
5. Refrigerated benchtop centrifuge Eppendorf 5810A.
6. Refrigerated benchtop centrifuge Eppendorf 5702R.
7. Rotating wheel.
8. Magnetic stands suitable for microfuge or larger tubes.
9. Aspirator.

2.3 GST Binding Assay

2.3.1 Solutions

1. NET-N+ buffer: 50 mM Tris–HCl, pH 7.4, 150 mM NaCl, 5 mM EDTA, 0.5% NP-40, and a protease inhibitor cocktail complete (Roche).
2. NET-N+ washing buffer: 50 mM Tris–HCl, pH 7.4150 mM NaCl, 5 mM EDTA, 0.5% (v/v) NP-40, and protease inhibitor cocktail complete (*see* **Note 2**).

3. Ponceau S staining solution: 0.5% (w/v) Ponceau S and 1% (v/v) acetic acid.
4. TNT® Quick Coupled Transcription/Translation System (Promega).
5. EasyTag L-[^{35}S]-methionine 500 μCi (18.5 MBq) in stabilized aqueous solution (hot methionine).

2.3.2 Tools

1. Heat block.
2. Apparatus suitable for conventional (denaturing) sodium dodecyl sulfate polyacrylamide gel electrophoresis (SDS-PAGE) and Western blotting. Use 10–12% SDS-polyacrylamide gels for optimal separation of small- to large-sized proteins.
3. iBlot Dry Blotting System (ThermoFisher).
4. PCR machine.
5. Kodak BioMax XAR Films (Sigma–Aldrich).
6. Intensifying screen (SIGMA).

2.4 Phosphorylation Assay

2.4.1 Solutions

1. Kinase buffer: 20 mM HEPES pH 7.5, 5 mM $MgCl_2$, and 1 mM DTT.
2. Kinase buffer + ATP: 0.1 mM adenosine 5′-triphosphate (ATP) in kinase buffer.
3. EasyTide® Adenosine 5′-triphosphate [gamma-32P], 250 μAd (9.25 MBq) in 50 mM tricine buffer (pH 7.6) with green dye (hot ATP).

2.4.2 Tools

1. Eppendorf thermomixer compact.

2.5 His-GFP::CHMP4C Purification

2.5.1 Solutions

1. 1 mM PMSF (AppliChem).
2. 10 μg/ml DNase and 1 mM phenylmethylsulfonyl fluoride (PMSF) (AppliChem) (*see* **Note 3**).
3. HisTrap Ni Sepharose High Performance (HP) affinity resin column.
4. Binding buffer: 20 mM Tris–HCl pH 8, 750 mM NaCl, 35 mM imidazole, and 2 mM 2-mercaptoethanol.
5. Wash buffer: 20 mM Tris–HCl pH 8, 1 M NaCl, 50 mM KCl, 10 mM $MgCl_2$, 2 mM ATP, 35 mM imidazole, and 2 mM 2-mercaptoethanol.
6. Elution buffer: 20 mM Tris–HCl pH 8, 750 mM NaCl, 400 mM imidazole, and 2 mM 2-mercaptoethanol.
7. Dialysis buffer: 20 mM Tris–HCl pH 8, 750 mM NaCl, and 2 mM DTT.
8. Size exclusion buffer: 20 mM Tris–HCl pH 8, 200 mM NaCl, 2 mM DTT.

2.5.2 Tools

1. Cloning vector pET-His6-msfGFP (Addgene).
2. Baffled flask.
3. Sonicator (Vibra-Cell, Sonics & Materials).
4. Heat block.
5. 10 kDa molecular mass cutoff cellulose concentrator (Millipore).
6. Spectra/Por dialysis tubing with a molecular mass cut-off of 6–8 kDa (Spectrum).
7. Superdex 200 increase 10/300 GL column (GE Healthcare).
8. Apparatus suitable for conventional (denaturing) sodium dodecyl sulfate polyacrylamide gel electrophoresis (SDS-PAGE) and Western blotting. Use 10%–12% SDS- polyacrylamide gels for optimal separation of small to large size proteins.
9. Beckman Avanti JXN-26, JLA 9.1000 rotor.
10. Nanodrop.

2.6 Atomic Force Microscopy (AFM)

2.6.1 Solutions

1. Avanti Polar Lipids (Alabaster, AL, USA) in chloroform, for the present application we used: 70% L-α-phosphatidylcholine (PC), 15% 1,2-dioleoyl-*sn*-glycero-3-phosphoethanolamine (DOPE), and 15% 1,2-dioleoyl-*sn*-glycero-3-phosphoserine (DOPS).
2. Imaging buffer: 20 mM Tris–HCl, pH 8, 100 mM NaCl, and 1 mM $MgCl_2$.
3. Kinase buffer: 20 mM HEPES pH 7.5, 5 mM $MgCl_2$, and 1 mM DTT.

2.6.2 Tools

1. AFM Bruker Dimension FastScan instrument (Bruker, Billerica, Massachusetts, USA).
2. Sonicator (Soniprep 150, MSE, Lower Sydenham, UK).
3. Mica sheet (Agar Scientific, Stansted, UK).
4. 12 mm metal SPM Specimen Discs 12mm (Agar Scientific) stub (Agar Scientific, Stansted, UK).
5. Nanoscope Analysis software, v 1.5 (Bruker, Billerica, Massachusetts, USA).

3 Methods

3.1 GST-Tagged Protein Expression

3.1.1 Generation of Plasmids with GST Fusion Protein

Gateway Technology® (Invitrogen) was used in all cloning procedures.

DNA fragments encoding for CHMP4C full-length, N-, or C-terminal fragments (both from *Drosophila melanogaster* and human genes) were amplified by PCR using Accuprime polymerase (Invitrogen) and primers containing *att*B flanking sequences for Gateway® Technology. The PCR product was purified and introduced into a pDONR221 plasmid (Invitrogen) to generate CHMP4C entry vectors. Then CHMP4C entry clone was recombined into pDEST15 using Gateway® Technology to express N-terminal GST-tagged polypeptides in *Escherichia coli*.

To generate CHMP4C mutant variants we used QuikChange® Lightning Site-Directed Mutagenesis Kit (Agilent Technologies).

3.1.2 Expression of GST-Tagged Proteins in *E. coli*

We used BL21AI bacteria strain to grow our GST-tagged protein.

1. Transform the BL21AI using plasmids containing GST fusion proteins.
2. Grow 5 ml culture in Luria broth (LB) with antibiotics overnight.
3. Inoculate 1/40 of the culture in 100 ml LB with antibiotics and grow at 37 °C for around 90 min or until they reach OD_{600} 0.6.
4. When the correct OD_{600} is reached, induce protein expression using 20% (v/v) of L(+)arabinose stock solution to a final concentration of 0.2% (v/v) L(+)arabinose. Before adding the L(+)arabinose, take 1 ml of culture of uninduced cells (not induced) (Fig. 1a).
5. Incubate at 30 °C for 5 h with shaking.
6. Collect 1 ml of culture by centrifugation for 5 min at 17,000 × *g* to control the expression of the GST-fused protein before purification (induced) (Fig. 1a). Collect all bacterial cultures by centrifugation for 15 min at 6000 × *g* and freeze the pellets at −80 °C.

3.1.3 Purification of GST-Tagged Proteins

This method has been designed according to the guidelines provided by the glutathione-Sepharose 4B beads supplier with slight modifications.

1. Resuspend the bacterial pellet in 5 ml of lysis buffer on ice using a 10 ml pipette with an up-and-down movement of the liquid buffer.
2. Sonicate the lysate on ice three times at 50% power (15 μA) for 30 s each.

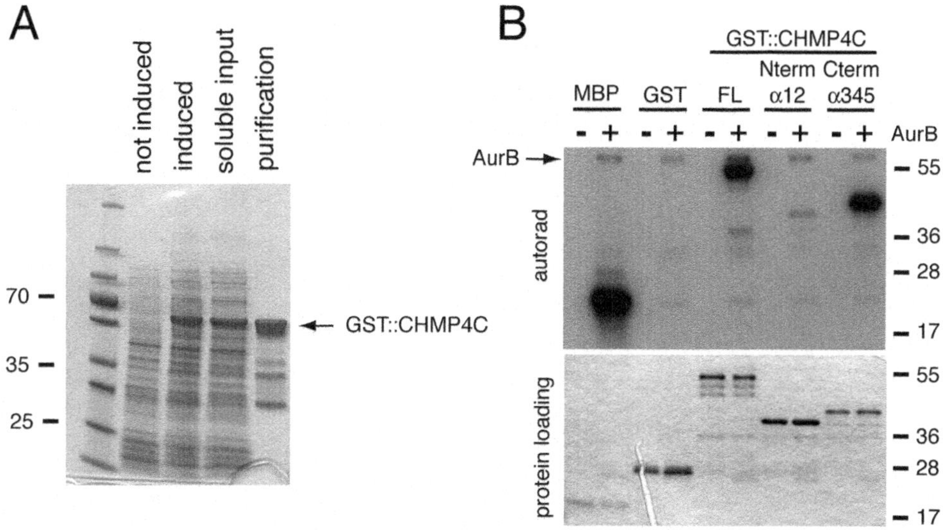

Fig. 1 Aurora B phosphorylates CHMP4C in vitro. (**a**) The Coomassie Blue-stained SDS-PAGE gel for the purification of GST-tagged full-length CHMP4C (FL) is shown. The lanes show the different steps of purification from the growing BL21 bacteria expressing the protein to the purified protein. The numbers on the left indicate the sizes in kDa of the molecular mass markers. (**b**) GST-tagged CHMP4 proteins, GST alone, or the positive control myelin basic protein (MBP) were incubated with (+) or without (−) recombinant Aurora B in the presence of [γ-^{32}P] ATP; α12 correspond to the half N term of CHMP4 (1–121aa) tagged with GST, whereas α345 correspond to the half C term of CHMP4 (122–233 aa) tagged with GST. The reactions were then separated by SDS-PAGE and gels were stained with Coomassie Blue, dried and exposed to film at −80 °C. The Coomassie Blue staining of the protein loading is shown at the bottom. The numbers on the right indicate the sizes in kDa of the molecular mass markers. Panel **b** is adapted from [11]

3. Incubate the bacterial cell extract on a rotating wheel set at low speed for 20 min at 4 °C.
4. Centrifuge at 10,000 × *g* at 4 °C in a refrigerated Eppendorf benchtop centrifuge for 20 min.
5. Transfer the supernatant to a new 15 ml Falcon tube and then add another 5 ml of lysis buffer with Triton.
6. Place 1 ml glutathione-Sepharose 4B beads in a 15 ml Falcon tube and wash with 10 ml of wash buffer—100 mM NaCl.
7. Centrifuge at 1500 × *g* for 1 min at 4 °C.
8. Mix the supernatant of the bacterial extract with the washed glutathione-Sepharose beads and incubate for 90 min at 4 °C on a rotating wheel (*see* **Note 4**).
9. Pellet the glutathione-Sepharose beads at 1500 × *g* for 1 min and remove the bacterial cell extract containing unbound proteins. Take a small fraction of bacterial extract after binding for subsequent Western blot analysis (soluble unbound).
10. Wash the beads three times with 10 ml of wash buffer—1 M NaCl for 5 min on a rotating wheel at 4 °C.

11. Wash the beads twice with 10 ml of wash buffer—100 mM NaCl for 5 min on a rotating wheel at 4 °C.
12. Wash once with 10 ml PBS 1× (*see* **Note 5**) for 5 min on a rotating wheel at 4 °C. Centrifuge the glutathione-Sepharose beads at 1500 × *g* for 1 min.
13. Resuspend the beads with 2 ml of PBS 1 × .
14. Freeze half of the beads with 10% glycerol at −80 °C and keep the rest at 4 °C for Western analysis.

3.2 Phosphorylation Assay

CHMP4C was phosphorylated in vitro using recombinant Aurora B.

1. Mix His::CHMP4C purified protein or GST-tagged CHMP4C (full-length or N- or C-terminal protein fragment) with the kinase buffer containing 0.1 mM unlabeled (cold) adenosine 5′-triphosphate (ATP).
2. Add 190 ng of recombinant Aurora B in a final reaction volume of 15 μl (omit this reagent in control reactions).
3. Finally add 5 μCi of [γ-32P] ATP (hot ATP) (*see* **Note 6**).
4. Incubate the reaction mixes at 30 °C for 30 min with constant agitation.
5. Add 15 μl of 2× Laemmli sample buffer to stop the reaction.
6. Boil the samples for 10 min at 90 °C and run on a 4–20% Tris-Glycine precast gel.
7. Stain gels with Quick Coomassie Stain® to check the protein loading (*see* **Note 1**).
8. Transfer the gel onto a nitrocellulose membrane using the iBlot Dry Blotting System.
9. Expose the membranes to Kodak BioMax XAR films at −80 °C with intensifying screen (Fig. 1b).

3.3 GST In Vitro Binding Assay

3.3.1 Production of [^{35}S] Methionine-Labeled Polypeptides with TnT® Quick Coupled Transcription/Translation Systems

1. Amplify PCR products encoding for CHMP4C, full-length, or N- or C-terminal fragments both from *Drosophila melanogaster* and human genes using primers containing the T7 promoter sequence.
2. Gel-purify the PCR products.
3. PCR products are transcribed and translated in vitro using the TNT® Quick Coupled Transcription/Translation System in the presence of [35S] methionine (*see* **Note 6**).

3.3.2 GST In Vitro Binding Assay

The binding reaction is performed in the NET-N+ buffer containing 150 mM NaCl. Subsequent washes use salt concentrations from 150 mM to 1 M NaCl according to the protein (*see* **Note 2**).

1. Prepare 25 μl glutathione-Sepharose beads containing purified GST-tagged proteins and 300 μl NET-N+ buffer in a 1.5 ml tube.

2. Add 10 μl of [^{35}S]methionine-labeled polypeptides.
3. Incubate the mixture at 4 °C for 30 min on a thermomixer with agitation.
4. Spin 500 × *g* for 1 min and remove the supernatant carefully.
5. Add 500 μl of the Net-N+ buffer and mix gently.
6. Spin 500 × *g* for 1 min and remove the supernatant carefully. Repeat this wash five times.
7. Add 300 μl of the PBS 1× and mix gently.
8. Spin 500 g for 1 min and remove the supernatant carefully (*see* **Note 7**).
9. Resuspend the glutathione-Sepharose beads in 25 μl of 2× Laemmli SDS-PAGE sample buffer and boil the samples.
10. Typically run 15 μl of the sample on a 4–20% gradient Tris-glycine gel.
11. Transfer the proteins onto a nitrocellulose membrane using a dry blotting system and expose the membrane to X-ray films at −80 °C with intensifying screens (*see* **Note 8**).

3.4 Purification His-GFP-Tagged CHMP4C Proteins

For purification of recombinant CHMP4C, its full-length Open reading frame (ORF) was cloned into a pET-His6-msfGFP cloning vector as a TEV-cleavable His-tagged protein.

3.4.1 Protein Expression

1. Transform the CHMP4C expression vector into *E. coli* BL21 Gold competent cells and plate 50 μl onto an LB media/ampicillin plate. Incubate overnight at 37 °C.
2. Next day resuspend a single colony from the plate into 5 ml of LB media/ampicillin media to prepare the starter culture. Grow the starter culture overnight at 37 °C with shaking.
3. Next day use 2.5 ml of the starter culture to inoculate 500 ml of LB/ampicillin media in a 2 l baffled flask. Grow at 37 °C with shaking.
4. Grow cells until OD_{600} ~0.8–1.2.
5. Induce cultures with 0.35 mM IPTG overnight at 18 °C.
6. Harvest cells by spinning cultures at 3320 × *g*, 4 °C for 10 min in Beckman Avanti JXN-26, JLA 9.1000 rotor.

3.4.2 Affinity Purification

1. Resuspend the pellet of bacterial cells with 100 ml of binding buffer. Add 1 complete EDTA-free tablet, 10 μg/ml DNase, and 1 mM PMSF (*see* **Note 3**).
2. Lyse the cells by sonication at 60% amplitude, 1 pulse on-1 pulse off for 5 min. Take 1 μl of total lysate for subsequent SDS-PAGE analysis (*see* **Note 9**).

3. Clarify the lysate by centrifugation at 22,000 × *g* for 50 min at 4 °C. Take 1 μl of the supernatant (soluble fraction) and the pellet (insoluble fraction) for SDS-PAGE analysis.
4. Meanwhile, wash a 5 ml HisTrap HP column connected to a peristaltic pump with 2 column volumes (cv) of distilled water followed by equilibration of the column with 2 cv of binding buffer.
5. Load the supernatant into the 5 ml HisTrap HP column at 4 °C and a flow rate of 1 ml/min.
6. Wash at 5 ml/min with 20 cv of binding buffer followed by 20 cv of wash buffer.
7. Elute at 5 ml/min with elution buffer collecting fractions every 5 ml. Check the protein concentration of each elution fraction with Nanodrop. Collect 1 μl of each fraction for SDS-PAGE analysis.
8. Run a 15% SDS-polyacrylamide gel of the purification and stain with Coomassie Blue. Pool the fractions containing pure protein (Fig. 2a).
9. Cleave the His-GFP tag of the protein by incubating the pooled fractions with 1 mg of TEV and dialyze overnight at 4 °C against the dialysis buffer, using Spectra/Por dialysis tubing with a molecular mass cut-off of 6–8 kDa. Run another 15% SDS-polyacrylamide gel to check for tag cleavage.
10. Concentrate the cleaved protein using a 10 kDa molecular mass cut-off cellulose concentrator (*see* **Note 10**).

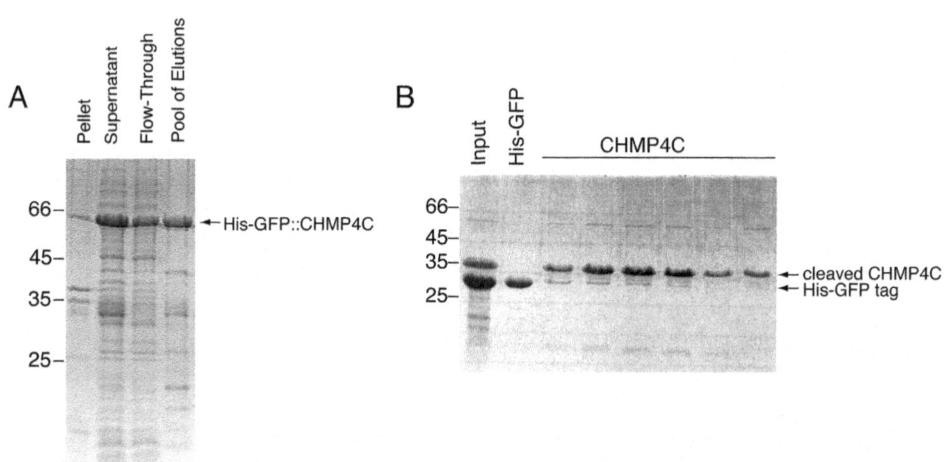

Fig. 2 (**a**) Representative SDS-PAGE of His-GFP-CHMP4C affinity purification using a HisTrap HP column. The numbers on the left indicate the sizes in kDa of the molecular mass markers. (**b**) Representative SDS-PAGE of size exclusion chromatography fractions with cleaved His-GFP-CHMP4C. CHMP4C can be separated from the His-GFP tag by using the S200 increase 10/300 GL column. Input is cleaved CHMP4C and His-GFP tag. The numbers on the left indicate the sizes in kDa of the molecular mass markers

3.4.3 Size Exclusion Chromatography

To separate the CHMP4C protein without the tag, size exclusion chromatography has been employed.

1. Inject 500 μl of the concentrated protein from **step 10** (Subheading 3.4.2) onto a Superdex 200 increase 10/300 GL column equilibrated with the size exclusion buffer (flow rate: 0.5 ml/min) (*see* **Note 11**).
2. Analyze the purity of the sample by 15% SDS- polyacrylamide gel followed by staining with Coomassie Blue (Fig. 2b).
3. Pool the fractions accordingly and concentrate the protein with a 10 kDa molecular mass cut-off cellulose concentrator if needed (*see* **Note 12).**

3.5 Atomic Force Microscopy (AFM)

3.5.1 Lipid Preparation

1. Different lipids were prepared at a final concentration of 2 mg/ml using lipid chloroform stocks from Avanti Polar Lipids.
2. Mix lipid stocks in desired ratios in a glass tube.
3. Dry the lipid mixture under a gentle nitrogen stream for at least 30 min, or until the chloroform has completely evaporated. A white film should be visible at the bottom of the tube.
4. Hydrate dried lipids in BioPerformance Certified water (SIGMA-ALDRICH) overnight at a final concentration of 2 mg/ml.
5. Vortex rehydrated lipids vigorously for 1 min to produce small multilamellar vesicles.
6. Sonicate lipids at an amplitude of 10 μA for 5–10 s with a probe sonicator until the mixture becomes transparent, indicating the formation of small unilamellar vesicles.

3.5.2 Substrate Preparation for Atomic Force Microscopy Imaging

1. Punch mica discs out of a mica sheet.
2. Attach the mica disc onto a 12 mm metal SPM stub.
3. Reveal a flat, clean layer of mica using Scotch tape.

3.5.3 Sample Preparation for Atomic Force Microscopy: Fluid Imaging

1. Incubate 40 μl of lipid suspension with the protein of interest (or plain buffer in the case of control experiments) for 30 min.
2. Deposit the sample onto freshly cleaved mica and incubate at room temperature for 15 min.
3. Remove excess liposomes by gentle washing with 1 ml of imaging buffer. Repeat this step three times.
4. Immerse the lipid bilayer in 150 μl of buffer.
5. Place the sample in the fluid cell of the atomic force microscope. All imaging was conducted under fluid, using FastScan D cantilevers. Their resonant frequencies under fluid were 110–140 kHz, and the actual scanning frequencies were

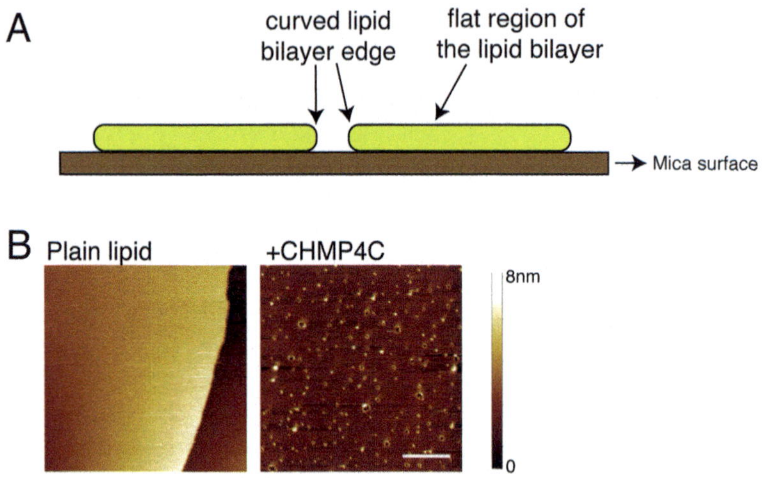

Fig. 3 CHMP4C binds to and remodels membranes in vitro. (**a**) Schematic representation of a lipid bilayer (yellow) assembled on a mica surface (brown). The central portion of the bilayer (indicated in yellow) is flat (~4 nm in thickness), while the edges of the bilayer are highly curved. (**b**) AFM image of a lipid bilayer without CHMP4C and one incubated with CHMP4C. A height bar is shown on the right. Scale bar, 20 nm. Panel **a** is adapted from Capalbo et al. [9]

approximately 5% below the maximal resonance peak. All imaging was performed at room temperature (Fig. 3).

6. AFM images were acquired at a rate of four frames/min. Acquired images need plane fitting to remove tilt and fit each scan line to a first-order equation.

3.5.4 Sample Preparation for Atomic Force Microscopy: Dry Imaging

We used imaging in air (dry) for our purified proteins without lipids and always on mica to visualize the molecular surface/structure.

1. Dilute the protein of interest to an appropriate concentration so that an even spread on the mica is achieved (*see* **Note 13**).
2. Deposit 45 μl of the sample onto freshly cleaved mica and incubate at room temperature for 10 min.
3. Remove excess protein by gentle washing with 1 ml of BPC water. Repeat this step five times.
4. Dry the sample under a gentle nitrogen stream.
5. Place the sample in sample holder of the atomic force microscope. Imaging was conducted in air, using FastScan A cantilevers. Scanning frequencies used were approximately 5% below the maximal resonance peak. All imaging was performed at room temperature (Fig. 4).

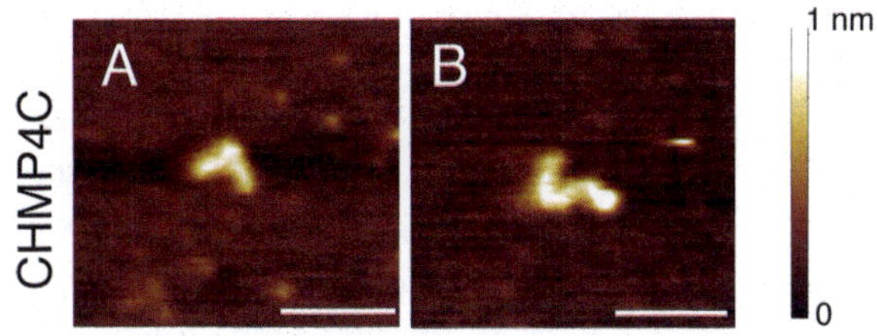

Fig. 4 CHMP4C can spontaneously polymerize into filaments. (**a**, **b**) High-magnification AFM images of dry mica incubated with CHMP4C. A height bar is shown on the right. Scale bar, 100 nm

3.5.5 Phosphorylation Assay for Atomic Force Microscopy

CHMP4C was phosphorylated in vitro using recombinant Aurora B.

1. Mix CHMP4C purified protein with the kinase buffer containing 0.1 mM cold adenosine 5′-triphosphate (ATP).
2. Add 190 ng of recombinant Aurora B in a final reaction volume of 15 μl (omit this reagent in control samples).
3. Incubate the reactions at 30 °C for 30 min with constant agitation and then incubate them with lipids for a further 30 min (*see* **step 1** in Subheading 3.5.3).

4 Notes

1. We use usually Quick Coomassie Stain® because it's easy, quick, and does not required destaining to visualize the protein bands on the SDS gel. The bands appear after 10 min in the reagent.
2. We use different NET-N + washing buffer solutions differing for NaCl concentration (between 150 mM and1 M NaCl) according the binding strengths of the interacting proteins. If the interaction is really strong, we use higher salt concentration.
3. Important to supplement the complete EDTA-free tablet with PMSF to avoid proteolytic degradation of CHMP4C protein.
4. After centrifugation in **step 4** (Subheading 3.1.3) take a small fraction of the clarified supernatant for subsequent Western blot analysis (soluble input) and resuspend the pellet in a volume of Laemmli sample buffer equal to the volume used for the fraction of the clarified supernatant and keep for Western Blot analysis (pellet).
5. The last wash in the GST protein purification protocol is done with filtered PBS 1× to eliminate all the detergent in the beads, also this step eliminates the bubbles of the detergent, making the removal of all the liquid easier so that final PBS 1× can be added at the end of the purification.

6. Handling of radioactive isotopes requires the use of filtered tips, double closure microfuge tubes, and disposal of all materials as solid radioactive residues waste. Please follow specific safety measures implemented in your research institute.
7. For the final **steps** (7 and **8**) of the GST binding assay, we add 300 μl of the filtered PBS 1× with a gentle mix to remove all the detergent and foam in the tube. After centrifugation, we remove the supernatant carefully and add Laemmli SDS-PAGE sample buffer.
8. At the end of the phosphorylation assay, to check the presence of the proteins in the gel, a quick staining of the gel with Quick Coomassie Stain® can be done before the transfer onto a membrane. The membrane can be stained using Ponceau S staining solution and proteins loading can be checked for.
9. Avoid heating up of the sample by sonication as it might lead to CHMP4C degradation.
10. At 750 mM NaCl, the protein can be concentrated up to around 20 mg/ml; at low salt concentration, solubility is much lower and protein tends to stick to the membrane of the concentrator.
11. In the tested buffer condition, purified CHMP4C elutes after the inclusion volume most likely due to its nonspecific interaction with the size exclusion column.
12. At 200 mM NaCl, CHMP4C protein can be concentrated up to around 1 mg/ml.
13. This step will need fine tuning depending on the sample; it is advised to prepare about three ten-fold dilutions initially, to help determine the appropriate protein concentration.

Acknowledgments

This work was supported by Cancer Research UK grant C12296/A12541 and Biotechnology and Biological Research Council grant BB/R001227/1 to PPD. IM and JME were supported by Biotechnology and Biological Research Council grant BB/J018236/1. AAJ is supported by a Well come Trust Senior Research Fellowship 202811.

References

1. Guizetti J, Gerlich DW (2012) ESCRT-III polymers in membrane neck constriction. Trends Cell Biol 22:133–140
2. Lata S et al (2009) Structure and function of ESCRT-III. Biochem Soc Trans 37:156–160
3. Williams RL, Urbe S (2007) The emerging shape of the ESCRT machinery. Nat Rev Mol Cell Biol 8:355–368
4. Stoten CL, Carlton JG (2018) ESCRT-dependent control of membrane remodelling

during cell division. Semin Cell Dev Biol 74:50–65

5. Christ L, Raiborg C, Wenzel EM, Campsteijn C, Stenmark H (2017) Cellular functions and molecular mechanisms of the ESCRT membrane-scission machinery. Trends Biochem Sci 42:42–56
6. Elia N, Sougrat R, Spurlin TA, Hurley JH, Lippincott-Schwartz J (2011) Dynamics of endosomal sorting complex required for transport (ESCRT) machinery during cytokinesis and its role in abscission. Proc Natl Acad Sci U S A 108:4846–4851
7. Wollert T, Wunder C, Lippincott-Schwartz J, Hurley JH (2009) Membrane scission by the ESCRT-III complex. Nature 458:172–177
8. Guizetti J et al (2011) Cortical constriction during abscission involves helices of ESCRT-III-dependent filaments. Science 331:1616–1620
9. Capalbo L et al (2016) Coordinated regulation of the ESCRT-III component CHMP4C by the chromosomal passenger complex and centralspindlin during cytokinesis. Open Biol 6:160248
10. Bhutta MS, McInerny CJ, Gould GW (2014) ESCRT function in cytokinesis: location, dynamics and regulation by mitotic kinases. Int J Mol Sci 15:21723–21739
11. Capalbo L et al (2012) The chromosomal passenger complex controls the function of endosomal sorting complex required for transport-III Snf7 proteins during cytokinesis. Open Biol 2:120070
12. Carlton JG, Caballe A, Agromayor M, Kloc M, Martin-Serrano J (2012) ESCRT-III governs the Aurora B-mediated abscission checkpoint through CHMP4C. Science 336:220–225

Chapter 16

Assessment of ESCRT Protein CHMP5 Activity on Client Protein Ubiquitination by Immunoprecipitation and Western Blotting

Francheska Son, Katharine Umphred-Wilson, Jae-Hyuck Shim, and Stanley Adoro

Abstract

The charged multivesicular body protein-5 (CHMP5) is a member of the endosomal-sorting complex required for transport (ESCRT) that controls membrane-scission events in eukaryotic cells. Recent studies have revealed novel functions of CHMP5 beyond its role in the ESCRT machinery, notably as a critical nonenzymatic regulator of the ubiquitination and subsequent degradation of proteins in immune cells. Here we describe an immunoprecipitation and western blot methodology for assessing CHMP5 activity on client protein ubiquitination in T lymphocytes.

Key words CHMP5, T lymphocytes, Posttranslational modification, Ubiquitination, Deubiquitinase

1 Introduction

Degradation of ubiquitin-tagged proteins by the proteasome is a major conserved pathway for stringently regulating cellular protein abundance in cells. This process involves not only the dynamic balance between the activity of E3-ligases that attach ubiquitin molecules to proteins and deubiquitinases that remove ubiquitin tags but also the non-catalytic proteins that function as adaptors to nucleate enzymes participating in client protein ubiquitination and degradation [1]. The plethora of human diseases that arise from defects in the protein degradation machinery is a testament to the critical role of the ubiquitination machinery in cellular homeostasis [2].

The ESCRT machinery is comprised of cytosolic protein complexes (ESCRT-0, -I, -II, and -III) that are sequentially assembled to initiate and execute membrane remodeling and scission. Through this membrane remodeling activity, the ESCRT machinery has been shown to regulate various cellular processes including

Emmanuel Culetto and Renaud Legouis (eds.), *The ESCRT Complexes: Methods and Protocols*, Methods in Molecular Biology, vol. 1998, https://doi.org/10.1007/978-1-4939-9492-2_16, © Springer Science+Business Media, LLC, part of Springer Nature 2019

multivesicular body biogenesis, cytokinesis, autophagy, cell polarity, and viral budding in eukaryotes [3, 4]. There is growing evidence that individual ESCRT proteins may directly regulate cellular homeostasis in ways that are independent of their membrane remodeling activity. For example, cell-death rescue studies in yeast identified the human ESCRT-III protein variant, VPS24β, which lacks the N-terminal of VPS24, as a direct suppressor of proapoptotic Bax activity [5]. In two immune cell subsets (T lymphocytes and osteoclasts), it was found that CHMP5 suppressed the ubiquitination of key client proteins. In osteoclasts, CHMP5 was essential to suppress IκBα protein stability such that CHMP5-deficient osteoclasts displayed marked NF-κB activation downstream of RANKL signals that caused elevated bone turnover rates [6]. In thymocytes, we have recently discovered that CHMP5 is essential during positive selection [7]. Despite its reported role in the activation of VPS4 whose ATPase activity catalyzes the final step of endosomal membrane scission [8, 9], CHMP5 deficiency did not alter surface T-cell receptor (TCR), prosurvival cytokine interleukin-7 (IL-7) receptor expression, or receptor signal transduction in thymocytes [7], suggesting that its activity during positive selection was likely independent of endocytic trafficking. Instead, upon positive selection TCR signaling in developing thymocytes, CHMP5 promoted posttranslational protein stabilization of pro-survival protein, Bcl-2, leading to the progression of T-cell development [7]. We thus proposed a noncanonical activity of CHMP5, whereby it promoted protein stability by recruiting cell-type-specific deubiquitinases to specific client proteins in a dose-dependent manner [6, 7]. Here we describe how we analyze CHMP5 protein levels and assess CHMP5's activity on client protein ubiquitination by immunoprecipitation and Western blotting. Although this protocol was applied to T-cells in our laboratory, we expect that it is applicable to analysis of other cell types.

2 Materials

Prepare all cell culture media using aseptic tissue culture techniques, filter-sterilize, and store at 4 °C. All buffers and solutions should be prepared with distilled deionized water (molecular biology grade) and stored at room temperature except otherwise specified.

2.1 Cell Culture

1. T-cell culture media (RPMI-10): Roswell Park Memorial Institute (RPMI) medium 1640 (*see* **Note 1**) supplemented with 10% (v/v) fetal bovine serum (FBS), 2 mM L-glutamine, 1 mM sodium pyruvate, 10 mM *N*-2-hydroxyethylpiperazine-*N*-2-ethane sulfonic acid, counter-ion: NaOH (HEPES), pH 7.4, 1 × Minimum essential medium (MEM) nonessential amino

acids (*see* **Note 2**), 50 IU/ml penicillin, 50 μg/ml streptomycin, and 55 μM β-mercaptoethanol (*see* **Note 3**).

2. HEK293T (obtained from ATCC) culture media (DMEM-10): Dulbecco's modified Eagle medium (DMEM) supplemented with 10% (v/v) fetal bovine serum, 2 mM L-glutamine, 50 IU/ml penicillin, and 50 μg/ml streptomycin.
3. All cells are cultured in 5% CO_2 incubators maintained at 37 °C.

2.2 Immunoblot of CHMP5 Protein

1. Cells: thymocytes, splenic T-lymphocytes, Jurkat T-cell lines, or HEK 293T cells.
2. Lysis buffer: 10 mM Tris–HCl, pH 8.0, 50 mM NaCl, 5 mM EDTA, 2 mM NaF, 30 mM sodium pyrophosphate, 100 mM Na_3VO_4, 0.5 mM PMSF, 1 μg/ml leupeptin, 5 μg/ml aprotinin, and 1% (v/v) Triton X-100 (*see* **Note 4**).
3. 4× LDS sample buffer: 106 mM Tris–HCl, 141 mM Tris base, 2% LDS, 10% glycerol, 0.51 mM EDTA, 0.22 mM SERVA Blue, G250 0.175 mM Phenol Red, and pH 8.5.
4. 10× Reducing agent: 500 mM dithiothreitol (DTT).
5. Phosphate buffered saline (PBS): 0.144 g/l potassium dihydrogen phosphate, 9 g/l sodium chloride, and 0.795 g/l disodium phosphate.
6. BCA Assay Kit: Reagent A: sodium carbonate, sodium bicarbonate, bicinchoninic acid, and sodium tartrate in 0.1 M sodium hydroxide. Reagent B: 4% cupric sulfate, albumin standard ampules: 2 mg/ml bovine serum albumin (BSA) in 0.9% saline and 0.05% sodium azide.
7. Standard 4–12% SDS-PAGE gradient gel and Immobilon-P polyvinylidene difluoride (PVDF) membrane for protein transfer (*see* **Note 5**).
8. SDS running buffer: 50 mM 3-(*N*-morpholino) propanesulfonic acid (MOPS), 50 mM Tris base, 0.1% SDS, 1 mM EDTA, and pH 7.7.
9. Transfer buffer: 25 mM bicine, 25 mM Bis-Tris (free base), 1 mM EDTA, and pH 7.2.
10. Tris-buffered saline (TBS): 137 mM NaCl, 20 mM Tris–HCl, and pH 7.6 in dH_2O (deionized water).
11. Blocking solution: 5% (w/v) non-fat dry milk and 0.1% (v/v) Tween-20 in Tris-buffered saline (TBS) (*see* **Note 6**).
12. Antibodies: mouse monoclonal anti-CHMP5 antibody (*see* **Note 7**) and goat anti-mouse IgG (heavy and light chain) horseradish peroxidase conjugated antibody.

2.3 Assessment of CHMP5 Activity on Protein Ubiquitination

1. Cells: HEK293T cells.
2. Reagents: Ni-NTA His-bind resin or Ni-NTA magnetic agarose beads, MG132 (proteasome inhibitor) stock solution: 1 M MG132 in DMSO.
3. Plasmids: His-tagged ubiquitin (*see* **Note 8**), c-terminal FLAG-tagged CHMP5 (*see* **Note 9**), hemagglutinin (HA)-tagged Bcl-xL (or any candidate client protein), and Myc-tagged Cbl (*see* **Note 10**).
4. Transfection reagents: lipofectamine 2000 or Effectene transfection reagent.
5. Denaturing lysis buffer: 8 M urea, 50 mM Tris–HCl pH 8.0, 1.0% (v/v) Triton X-100, 10 mM imidazole, and 10 mM β-mercaptoethanol.
6. Sample buffer, 4× LDS: 106 mM Tris–HCl, 141 mM Tris base, 2% LDS, 10% glycerol, 0.51 mM EDTA, 0.22 mM SERVA Blue, G250 0.175 mM Phenol Red, and pH 8.5. (Plus 10× reducing agent: *see* **item 4** in Subheading 2.2).
7. Standard 4–12% SDS-PAGE gradient gel and Immobilon-P PVDF membrane for protein transfer mentioned earlier (*see* **item** 7 in Subheading 2.2).
8. Buffers: *see* **items 8–11** in Subheading 2.2.
9. Antibodies: mouse monoclonal M2 anti-FLAG (*see* **Note 11**) and goat anti-rabbit IgG (heavy and light chain) horseradish peroxidase-conjugated antibody (*see* **item 12** in Subheading 2.2).
10. Probe sonicator.
11. Refrigerated benchtop centrifuge suitable for Eppendorf tubes.

3 Methods

It is essential to carry out all procedures as quickly as possible given the rapid turnover of CHMP5 in most cells as per our observation. Accordingly, also prechill lysis buffers before use and carry out all centrifugation steps at 4 °C.

3.1 Immunoblot of Endogenous CHMP5 Protein

1. Harvest cells from tissue culture or isolate immune cell type of interest from single cell suspensions of lymphoid organs. Pellet cells by low-speed centrifugation (2500 × *g*, 5 min at 4 °C).
2. Wash cells with ice-cold PBS; make sure to remove residual PBS from cell pellets. If desired, cell pellets can be stored at −80 °C until use.
3. Lyse the cells in 50–100 μl of lysis buffer and incubate on ice for 1 h. Spin down the lysate to remove cellular debris at

12,000 × *g* for 12 min and proceed to quantify the protein concentration of each lysate using a BCA assay (*see* **Note 12**).

4. Bring each sample lysate to the same concentration and volume with lysis buffer, then add 4× LDS sample buffer and 10× reducing agent to each and boil at 70 °C for 10 min.
5. Perform standard SDS-PAGE and transfer onto a PVDF membrane (*see* **item** 7 in Subheading 2.2). Blocking with non-fat dry milk in Tris-buffered saline (TBS) containing 0.1% (v/v) Tween (TBST) is recommended.
6. For optimal endogenous CHMP5 detection, it is essential to incubate with the recommended concentration of primary antibody in blocking solution on a rocker overnight at 4 °C.
7. Continue the next day by washing the membrane three times with TBST, each for 7 min.
8. After blocking for an additional 10 min, add the secondary antibody at the recommended concentration in blocking solution and incubate on a rocker for 1 h at room temperature.
9. Wash once with TBST and twice with TBS for 7 min each. Complete the procedure by visualizing with autoradiography or digital chemiluminescence imaging.

3.2 Assessment of CHMP5 Activity on the Ubiquitination of Client Proteins

1. Culture HEK293T cells to 70% confluency in 10-cm tissue culture dishes. Make sure to plate enough dishes for all the required conditions and appropriate controls.
2. Using desired lipid-based transfection reagent, transfect cells with plasmids encoding epitope-tagged client protein (HA-Bcl-xL in this case), His-tagged ubiquitin, and titrated doses of the CHMP5 encoding plasmid. To induce ubiquitination of Bcl-xL, we also co-transfected HEK293T cells with Cbl.
3. After 36 h, treat cells with 10 μM MG132 and incubate for additional 6 h (*see* **Note 13**). Harvest cells from culture and wash two times with 1 ml of ice-cold PBS and centrifuge at 2500 × *g* for 5 min at 4 °C to pellet cells.
4. Lyse cell pellet in 1 ml of denaturing lysis buffer. To reduce viscosity of the lysates sonication is recommended (*see* **Note 14**).
5. Save an aliquot of the lysate to run as "Pre-IP" lysate (Fig. 1), and add 30 μl of Ni-NTA His-bind resin to the remainder lysate.
6. After rotation overnight at room temperature, wash three times with 1 ml of the denaturing lysis buffer at room temperature. Centrifuge at 800 × *g* for 45 s at 4 °C to sediment Ni-NTA His-bind resin and discard supernatant between each wash.
7. Elute protein by resuspending in bead volume (*see* **Note 15**) of LDS sample buffer with reducing agent.

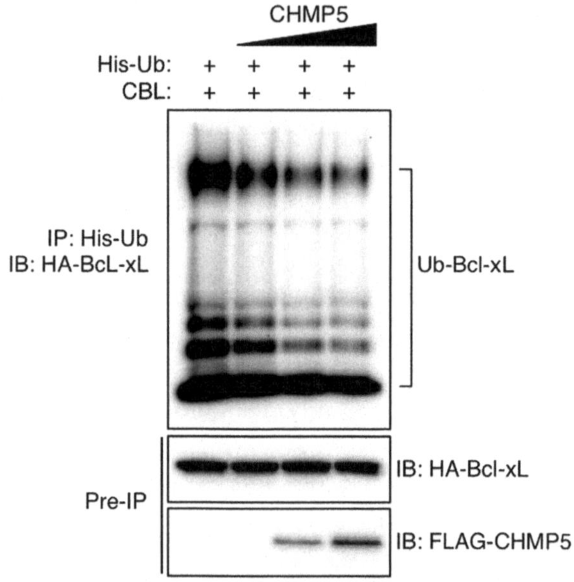

Fig. 1 Dose-dependent inhibition of Bcl-xL ubiquitination by CHMP5. Shown is a representative immunoblot of Ni-NTA eluates from HEK293 cells co-transfected with His-tagged ubiquitin (His-Ub), Cbl, HA-Bcl-xL, and increasing amounts of FLAG-CHMP5 plasmid. The top blot shows anti-HA antibody-probed membrane of the Ni-NTA eluates resolved by SDS-PAGE and transferred onto PVDF membranes. The bottom blots are anti-HA (Bcl-xL) and anti-FLAG (CHMP5) antibody-probed membranes of the HEK293 lysates before ("Pre-IP") they were subjected to Ni-NTA binding

8. Run eluted immunoprecipitates on SDS-PAGE, transfer onto a PVDF membrane, and proceed with standard primary and secondary antibody detection (*see* **steps 5–9** in Subheading 3.1).
9. As shown in Fig. 1, the ubiquitination of the T-lymphocyte prosurvival protein Bcl-xL was reduced when co-expressed with CHMP5 in a dose-dependent manner.

4 Notes

1. Based on our experience and others, RPMI medium was optimal and was routinely chosen for culturing primary thymocytes, splenic T-cells, and Jurkat suspension cell lines.
2. 1× MEM contains nonessential amino acids that are found in minimum essential medium. Supplementation of media with nonessential amino acids stimulates cell growth and prolongs cell viability. We routinely used HEPES (Sodium salt) available as a 1 M sterile solution from Gibco.

3. β-Mercaptoethanol is used in culture medium as a reducing agent to enhance cell growth by preventing toxic levels of oxygen-free radicals.
4. In our laboratory this lysis buffer formulation provided the best CHMP5 protein extraction and subsequent detection by western blotting.
5. In our hands, lysates resolved well in standard precast 4–12% Bis-Tris or Tris-glycine gels (ThermoFisher Scientific), and transfer onto a PVDF membrane (EMD Millipore) was adequate for probing our proteins of interest.
6. The use of Tween-20 in antibody-blocking solutions is recommended to prevent nonspecific binding of the antibody.
7. The anti-CHMP5 antibody (Santa Cruz Biotechnology, H-90, sc-67,230) used in our previous studies [7] has been discontinued by the manufacturer. For mouse or human CHMP5 detection, we have validated and recommend rabbit polyclonal anti-CHMP5 (Ab96273, Abcam) or rabbit polyclonal anti-CHMP5 (PA5-63303, ThermoFisher Scientific) as alternatives.
8. HA-tagged ubiquitin encoding plasmid (Addgene) may alternatively be used instead of His-tagged ubiquitin encoding plasmid.
9. We found that C-terminal epitope-tagged CHMP5 proteins maintained native activity compared to N-terminal epitope-tagged versions.
10. In this specific method, co-transfection of the E3-ligase Cbl is used to induce Bcl-xL ubiquitination.
11. We have validated and recommend the mouse monoclonal M2 anti-FLAG antibody from Sigma-Aldrich (F1804).
12. Perform the BCA assay as indicated in the protocol provided by Thermofisher in the Pierce BCA Protein Assay Kit. For optimal detection of CHMP5 protein, load at least 30 μg protein of lysate per well for each sample.
13. At this time-point, if necessary (when E3-ligase is not co-transfected), treat cells with predetermined inducer of client protein ubiquitination.
14. Sonication conditions and parameters should be determined and optimized based on instrument type and cell type being used. It is important to sonicate samples in the cold (e.g., place sample tubes in ice bucket) to minimize protein degradation.
15. Bead volume equals the volume of Ni-NTA resin used for immunoprecipitation. That is, elute with 30 μl LDS buffer if you use 30 μl of Ni-NTA resin for immunoprecipitation.

Acknowledgment

This work was supported in part by a Transition Career Development Award (K22-CA218467) from the National Cancer Institute of the US National Institutes of Health to S.A.

References

1. Mevissen TET, Komander D (2017) Mechanisms of deubiquitinase specificity and regulation. Annu Rev Biochem 86:159–192
2. Popovic D, Vucic D, Dikic I (2014) Ubiquitination in disease pathogenesis and treatment. Nat Med 20(11):1242–1253
3. Raiborg C, Stenmark H (2009) The ESCRT machinery in endosomal sorting of ubiquitylated membrane proteins. Nature 458 (7237):445–452
4. Rusten TE, Vaccari T, Stenmark H (2011) Shaping development with ESCRTs. Nat Cell Biol 14 (1):38–45
5. Khoury CM, Yang Z, Ismail S, Greenwood MT (2007) Characterization of a novel alternatively spliced human transcript encoding an N-terminally truncated Vps24 protein that suppresses the effects of Bax in an ESCRT independent manner in yeast. Gene 391(1-2):233–241
6. Greenblatt MB, Park KH, Oh H, Kim JM, Shin DY, Lee JM, Lee JW, Singh A, Lee KY, Hu D, Xiao C, Charles JF, Penninger JM, Lotinun S, Baron R, Ghosh S, Shim JH (2015) CHMP5 controls bone turnover rates by dampening NF-kappaB activity in osteoclasts. J Exp Med 212(8):1283–1301
7. Adoro S, Park KH, Bettigole SE, Lis R, Shin HR, Seo H, Kim JH, Knobeloch KP, Shim JH, Glimcher LH (2017) Post-translational control of T cell development by the ESCRT protein CHMP5. Nat Immunol 18(7):780–790
8. Azmi IF, Davies BA, Xiao J, Babst M, Xu Z, Katzmann DJ (2008) ESCRT-III family members stimulate Vps4 ATPase activity directly or via Vta1. Dev Cell 14(1):50–61
9. Stuchell-Brereton MD, Skalicky JJ, Kieffer C, Karren MA, Ghaffarian S, Sundquist WI (2007) ESCRT-III recognition by VPS4 ATPases. Nature 449(7163):740–744

Chapter 17

Purification of Plant ESCRT Proteins for Polyclonal Antibody Production

Julio Paez-Valencia and Marisa S. Otegui

Abstract

Most endosomal sorting complex required for transport (ESCRT)-III proteins are not fully functional when expressed as fusion of fluorescent or epitope tags, frequently making the use of specific antibodies the only available method for their detection. Heterologous expression of ESCRT-III proteins in bacteria often results in the formation of insoluble aggregates or inclusion bodies that interfere with their purification. However, inclusion bodies are usually pure protein aggregates with high antigenicity. In addition, since proteins within inclusion bodies are presented in a range of folding states, immunization with inclusion bodies can potentially result in antibodies with specificity for different folding states of the protein under study. We describe here a protocol to isolate bacterial inclusion bodies of plant ESCRT-III proteins for production of polyclonal antibodies. We also describe a nitrocellulose-based immunoaffinity purification method that allows the immobilization of ESCRT-III proteins and the subsequent isolation of specific antibodies from a crude serum.

Key words ESCRT, Antibodies, Inclusion bodies, Nitrocellulose-based immunoaffinity purification

1 Introduction

Endosomal sorting complex required for transport (ESCRT) proteins mediate membrane bending in reverse topology (away from the cytoplasm) and sorting of ubiquitylated protein in endosomal intralumenal vesicles. They have also been implicated in other membrane remodeling events such as retroviral budding [1], animal cytokinesis [2–5], plasma membrane repair [6], microautophagy and macroautophagy [7, 8], and nuclear envelope closure after mitosis [9, 10]. ESCRT proteins are organized in five molecular complexes, ESCRT-0, I, II, III, and the vacuolar protein sorting 4-vesicle trafficking 1 (VPS4-VTA1) complex (also suppressor of K^+ transport growth defect 1-LYST-interacting protein5 (SKD1-LIP5) complex in plants). ESCRT modules are present from archaea to eukaryotes. Some ESCRT subunits have undergone gene expansion in multicellular organisms, particularly in

Emmanuel Culetto and Renaud Legouis (eds.), *The ESCRT Complexes: Methods and Protocols*, Methods in Molecular Biology, vol. 1998, https://doi.org/10.1007/978-1-4939-9492-2_17, © Springer Science+Business Media, LLC, part of Springer Nature 2019

plant linages in which they are represented by multiple isoforms [11]. Plants contain putative orthologs for most of the ESCRT proteins originally identified in metazoans and fungi with the exception of the canonical ESCRT-0 [12, 13], and their interaction networks seem to be conserved [14–17].

ESCRT-III subunits occur in the cytoplasm in an autoinhibited monomeric state [18], and their recruitment to membrane involves deep conformational changes and oligomerization [19]. In addition, ESCRT-III function is dependent on the proper concentration and stoichiometry of ESCRT-III subunits [20]. Because of these unique features, the expression of functional tagged versions of ESCRT-III proteins is very challenging. Thus, the use of antibodies against native ESCRT-III proteins has been often the only reliable way to study their cellular dynamics and biochemical associations [15, 21]. When ESCRT-III proteins are expressed in bacteria, they tend to aggregate in inclusion bodies, complicating their subsequent use for biochemical applications. Interestingly, proteins within inclusion bodies are usually found in a range of folding states [22]; therefore, animal immunization with an emulsion of inclusion bodies can potentially lead to the production of antibodies with specificity for different conformations. We present here a protocol to express and purify plant ESCRT-III proteins from bacteria for their use in the production of polyclonal antibodies and a nitrocellulose-based immunoaffinity purification method to isolate specific antibodies from a crude serum.

2 Materials

2.1 Protein Expression and Isolation of Inclusion Bodies from Bacteria

Isolation of protein inclusion bodies from bacteria.

1. Plasmid-bearing recombinant gene of interest.
2. *Escherichia coli* host strain (e.g., BL21).
3. Lysogeny broth (LB) growth media.
4. Isopropyl β-D-1-thiogalactopyranoside (IPTG) stock solution: 1 M IPTG in deionized distilled water (ddH_2O).
5. Ampicillin stock solution (1000×): 100 mg/mL ampicillin in ddH_2O sterilized by filtration (filter pore size 0.20 μm).
6. Phenylmethane sulfonyl fluoride or phenylmethylsulfonyl fluoride (PMSF) stock solution: 100 mM PMSF in isopropanol (store at −20 °C).
7. Lysis buffer: 50 mM Tris–HCl, pH 8.0, 1 mM EDTA, 50 mM NaCl, 1 mM PMSF (*see* **Note 1**).
8. Lysozyme stock solution: 50 mg lysozyme from chicken egg white/mL in ddH_2O.

9. DNase stock solution: 50 mg/mL DNase from bovine pancreas in ddH_2O.
10. RNase stock solution: 50 mg/mL RNase from bovine pancreas in ddH_2O.
11. 5% Nonidet P-40 in lysis buffer. Dissolve 5 μL of Nonidet-P40 in 100 μL of lysis buffer.
12. Inclusion bodies washing buffers: 1, 2, 3, or 4 M urea lysis buffers with 0.5% Triton X-100.
13. Ultrasonic homogenizer for bacterial cell disruption (e.g., Q125 Sonica™).

2.2 SDS-PAGE and Western Blotting

1. Precast sodium dodecyl sulfate-polyacrylamide gel electrophoresis (SDS-PAGE) gel: 4–10% polyacrylamide gradient gels for proteins larger than 100 kD and 12–20% for proteins smaller than 100 kD.
2. 5× SDS-PAGE sample buffer: 60 mM Tris–HCl, pH 6.8, 25% (v/v) glycerol, 2% (w/v) SDS, 1 mM dithiothreitol (DTT), and 0.1% (w/v) bromophenol blue. Store in 10 mL aliquots at −20 °C.
3. Coomassie blue stain solution: 10% (v/v) glacial acetic acid, 45% (v/v) methanol, and 0.1% (w/v) Coomassie Blue R-250 in ddH_2O. Filter using Whatman # 1 filter paper, and store at room temperature.
4. Destaining solution: 10% (v/v) glacial acetic acid and 10% (v/v) methanol in ddH_2O.
5. Ponceau-S solution: 0.2 g of Ponceau-S in 100 mL of 3% (w/v) trichloroacetic acid (TCA) in ddH_2O. Prepare TCA solution by adding ddH_2O to 3 g of TCA to a final volume of 100 mL. This solution can be reused.
6. Nitrocellulose membrane: nitrocellulose blotting membrane of 0.2 or 0.45 μm pore size. It may be necessary to identify the optimal membrane suitable for the immobilization of a specific protein (*see* **Note 2**).
7. Protein transfer buffer: 25 mM Tris base, 192 mM glycine, pH 8.3, and 10% (v/v) methanol in ddH_2O.

2.3 Nitrocellulose-Based Immunoaffinity

1. Tris buffered saline-Tween (TBST) buffer: 10 mM Tris base, pH 8.0, 150 mM NaCl, and 0.05% (v/v) Tween-20. Dissolve 1.2 g Tris base and 8.76 g of NaCl in 800 mL of ddH_2O. Adjust pH to 8.0 with 1 N HCl and make up to 1 L with ddH_2O.
2. Blocking solution: 5% (w/v) nonfat powdered milk in TBST.
3. Glycine elution buffer, pH 2: 0.1 M glycine, 0.5 M NaCl, and 0.05% (v/v) Tween-20. Weigh 0.75 g glycine and 2.9 g NaCl and add ddH_2O to a volume of 80 mL. Mix and adjust pH with

3 mL 1 N HCl. Add 50 mL Tween-20. Make up to a final volume of 100 mL in ddH_2O. Precool at 4 °C before use.

4. 1 M Tris buffer, pH 8.1. Weigh 12.1 g Tris base, 40 mL 1 N HCl, and add ddH_2O to a final volume of 100 mL.
5. Bovine serum albumin (BSA) solution: 10% BSA (w/v) (IgG-free fraction) in ddH_2O.
6. Sodium azide solution: 5% (w/v) sodium azide in ddH_2O.

2.4 Protein Silver Staining

1. Silver stain fixing solution: 50% (v/v) methanol and 5% (v/v) acetic acid in ddH_2O.
2. Silver stain washing solution: 50% (v/v) methanol in ddH_2O.
3. Silver stain-sensitizing solution: 0.02% (w/v) sodium thiosulfate ($Na_2S_2O_3$) in ddH_2O.
4. Silver nitrate solution: 0.1% (w/v) silver nitrate ($AgNO_3$) in ddH_2O. Alternatively, a 200× stock solution can be made by adding 2 g of silver nitrate in 10 mL of ddH_2O and store in a dark glass tube.
5. Silver developer solution: add 0.05% (v/v) formaldehyde and 2% sodium carbonate in ddH_2O. Mix 60 μL of 37% formaldehyde and 1 g sodium carbonate in ddH_2O to a total volume of 50 mL.
6. Silver stop solution: 5% (v/v) acetic acid in ddH_2O.
7. Silver stain storage solution: 0.3% (w/v) sodium carbonate in ddH_2O.

3 Methods

3.1 Isolation of Protein Inclusion Bodies from Bacteria

1. Inoculate a single bacterial colony transformed with plasmid containing gene of interest into 2 mL of LB medium supplemented with appropriate antibiotic for selection. Incubate the culture overnight at 37 °C with vigorous shaking (*see* **Note 3**).
2. Inoculate a small amount of the overnight culture into 20 mL of LB medium with selective antibiotic to an optical density at 600 nm (OD_{600}) of 0.02–0.05 (e.g., 1:100 dilution of an overnight of *E. coli* culture grown in LB medium under standard conditions).
3. Monitor culture growth. When the culture reaches an OD_{600} of 0.2–0.5 (mid-log phase growth), induce gene expression by the addition of IPTG to a final concentration 1 mM (*see* **Note 4**).
4. Grow culture for at least 2 h after induction.
5. Harvest bacterial cells by centrifugation at 5000 × *g* for 5 min at 4 °C.

6. Suspend cells in 20 mL of lysis buffer, transfer the suspension to a conical tube (e.g., 50 mL falcon tube), and freeze suspension at −80 °C overnight.
7. Thaw cells and add 100 μL from the 100 mM PMSF stock. Check pH; it should be between 7 and 8. If necessary, adjust pH by adding drops of 0.5 M NaOH with vigorous mixing. Add PMSF stock solution to lysis buffer right before use.
8. Add lysozyme to a final concentration of 1 mg/mL (400 μL of the 50 mg/mL lysozyme stock) and stir the suspension for 2 h at 37 °C. Lysozyme concentration as high as 10 mg/mL may be used to reduce incubation times to 15 min.
9. Add 100 μL of Triton X-100 to a final concentration of 0.5%. Stir at room temperature for 1 h.
10. Cool down the suspension in ice and sonicate in an ultrasonic homogenizer for 10 min at 40% amplitude under 10 s cycles followed by 30 s cooling intervals (*see* **Note 5**).
11. After sonication, inoculate 100 μL of the suspension on LB plates with the corresponding selective antibiotic (e.g., ampicillin) and incubate at 37 °C (*see* **Note 6**).
12. Add 100 μL of 5% Nonidet-P40 in lysis buffer to the suspension. Incubate at 4 °C for 1 h under agitation.
13. Add DNAse and RNase to a concentration of 10 mg/mL supplemented with 10 mM $MgSO_4$ (DNAse cofactor). Stir the solution for 45 min at 37 °C to remove nucleic acids.
14. Centrifuge the sample at 15,000 × *g* for 15 min at 4 °C, save the supernatant (soluble fraction) for further analysis, and wash pellet (insoluble fraction) containing inclusion bodies with 1 mL of lysis buffer with 0.5% (v/v) Triton X-100.
15. Centrifuge at 15,000 × *g* for 15 min at 4 °C and resuspend pellet in 500 μL of lysis buffer. Prepare 50 μL aliquots and store at −80 °C.
16. Check protein profiles of all the steps by running samples by SDS-PAGE.

3.2 Purification of Inclusion Bodies

Contaminating proteins can be present in the insoluble fraction containing the inclusion bodies. By washing with detergent and urea, it is possible to remove contaminant proteins and isolate purified inclusion bodies (*see* **Note** 7). It is advisable to carry out a small-scale trial to optimize washing conditions in which the protein of interest is not fully solubilized.

1. Resuspend 100 μL of the pelleted insoluble fraction in a range of test solutions of lysis buffer containing 0.5% Triton X-100 and 1, 2, 3, or 4 M urea. Then mix and incubate for 10 min at room temperature. Centrifuge at 12,000 × *g* for 10 min at

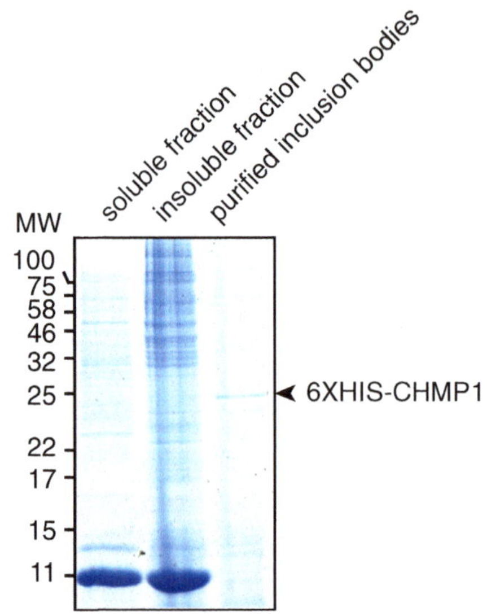

Fig. 1 Purification of inclusion bodies containing the Arabidopsis ESCRT-III subunit CHMP1 fused to a 6XHIS tag (band of 23 KDa as indicated) isolated from bacteria. Protein gel electrophoresis under denaturing conditions stained with Coomassie blue. *MW* molecular weight markers

4 °C, discard the supernatant, and resuspend the isolated inclusion bodies in 100 μL of H_2O.

2. Add 5× SDS-PAGE sample buffer to the saved supernatant (soluble fraction), insoluble fraction, and resuspended inclusion bodies. Analyze samples to detect protein of interest using SDS-PAGE (*see* **Note 8**).
3. Incubate gel with Coomassie blue stain for 2–16 h with slow agitation.
4. Incubate gel in destain solution with slow shaking until protein bands are clearly visible (*see* **Note 9**). The best washing buffer will result in an inclusion body fraction enriched in the protein of interest and free of contaminants (Fig. 1).
5. Scale this procedure up, and wash the inclusion bodies twice with buffer containing the optimum concentration of urea.
6. For animal immunization purposes, emulsified inclusion bodies can be used as immunogens since particulate antigens are highly immunogenic. Sonication of the inclusion bodies into smaller particles is recommended.

3.3 Nitrocellulose Immunoaffinity-Purification of Antibodies

To purify antibodies, we first immobilize the corresponding antigen to a solid-phase matrix (*see* **Note 10**). Nitrocellulose immunoaffinity purification requires resolving proteins by SDS-PAGE and their subsequent transfer to a nitrocellulose membrane. The inclusion of Tween-20 in all buffers used in this procedure allows for

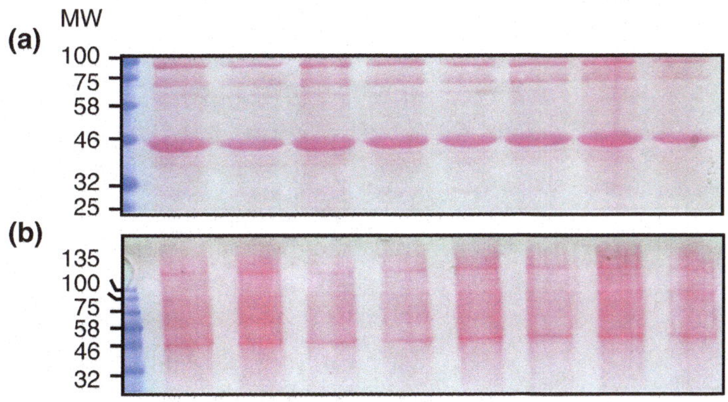

Fig. 2 Preparation of nitrocellulose immunoaffinity membranes. (**a**) Immobilized GST-CHMP1A on nitrocellulose membrane after denaturing gel electrophoresis and stained with Ponceau S. (**b**) Immobilized GST-CHMP1A on nitrocellulose membrane after native gel electrophoresis and stained with Ponceau S. *MW* molecular weight markers

more than 80% antibody recovery. We have used immunoblot affinity purification to isolate specific antibodies for different plant ESCRT proteins, such as Arabidopsis SNF7.1 [21] and CHMP1A. The isolated antibodies can be used for Western blots, immunoprecipitation, immunoelectron microscopy, and immunohistochemistry.

1. Load the first well of a SDS-PAGE gel with pres-stained molecular markers and the rest of the wells with 10 μL of resuspended inclusion bodies (*see* **Note 11**).
2. Run the gel in an electrophoresis apparatus. If antibodies specific to either native or denatured protein states want to be purified separately, samples can be run under both native and denaturing conditions (*see* **Note 12**). When the dye front of the sample buffer reaches the bottom of the plate, remove the gel.
3. Set the western blot transfer system following manufacturer's instructions.
4. To locate the protein band of interest, stain the membrane with Ponceau S solution for 5 min, cut a membrane strip containing the recombinant protein band (Fig. 2), and destain with distilled water. Avoid drying the membrane.
5. Block nitrocellulose membrane strip with 5% low-fat milk in TBST for 30 min at room temperature with mild agitation.
6. Wash the membrane with TBST three times for 10 s.
7. Add 1 mL of antiserum to 1 mL of TBST with 50 μL of 10% IgG-free BSA and 50 μL of 5% sodium azide. Incubate the nitrocellulose membrane strip in small container (e.g., 15 mL falcon tube or similar) overnight at room temperature with mild agitation.

8. Wash the nitrocellulose membrane strip three times with 10 mL of TBST for 5 min with mild agitation to remove unbound antibodies.
9. Elute the antibodies bound to the antigen in the membrane strips by incubating with 1 mL of precooled glycine elution buffer for 3 min over ice with gentle agitation.
10. Transfer immediately the elution buffer containing the antibodies to a small plastic tube containing 150 μL 1 M Tris base, pH 8.1 for neutralization.
11. Repeat **steps 9** and **10** once again. Pull together the two eluted samples (approximately 2.3 mL; take a 50 μL aliquot of each elution step for further analysis by SDS-PAGE and silver staining) and mix them with 250 μL of 10% BSA and 75 μL of 5% sodium azide. Prepare aliquots of 250 μL and store them at 4 °C.

3.4 Evaluation of Purified Antibodies

There are different procedures to visualize the reduced IgG fraction separated by SDS-PAGE. Two commonly used techniques are staining gels with either Coomassie blue or silver nitrate. We recommend silver staining of proteins because of its higher sensitivity [23].

1. After running samples by SDS-PAGE, incubate gel in 50 mL of the silver stain fixing solution for 20 min. Discard the solution and rinse gel with 50 mL of silver stain washing solution.
2. Rinse with 50 mL of ddH_2O for 1 h.
3. Incubate with silver stain-sensitizing solution for 1 min. Rinse twice with ddH_2O for 1 min each.
4. Incubate gel with precooled silver nitrate solution for 20 min at 4 °C. Rinse twice with ddH_2O for 1 min each.
5. Incubate with 50 mL of silver stain developing solution monitoring the color change of the gel. Change the solution if the developer turns yellow. Stop the developing reaction when protein bands are clearly visible by adding 50 mL of silver stain stop solution and incubate for 5 min (Fig. 3).
6. Keep gels immersed in silver stain storage solution.

The titer and specificity of the purified antibodies can be determined by western blotting, resolving three different concentrations of crude extracts and different dilution of the purified antibodies (Fig. 4).

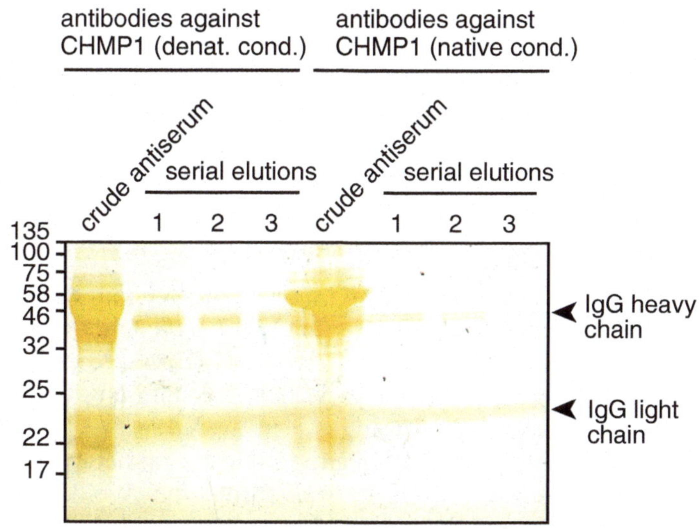

Fig. 3 Purification of anti-CHMP1A. Silver-stained gels of crude antiserum and eluted antibody fractions immunopurified from nitrocellulose membranes with GST-CHMP1 resolved under either denaturing or native conditions. *MW* molecular weight markers

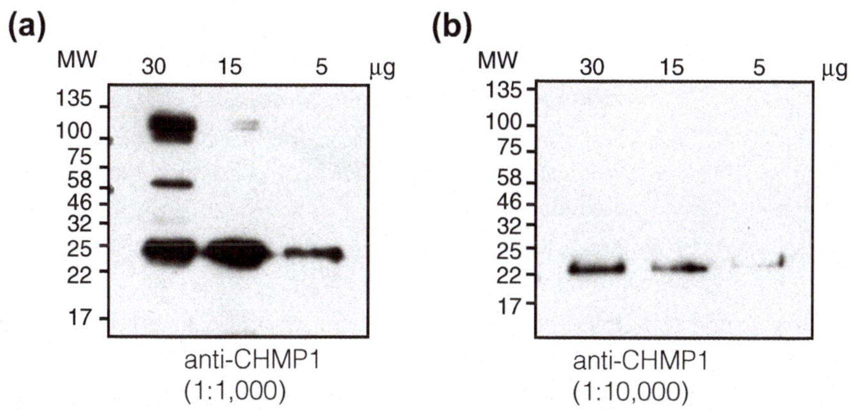

Fig. 4 Titration and evaluation of cross-reactivity of the antibodies analyzed by western blot. Two different concentrations of purified antibodies (**a**) 1:1000 and (**b**) 1:10,000 were used to detect CHMP1 from three different concentrations of *Arabidopsis thaliana* crude protein extract (30 μg, 15 μg, 5 μg of total protein). *MW* molecular weight markers

4 Notes

1. The lysis buffer for enzymatic digestion is critical. Lysozyme from chicken egg has a pH optimum between 7.0 and 8.6 and works better in ionic strength of 0.05 M.
2. Alternatively, polyvinylidene difluoride (PVDF) membrane can be used.

3. To ensure adequate aeration of the cell culture, the tube should be loosely capped, its volume should be at least four times greater that the volume of bacterial culture, and should be incubated under vigorous agitation.
4. IPTG is unstable at room temperature. Only thaw the working stock prior to the induction of the culture. It is recommendable to grow and harvest a non-induced culture control for comparison to induced cultures.
5. Longer sonication pulses of up to 5 min can be used to ensure complete cell lysis pure inclusion bodies [24, 25]. Optimal sonication times as well as lysozyme concentration must be determined for each case. To minimize froth formation, we recommend using 1.5 mL plastic tubes containing 1 mL of culture and placing the sonicator tip between the surface of the sample and the bottom of the tube. Avoid contact between the tip and the tube walls.
6. Since complete bacterial lysis is critical for purification of inclusion bodies, this control allows to test whether bacterial cells remain intact after the lysis step.
7. The pellet containing inclusion bodies is washed with buffer containing chaotropic agents (urea) and detergents (e.g., 1% Triton X-100) to remove contaminants, especially proteins (proteases), that may have absorbed onto the hydrophobic inclusion bodies during processing.
8. For mini-gels such as those for Bio-Rad Mini Protean systems, we recommend using 5 μL of cell lysate extracts and 1 μL of inclusion bodies.
9. To accelerate the destaining process, place a sponge in the same container to absorb excess dye.
10. Recombinant proteins are usually expressed with a tag (e.g., 6XHIS, glutathione S-transferase (GST)); therefore, the antisera against a tagged protein can also contain a fraction of anti-tag antibodies. In fact, about half of the antibodies reacting with the fusion protein may be directed against the tag [26]. We recommend performing blot-affinity purification with a recombinant protein with either no tag or fused to a different tag to the one that was used for immunization. This will avoid the binding of anti-tag antibodies to the matrix and their elimination from the purified antibody fraction.
11. One cm^2 of nitrocellulose binds approximately 300 μg of protein. The amount of IgG that can bind to this amount of immobilized protein is in the same range. Assuming 10 mg/mL of total IgG in hyperimmune sera and 1% of specific antibodies, 100 μg of membrane-immobilized protein could bind antibodies in up to 1 mL of antiserum.

12. Although inclusion bodies have traditionally been described as biologically inert protein aggregates, they can also contain proteins in native-like conformation [22]. Thus, protein species in fully folded to unfolded states can coexists in inclusion bodies. For non-denaturing conditions, avoid using SDS in the loading and migration buffers and boiling samples. It is also important to keep in mind that some antibodies only recognize proteins in their non-reduced or oxidized forms. If such antibodies need to be isolated, reducing agents such as β-mercaptoethanol or DTT should not be included in the buffers.

Acknowledgments

This work was supported by NSF grant MCB1614965 and funds from UW-Madison (Vilas Research Associate Award) to M.S.O.

References

1. Carlton JG, Martin-Serrano J (2007) Parallels between cytokinesis and retroviral budding: a role for the ESCRT machinery. Science (New York, NY) 316:1908–1912
2. Guizetti J, Schermelleh L, Mäntler J, Maar S, Poser I, Leonhardt H, Müller-Reichert T, Gerlich DW (2011) Cortical constriction during abscission involves helices of ESCRT-III–dependent filaments. Science (New York, NY) 331:1616–1620
3. Konig J, Frankel EB, Audhya A, Muller-Reichert T (2017) Membrane remodeling during embryonic abscission in *Caenorhabditis elegans*. J Cell Biol 216:1277–1286
4. Elia N, Sougrat R, Spurlin TA, Hurley JH, Lippincott-Schwartz J (2011) Dynamics of endosomal sorting complex required for transport (ESCRT) machinery during cytokinesis and its role in abscission. Proc Natl Acad Sci U S A 108:4846–4851
5. Lafaurie-Janvore J, Maiuri P, Wang I, Pinot M, Manneville JB, Betz T, Balland M, Piel M (2013) ESCRT-III assembly and cytokinetic abscission are induced by tension release in the intercellular bridge. Science (New York, NY) 339:1625–1629
6. Jimenez AJ, Maiuri P, Lafaurie-Janvore J, Divoux S, Piel M, Perez F (2014) ESCRT machinery is required for plasma membrane repair. Science (New York, NY) 343:1247136
7. Spitzer C, Li F, Buono R, Roschzttardtz H, Chung T, Zhang M, Osteryoung KW, Vierstra RD, Otegui MS (2015) The endosomal protein CHARGED MULTIVESICULAR BODY PROTEIN1 regulates the autophagic turnover of plastids in Arabidopsis. Plant Cell 27:391–402
8. Liu XM, Sun LL, Hu W, Ding YH, Dong MQ, Du LL (2015) ESCRTs cooperate with a selective autophagy receptor to mediate vacuolar targeting of soluble cargos. Mol Cell 59:1035–1042
9. Vietri M, Schink KO, Campsteijn C, Wegner CS, Schultz SW, Christ L, Thoresen SB, Brech A, Raiborg C, Stenmark H (2015) Spastin and ESCRT-III coordinate mitotic spindle disassembly and nuclear envelope sealing. Nature 522:231–235
10. Olmos Y, Hodgson L, Mantell J, Verkade P, Carlton JG (2015) ESCRT-III controls nuclear envelope reformation. Nature 522:236–239
11. Paez Valencia J, Goodman K, Otegui MS (2016) Endocytosis and endosomal trafficking in plants. Ann Rev Plant Biol 67:309–335
12. Winter V, Hauser M-T (2006) Exploring the ESCRTing machinery in eukaryotes. Trends Plant Sci 11:115–123
13. Leung KF, Dacks JB, Field MC (2008) Evolution of the multivesicular body ESCRT machinery; retention across the eukaryotic lineage. Traffic 9:1698–1716
14. Shahriari M, Richter K, Keshavaiah C, Sabovljevic A, Huelskamp M, Schellmann S (2011) The *Arabidopsis* ESCRT protein-protein interaction network. Plant Mol Biol 76:85–96
15. Spitzer C, Reyes FC, Buono R, Sliwinski MK, Haas TJ, Otegui MS (2009) The ESCRT-

related CHMP1A and B proteins mediate multivesicular body sorting of auxin carriers in *Arabidopsis* and are required for plant development. Plant Cell 21:749–766

16. Haas TJ, Sliwinski MK, Martínez DE, Preuss M, Ebine K, Ueda T, Nielsen E, Odorizzi G, Otegui MS (2007) The *Arabidopsis* AAA ATPase SKD1 is involved in multivesicular endosome function and interacts with its positive regulator LYST-INTERACTING PROTEIN5. Plant Cell 19:1295–1312
17. Spitzer C, Schellmann S, Sabovljevic A, Shahriari M, Keshavaiah C, Bechtold N, Herzog M, Muller S, Hanisch FG, Hulskamp M (2006) The *Arabidopsis elch* mutant reveals functions of an ESCRT component in cytokinesis. Development 133:4679–4689
18. Bajorek M, Schubert HL, McCullough J, Langelier C, Eckert DM, Stubblefield W-MB, Uter NT, Myszka DG, Hill CP, Sundquist WI (2009) Structural basis for ESCRT-III protein autoinhibition. Nat Struct Mol Biol 16:754–762
19. Tang S, Henne WM, Borbat PP, Buchkovich NJ, Freed JH, Mao Y, Fromme JC, Emr SD (2015) Structural basis for activation, assembly and membrane binding of ESCRT-III Snf7 filaments. elife 4:e12548
20. Arlt H, Perz A, Ungermann C (2011) An overexpression screen in *Saccharomyces cerevisiae* identifies novel genes that affect endocytic protein trafficking. Traffic 12:1592–1603
21. Buono RA, Leier A, Paez-Valencia J, Pennington J, Goodman K, Miller N, Ahlquist P, Marquez-Lago TT, Otegui MS (2017) ESCRT-mediated vesicle concatenation in plant endosomes. J Cell Biol 216:2167–2177
22. Peternel Š, Komel R (2011) Active protein aggregates produced in *Escherichia coli*. Int J Mol Sci 12:8275–8287
23. Chevallet M, Luche S, Rabilloud T (2006) Silver staining of proteins in polyacrylamide gels. Nat Protoc 1:1852–1858
24. Rodriguez-Carmona E, Cano-Garrido O, Seras-Franzoso J, Villaverde A, Garcia-Fruitos E (2010) Isolation of cell-free bacterial inclusion bodies. Microb Cell Factories 9:71
25. Menzella HG, Gramajo HC, Ceccarelli EA (2002) High recovery of prochymosin from inclusion bodies using controlled air oxidation. Protein Expr Purif 25:248–255
26. Greenbaum JA, Andersen PH, Blythe M, Bui HH, Cachau RE, Crowe J, Davies M, Kolaskar AS, Lund O, Morrison S, Mumey B, Ofran Y, Pellequer JL, Pinilla C, Ponomarenko JV, Raghava GP, van Regenmortel MH, Roggen EL, Sette A, Schlessinger A, Sollner J, Zand M, Peters B (2007) Towards a consensus on datasets and evaluation metrics for developing B-cell epitope prediction tools. J Mol Recognit 20:75–82

Chapter 18

Genetic and Cytological Methods to Study ESCRT Cell Cycle Function in Fission Yeast

Imane M. Rezig, Shaun K. Bremner, Musab S. Bhutta, Ian P. Salt, Gwyn W. Gould, and Christopher J. McInerny

Abstract

The fission yeast *Schizosaccharomyces pombe,* an ascomycete fungus, is an established model organism for studying eukaryotic molecular and cellular events such as the cell cycle due to its powerful genetics, a sequenced genome, and the ease of molecular manipulation (Wood et al., Nature 415:871–880, 2002; Hoffman et al., Genetics 201:403–423, 2015). This chapter describes genetic and cytological methods to study endosomal sorting complex required for transport (ESCRT) function during the cell cycle in fission yeast. These include tetrad analysis to allow the creation of double mutants to test for genetic interactions by synthetic phenotype characterization, such as cellular growth and the analysis of division septa by calcofluor-white staining.

Key words Fission yeast, *Schizosaccharomyces pombe*, Tetrad analysis, Genetic interaction, Synthetic phenotype, Septation, Calcofluor-white, ESCRT, Anillin, Polo kinase

1 Introduction

1.1 Tetrad Analysis: A Method to Identify Double Mutants and Genetic Interactions in Fission Yeast

Schizosaccharomyces pombe was first developed as a genetic organism by Urs Leupold in the 1950s when he isolated a self-mating homothallic strain h^{90} and two heterothallic strains derived from the h^{90} strain, h^{+} and h^{-}, that can mate with each other [1–5]. In complete nutritional medium, cells grow by medial fission, but upon nitrogen starvation, they arrest at G1 phase; if h^{+} and h^{-} cells are in proximity, conjugation takes place resulting in sexual reproduction consisting of mating, meiosis, and the formation of asci containing four ascospores (tetrads) [6].

Tetrad analysis is used in *S. pombe* to analyze the phenotypes of meiotic products to detect combinatorial mutants and enable the identification of genetic interactions. Here, as an example, we

---

Imane M. Rezig and Shaun K. Bremner contributed equally to this work.

Emmanuel Culetto and Renaud Legouis (eds.), *The ESCRT Complexes: Methods and Protocols*, Methods in Molecular Biology, vol. 1998, https://doi.org/10.1007/978-1-4939-9492-2_18, © Springer Science+Business Media, LLC, part of Springer Nature 2019

describe a synthetic phenotype generated by mating two chromosomal deletion mutation strains, one a deletion of the ESCRT gene *vps4*$^+$ (*vps4Δ*) and the other a deletion of the anillin gene *mid1*$^+$ (also called *dmf1*$^+$) (*mid1Δ*). Anillin proteins have important roles in cytokinesis controlling the placement of the actin ring that predicts the site of cell separation [7]. Single deletion mutants of *vps4Δ* and *mid1Δ* are viable, though in both cases cytokinesis is impaired [7]. The observed impaired growth in the *vps4Δ mid1Δ* double mutant demonstrates a genetic interaction between *vps4*$^+$ and *mid1*$^+$ and further suggests an interaction between the encoded Vps4p and Mid1p proteins, which can be confirmed using biochemical methods [7].

Tetrad analysis is based on the sexual life cycle of *S. pombe*. Two haploid yeast strains of opposite mating type (h^+ and h^-) are mated to produce diploid cells. The diploid cells undergo sporulation and, in turn, produce asci, with each ascus containing four haploid ascospores protected by a membrane; this structure is collectively called a tetrad. Tetrads are incubated at 37 °C to dissolve the membrane and the individual spores manually separated using a micromanipulator and incubated at 25–30 °C for 3–5 days. Following growth, selective media are used to determine the genotypes of each spore [3, 8]. Genotype screening is completed using replica plating where yeast colonies are transferred by imprinting on velvet material in the same arrangement to different selective media or grown at different temperatures [9]. As the four progeny from one tetrad undergo classical Mendelian segregation, with gene alleles in a haploid organism segregating in a 2:2 ratio, the phenotypes and genotypes of viable and unviable colonies can be assigned.

1.2 Staining of Division Septa in Fission Yeast Using Calcofluor-White

Calcofluor-white is a dye that binds to β(1–4) polysaccharides such as cellulose and chitin, which, when excited by ultraviolet light, emits a blue fluorescence [10]. In fission yeast, calcofluor-white can be used to rapidly stain the division septum and characterize septation (cytokinesis) mutations [9]. We illustrate this by showing the septation phenotypes of wild-type fission yeast and strains containing chromosomal deletions of several ESCRT genes and a temperature-sensitive Polo kinase mutant *plo1-ts35*. Altered septation phenotypes of double mutant strains compared with the single mutant strains suggest interactions between the ESCRT and Polo kinase proteins in the control of cytokinesis.

2 Materials

2.1 Fission Yeast Media and Solutions

1. All media autoclaved and stored at room temperature in 500 ml bottles.

2. Complete medium. Yeast extract (YE) per liter of distilled H_2O: 30 g D-glucose, 5 g Bacto yeast extract, 225 mg adenine, and 255 mg uracil (*see* **Note 1**).
3. Mating medium (limiting for nitrogen). Malt extract (ME) per liter of distilled H_2O: 30 g Bacto malt extract (*see* **Note 1**).
4. Selective medium. Edinburgh Minimal Medium (EMM) per liter of distilled H_2O: 20 g D-glucose, 5 g NH_4Cl, 0.1 g Na_2SO_4, 0.1 g $MgCl_2$, 15 mg $CaCl_2$, 3 g $C_8H_5KO_4$ (potassium hydrogen phthalate), 1.8 g Na_2HPO_4, 1 ml vitamins (per liter of distilled H_2O: 10 g inositol, 10 g nicotinic acid, 1 g calcium pantothenate, and 10 mg biotin; stored at −20 °C), and 0.1 ml trace minerals (per liter of distilled H_2O: 5 g boric acid, 5.2 g $MnSO_4$, 4 g $ZnSO_4$, 2 g $FeCl_3$, 1.44 g molybdic acid, 0.4 g $CuSO_4$, 10 g citric acid, and 0.1 g KI; stored at 4 °C) (*see* **Note 1**).
5. Solid media prepared by adding 20 g Bacto agar (*see* **Note 1**) to 1 liter of medium (YE, ME, or EMM) prior to autoclaving. Medium melted by heating in a microwave oven, poured into 90 mm Petri dishes, and allowed to set. For standard solid medium, pour ~20 ml per Petri dish; for "thin" solid YE medium for micromanipulation of tetrads, pour ~10 ml per Petri dish.
6. Supplements for EMM, per liter of yeast medium: 375 mg adenine, 375 mg uracil, and 187.5 mg leucine (*see* **Note 1**). These are added when required to liquid or solid media prior to autoclaving, or by spreading 400 μl of a 50× stock solution (750 mg 100 ml^{-1}, autoclaved and stored at room temperature) on 20 ml of solid medium in a Petri dish and air dried for 10–30 min.
7. Geneticin (G418) (*see* **Note 1**) may be added to solid YE medium after autoclaving to a final concentration of 100 μg ml^{-1}. This is achieved by spreading 40 μl of a 500× stock solution (50 mg ml^{-1} distilled water, stored at 4°C in dark) on 20 ml of solid medium in a Petri dish and air-dried for 10–30 min.
8. Phosphate buffer saline (PBS) 10× stock: 8 g NaCl, 0.2 g KCl, 1.42 g Na_2HPO_4, and 0.24 g KH_2PO_4. Add distilled H_2O to 900 ml and adjust pH to 7.4 with concentrated HCl; this solution is not autoclaved. Increase to 1 liter with distilled H_2O and store at room temperature. Dilute with distilled H_2O to 1× prior to use.
9. Calcofluor-white (1 g l^{-1} distilled H_2O) (*see* **Note 1**): stored in the dark at room temperature.
10. 10% (w/v) KOH in distilled H_2O. Stored at room temperature.

2.2 Fission Yeast Strains

1. Strains used to illustrate synthetic phenotypes in double mutants:

   GG 1554: h^+ *mid1Δ* (*dmf1::kanMX4 ura4-D18 leu1-32 ade2-216*).

   GG 1622: h^- *vps4Δ* (*vps4::ura4⁺ ura4-D18 leu1-32*).

   The GG numbers refer to the Glasgow laboratory collection annotation.

   Yeast strains are either obtained directly from the research laboratory that created and published them or from the Japanese National Bio Resource Project (http://yeast.nig.ac.jp/yeast/). The strains used here have the chromosomal genes *mid1⁺* and *vps4⁺* replaced by the *kanMX4* and *ura4⁺* genes, respectively; mutations in the *ura4⁺*, *leu1⁺*, and *ade2⁺* nutritional genes, which are used for genetic procedures.

2.3 Equipment

1. Replica-plating device and sterile velvet material (cut to appropriate size ~10 cm^2 and autoclaved) (*see* **Note 2**).
2. Scalpel blade, spreader, ethanol, and autoclaved sterile distilled H_2O.
3. Sterile disposable toothpicks and 90 mm plastic Petri dishes.
4. Singer MSM Ascus Dissector. The Singer website illustrates the structure and components of the ascus dissector, which consists of a standard light microscope, a grid-based stage where a Petri dish containing mated yeast is placed, and a microneedle used for micromanipulation of tetrads.
5. Sterile 150 ml Erlenmeyer flasks and shaking water bath with adjustable temperature.
6. 1.5 ml micro-centrifuge tubes.
7. Phase contrast microscope with 20× and 40× objective lenses.
8. Glass microscope slides 76 × 26 mm and 22 × 22 mm (No. 1; 0.13–0.16 mm thickness) glass coverslips.
9. Clear nail polish.
10. Bunsen burner.
11. Zeiss Axiovert 135 fluorescent microscope with a Zeiss 63× Plan-APOCHROMAT oil-immersion objective lenses and 4′,6-diamidino-2-phenylindole (DAPI) filters.

3 Methods

3.1 Glycerol Stocks for Long-Term Storage of Fission Yeast

1. Grow yeast cells in liquid YE medium at 25–30 °C (temperature depending on the genotype) for 2–3 days.
2. Add 30% (v/v) sterile glycerol, mix and place at −70 °C (*see* **Note 3**).

3. To recover cells use a sterile toothpick to scrape some glycerol stock; complete the procedure as rapidly as possible to prevent thawing of stock cells (*see* **Note 4**).
4. Streak on solid YE medium and incubate at 25–30 °C (temperature depending on the genotype) for 2–3 days. For example, strain GG 1554 was incubated at 30 °C and required 2 days to form colonies.

3.2 Mating Fission Yeast Strains

- On solid ME medium, place 1–2 mm patches of freshly growing h^+ and h^- yeast cells (*see* **Note 5**). Add 30 μl sterile distilled H_2O and mix the two patches with a toothpick in an area of about 1 cm^2.
- Allow the mating mixture to air dry (5–10 min) and incubate at 25 °C for up to 3 days (*see* **Note 6**). Monitor the mixture on day 2 and day 3 for the formation of asci tetrads containing four ascospores using light microscopy (Fig. 1).

3.3 Tetrad Analysis of Fission Yeast

1. When asci tetrads are fully formed, judged by the presence of four ascospores (Fig. 1), transfer a small amount of the mating mixture to solid "thin" YE medium along one side of the Petri dish in the "inoculum" area (Fig. 2 and *see* **Note 7**).
2. Invert the Petri dish and place on the Singer MSM Ascus Dissector stage. Microscopically identify intact asci tetrads in the mating mixture through the Petri dish base and medium, and with the micromanipulator needle pick and place up to eight intact asci tetrads in an adjacent vertical line at positions

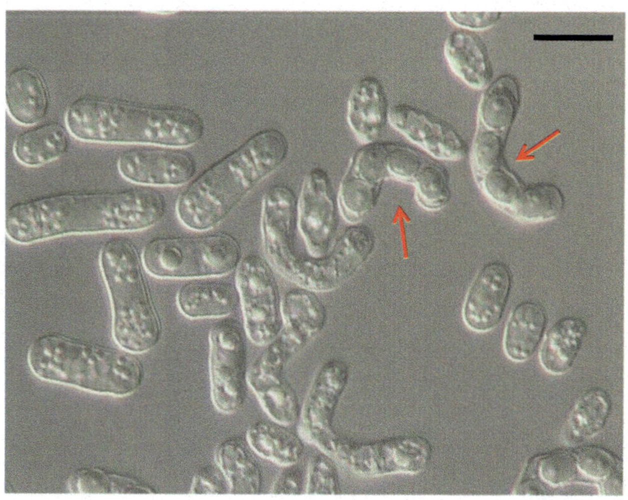

Fig. 1 Fission yeast mating mixture viewed by differential interference contrast microscopy. Cells of opposite mating type (h^+ and h^-) were mixed and grown on solid ME medium for 2 days at 25 °C. Fully formed asci tetrads containing four ascospores indicated by arrows. Scale bar 10 μm

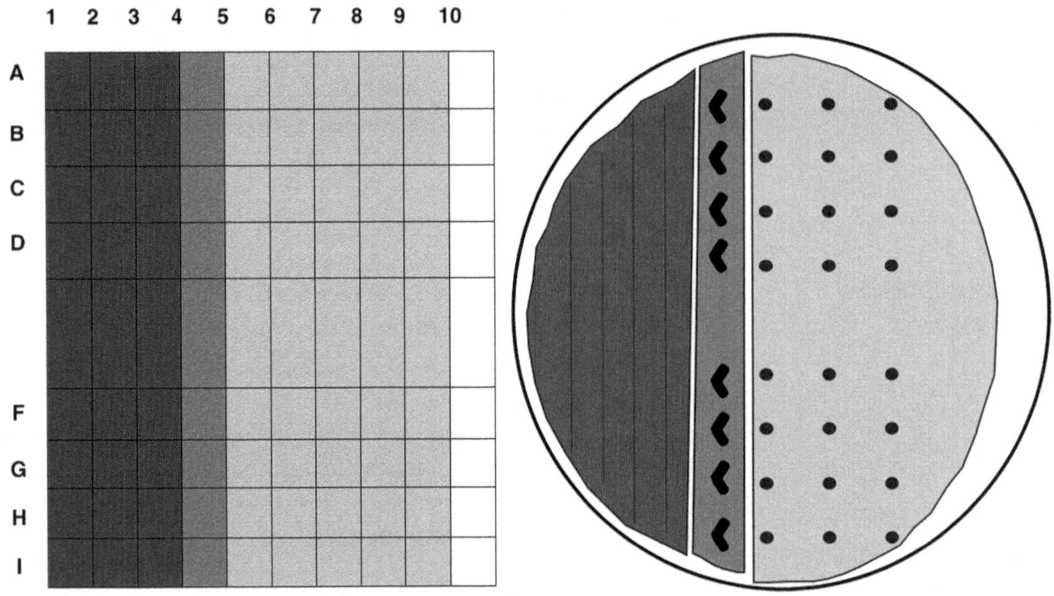

Fig. 2 Schematic representation of the grid-based stage of the Singer MSM Ascus Dissector (left), and its transfer to cells on solid medium in a 90 mm Petri dish (right). The dark grey area (named the "inoculum" by Singer MSM) represents the area on the solid "thin" YE medium in a Petri dish where the mating mixture is spread. The mid-grey area represents the location of the eight complete tetrads placed in a vertical line at positions A4, B4, C4 to I4. The pale grey area represents the location of three spores separated from each tetrad and placed at horizontal positions A5-7, B5-7, etc., with one spore left at position A4, etc. Positions 8, 9, and 10 can also be used for spore placement, if required

A4, B4, C4, etc. on the same solid "thin" YE medium Petri dish (Fig. 2).

3. Mark the position of A4 by piercing the surface of the solid medium nearby with the micromanipulator needle to facilitate relocating the tetrad in subsequent steps.
4. Incubate the asci tetrads at 37 °C for 3–5 h to induce membrane dissolution (*see* **Note 8**).
5. Using the micromanipulator needle, separate the four spores and place one spore each at horizontal three positions A5-7, B5-7, etc. (Fig. 2), with one spore left at position A4, B4, etc.
6. Incubate at 25–30 °C (temperature depending on the genotype) for 3–5 days until the spores form colonies, 1–2 mm in size.

3.4 Genotype Screening of Fission Yeast

1. Using a sterile scalpel, carefully remove the solid YE medium area (the "inoculum") containing the grown yeast mating mixture (Fig. 2).
2. Assemble a sterile velvet material over the replica-plating device.

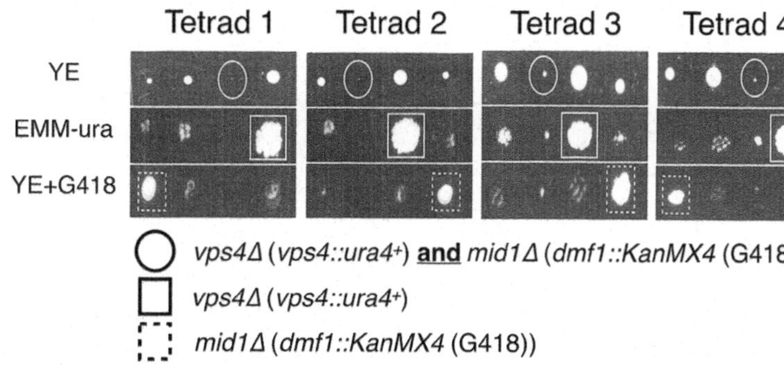

Fig. 3 Tetrad analysis of h$^-$ *vps4Δ* (*vps4::ura4$^+$*) (GG 1622) mated with h$^+$ *mid1Δ (dmf1::KanMX4)* (GG 1554) to identify *vps4Δ mid1Δ* double mutants that show a synthetic impaired growth phenotype. Tetrads were formed by mating the two strains on solid ME medium at 25 °C for 2–3 days. Four tetrads were then dissected using a micromanipulator and spores allowed to grow on solid YE medium at 30 °C until colonies formed. Colonies were replicated to solid YE + G418 and EMM-ura (for these genotypes EMM without the supplement uracil, but with adenine and leucine added) media and incubated at 30 °C to identify phenotypes and double mutants. The double mutants can be used for future experiments to study ESCRT/anillin function; and stored in 30% glycerol at −70 °C (*see* above)

3. Invert the YE medium Petri dish, remove the lid and place on the sterile velvet, and press gently and evenly to transfer the yeast colonies.
4. Carefully remove the YE medium Petri dish (*see* **Note 9**). Repeat **step 3** with Petri dishes containing selective solid medium, either EMM or YE with antibiotics (such as G418) or at restrictive temperature, to reveal single phenotypes (and genotypes) in each case (*see* **Note 10**). Incubate cells on Petri dishes at appropriate temperature(s) depending on the genotypes for 1–2 days.
5. Score phenotypes for each tetrad anticipating 2:2 segregation, where appropriate.
6. Figure 3 illustrates genotype screening for *vps4Δ mid1Δ* double deletion mutants following tetrad analysis. The double mutant reveals a synthetic impaired growth phenotype demonstrating a genetic interaction between the *vps4$^+$* and *mid1$^+$* genes and further suggests interaction between the encoded ESCRT Vps4p and anillin Mid1p proteins in fission yeast.

3.5 Fission Yeast Liquid Culture

1. Inoculate a single colony to 50 ml of liquid YE medium in a 150 ml Erlenmeyer flask and incubate in a shaking water bath overnight at 25–30 °C (temperature depending on the genotype).

2. Examine cells microscopically by phase contrast for the presence of division septa to confirm that they are undergoing the cell cycle and have not entered stationary phase (*see* **Note 11**).

3.6 Fission Yeast Calcofluor-White Staining

1. Transfer 1 ml of liquid yeast culture to a 1.5 ml microcentrifuge tube.
2. Pellet the cells by centrifugation at 17,530 × *g* for 30 s.
3. Remove the supernatant and resuspend the pellet in 1 ml PBS.
4. Add 30 μl of calcofluor-white followed by 30 μl of 10% KOH (*see* **Note 12**). Mix by inverting and incubate in the dark for 5 min.
5. Pellet the cells by centrifugation at 17,530 × *g* for 30 s.
6. Aspirate the supernatant and resuspend the pellet in 20–100 μl PBS (*see* **Note 13**).
7. Transfer 10 μl of the stained cells onto a glass microscope slide and cover with a glass coverslip.
8. With a tissue, dab the excess fluid from the edges of the coverslip.
9. Coverslips can be sealed by passing the microscope slide briefly through the flame of a Bunsen burner or by applying clear nail polish to the edges of the coverslip (*see* **Note 14**).

3.7 Visualization and Classification of Septation in Fission Yeast Cells

1. Cells are visualized using a Zeiss Axiovert 135 fluorescent microscope with a Zeiss 63× Plan-APOCHROMAT oil-immersion objective lenses.
2. Calcofluor-white is excited at 370 nm and emits at 440 nm allowing 4′,6-diamidino-2-phenylindole (DAPI) filters to be used. Parallel bright field images should also be captured to allow cell outlines to be determined.
3. Digital images are collected and processed using Microsoft PowerPoint and Adobe Photoshop.
4. Septation phenotypes are characterized in wild-type (*WT*) fission yeast and mutant strains containing either single chromosomal deletion mutants of various ESCRT genes or the temperature-sensitive Polo kinase mutant *plo1-ts35* (Fig. 4). Septation phenotypes were classified and labeled A–F (*see* **Note 15**).
5. The septation phenotype of 400 cells were analyzed across three replicates stained separately and repeated three times. Differences in septation between wild-type and ESCRT mutant cells suggest that ESCRT genes are required for septation in fission yeast (Fig. 4).
6. Double mutant strains were next created by tetrad analysis (*see* Subheading 3.3) of fission yeast strains containing

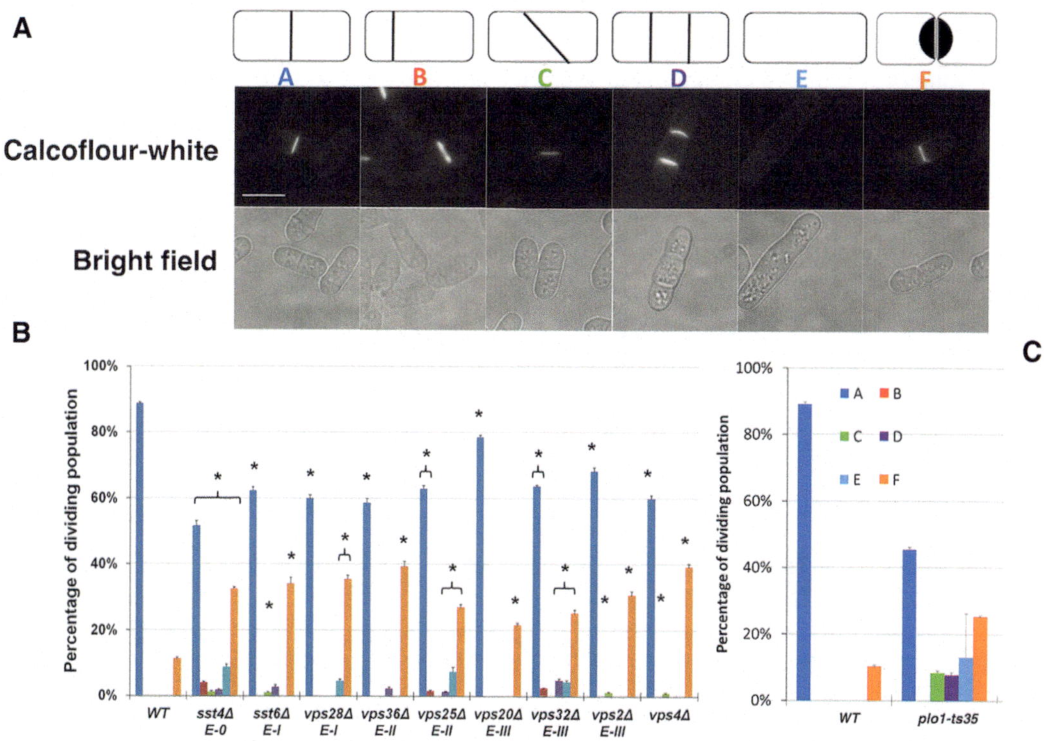

Fig. 4 Classification and quantification of septation in wild-type fission yeast and mutant strains containing either individual ESCRT chromosomal deletion mutants or the temperature-sensitive Polo kinase mutant *plo1-ts35.* Wild type (*WT*) and strains containing individual chromosomal deletions of ESCRT genes or *plo1-ts35* were grown at 25 °C in liquid YE medium to mid-exponential phase and harvested. Cells were stained with calcofluor-white and visualized by fluorescent microscopy. **(a)** Panels A–F show representative cells illustrating observed septation phenotypes. Schematic diagrams above panels represent each phenotype: phenotype (A) represents wild-type septa formation, (B) and (C) misaligned septa, (D) cells with more than two septa, (E) absent septa, and (F) delayed separation following septation. Quantitative analysis of the frequency of septation phenotypes A–F in strains containing single ESCRT chromosomal deletions (**b**) or *plo1-ts35* (**c**) in comparison to wild type (*WT*). In each case, 400 cells were counted in triplicate. An asterisk (*) indicates a *p* value <0.05 indicating a significant difference to wild type; $n = 3$. Scale bar 10 μm. Each of the ESCRT gene labels is accompanied by its respective ESCRT complex identification (E-0, E-I, E-II, and E-III). Image reproduced from Bhutta et al. [7] with permission from PLOS through the Creative Commons Attribution (CC BY) license

chromosome deletions of ESCRT genes and the temperature-sensitive Polo kinase mutant *plo1-ts35*, and their septa were characterized as described in **step 5**. Differences in septation phenotypes between the single mutants and ESCRT *plo1-ts35* double mutants suggest genetic interactions between the ESCRT and Polo kinase genes, and that both genes are required for septation (Fig. 5).

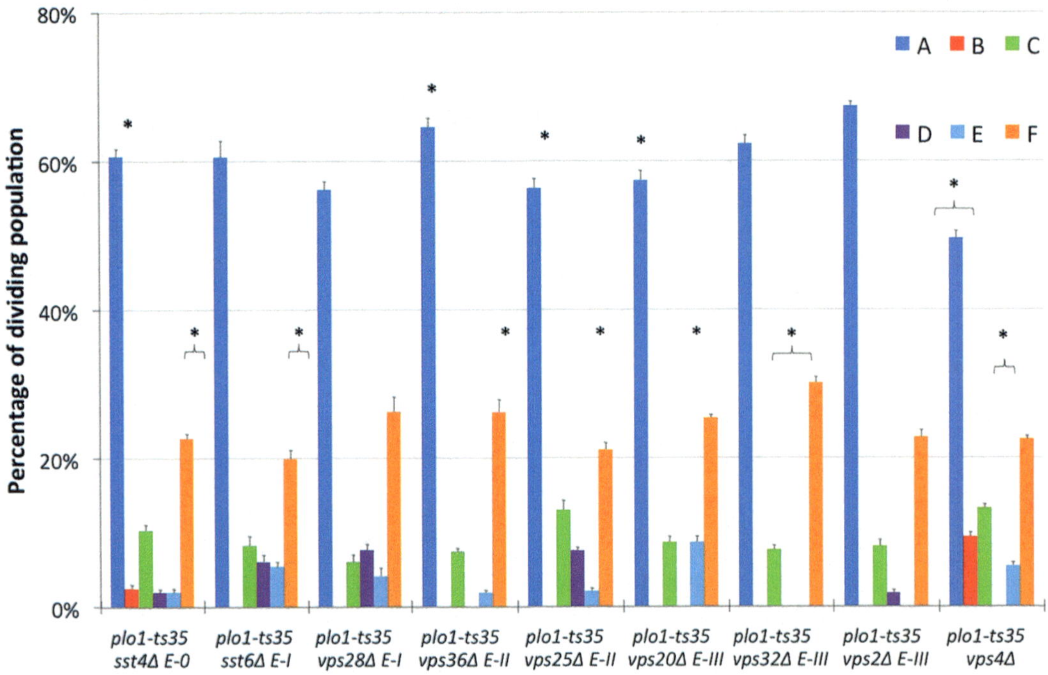

Fig. 5 Classification and quantification of septation in fission yeast double mutant strains containing individual ESCRT chromosomal deletion mutants in combination with the temperature-sensitive Polo kinase mutant *plo1-ts35*. Strains containing deletions of ESCRT genes in combination with *plo1-ts35* were grown at 25 °C in liquid YE medium to mid-exponential phase and harvested. Cells were stained with calcofluor-white and visualized by fluorescent microscopy. Quantitative analysis of the frequency of septation phenotypes A–F in strains in comparison to wild type (*WT*), as described in Fig. 4. In each case, 400 cells were counted in triplicate. An asterisk (*) indicates a *p* value <0.05 indicating a significant difference to wild-type; $n = 3$. Each of the ESCRT gene labels is accompanied by its respective ESCRT complex identification (E-0, E-I, E-II, and E-III). Image reproduced from Bhutta et al. [7] with permission from PLOS through the Creative Commons Attribution (CC BY) license

4 Notes

1. Bacto yeast media powder was purchased from BD Biosciences and all other chemical reagents were bought from Sigma/Aldrich. Dissolved adenine can precipitate and so require heating in a microwave oven to redissolve.
2. For example, the colony replica-plating device for Petri dishes from Bel-Art.
3. Glycerol stocks preserve the viability of cells for many years. To facilitate pipetting viscous glycerol, heat it in a microwave oven for 5–10 s at a low setting.
4. When recovering cells from glycerol stocks, ensure that the stock remains frozen and is not thawed; this can be done by holding stock tubes on dry ice.

5. For successful mating it is essential that cells are freshly growing: 1–2 days incubation on solid YE medium at a suitable temperature.
6. The mating mixture should be incubated below 30 °C as conjugation is impaired above this temperature.
7. To allow identification of tetrads and ascospores when using the Ascus Dissector microscope on looking through the base of the Petri dish and medium use "thin" solid YE medium to reduce optical refraction.
8. Some tetrad membranes may not readily break down after incubation at 37 °C. Isolate more tetrads than are needed and discard recalcitrant tetrads. Ideally, to isolate three double mutants, up to eight complete tetrads should be isolated.
9. When replica plating, the Petri dish should be lifted vertically from the velvet to avoid smudging colonies during transfer.
10. When replica plating, use the last replica Petri dish as a reference to ensure that all the colonies are transferred from the velvet and to exclude insufficient transfer if a colony fails to grow.
11. Cells entering stationary phase round up and become small, showing no septa.
12. The addition of KOH can reduce background fluorescence.
13. The volume of PBS used to resuspend the pellet depends on the density of the cell culture. An appropriate volume is required not only to prevent cells from clumping but also to enable microscopic visualization of individual cells.
14. For inverted microscopes, sealing coverslips with nail polish is preferable, but heat sealing is sufficient for upright microscopes.
15. Additional septation phenotypes may be identified by analysing other mutant strains.

Acknowledgments

This work is supported by a University of Glasgow Doctoral Training Programme to SKB and a BBSRC Studentship to MSB.

References

1. Wood V, Gwilliam R, Rajandream MA, Lyne M et al (2002) The genome sequence of *Schizosaccharomyces pombe*. Nature 415:871–880
2. Hoffman C, Wood V, Fantes P (2015) An ancient yeast for young geneticists: a primer on the *Schizosaccharomyces pombe* model system. Genetics 201:403–423
3. Moreno S, Klar A, Nurse P (1991) Molecular genetic analysis of fission yeast *Schizosaccharomyces pombe*. Methods Enzymol 194:795–823

4. Thon G, Klar A (1993) Directionality of fission yeast mating-type interconversion is controlled by the location of the donor loci. Genetics 134:1045–1054
5. Egel R (ed) (2004) The molecular biology of *Schizosaccharomyces pombe*. Springer-Verlag, Berlin and Heidelberg
6. Hayles J, Nurse P (1992) Genetics of the fission yeast *Schizosaccharomyces pombe*. Annu Rev Genet 26:373–402
7. Bhutta MS, Roy B, Gould GW, McInerny CJ (2014) A complex network of interactions between mitotic kinases, phosphatases and ESCRT proteins regulates septation and membrane trafficking in *S. pombe*. PLoS One 9: e111789
8. Forsburg S, Rhind N (2006) Basic methods for fission yeast. Yeast 23:173–183
9. Lederberg J, Lederberg E (1951) Replica plating and indirect selection of bacterial mutants. J Bacteriol 63:399–406
10. Herth W, Schnepf E (1980) The fluorochrome, calcofluor white, binds oriented to structural polysaccharide fibrils. Protoplasma 105:129–133

Chapter 19

ESCRT Mutant Analysis and Imaging of ESCRT Components in the Model Fungus *Ustilago maydis*

Carl Haag, Thomas Klein, and Michael Feldbrügge

Abstract

The ESCRT machinery (endosomal sorting complex required for transport) is an evolutionarily highly conserved multiprotein complex involved in numerous cellular processes like endocytosis, membrane repair, or endosomal long-distance transport. In fungal hyphae, endocytosis and long-distance mRNA transport are tightly linked, as endocytotic vesicles are also the key carrier vehicles for mRNAs. Studying the regulatory component Did2 (CHMP1) in the plant pathogen *Ustilago maydis* revealed that loss of Did2 resulted in disturbed endosomal maturation, thereby causing defects in microtubule-dependent transport of early endosomes. Here, we describe methods and protocols that allow studying the role of ESCRT components during endosomal transport. We present experimental strategies to analyze *U. maydis* ESCRT mutant phenotypes and test complementation with heterologous components, such as ESCRT regulators from *Drosophila melanogaster*.

Key words CHMP, Early endosomes, ESCRT regulator, Microtubules, mRNA transport, Small Rab-type GTPases

1 Introduction

The ESCRT machinery is a multipurpose complex that functions in membrane remodeling during various cellular processes such as endocytosis, cytokinesis, viral budding, or nuclear envelope repair [1, 2]. During endocytosis, for example, transmembrane proteins such as signaling receptors are sorted for degradation in the vacuolar/lysosomal compartment. A critical part of this highly regulated process is the step-wise assembly of defined ESCRT subcomplexes, resulting in the formation of intraluminal vesicles within endosomes. Thereby, sorting cargo is transferred to the inside of endosomes and hence insulated from the cytoplasm [3, 4].

A key step in ESCRT function is the polymerization of Snf7p (Vps32p; CHMP4) into membrane-associated spirals [1]. Its activity is regulated by the ATPase Vps4p (VPS4) that is needed for disassembly and recycling [5]. Vps4p function in turn is modulated

Emmanuel Culetto and Renaud Legouis (eds.), *The ESCRT Complexes: Methods and Protocols*, Methods in Molecular Biology, vol. 1998, https://doi.org/10.1007/978-1-4939-9492-2_19,

by regulatory factors such as Vps60p and Did2p (CHMP5 and CHMP1, respectively) [6, 7].

Such mechanistic insights into the function of the ESCRT machinery have been obtained by investigating mutants in *Saccharomyces cerevisiae*. Here, loss of ESCRT function causes defects in vacuolar protein sorting (*vps* mutants) [8, 9]. Two commonly studied cargoes are the vacuolar proteases Prc1p and Cps1p that are sorted in an ESCRT-dependent manner. In ESCRT mutants, Cps1p is no longer targeted to the vacuolar lumen but retained in the vacuolar membrane, while Prc1p still reaches the vacuole, but a substantial fraction is also erroneously exported from the cell [10, 11].

In highly polarized cells, such as fungal hyphae of *Aspergillus nidulans* or *Ustilago maydis,* long-distance transport of early endosomes was observed to involve ESCRT components, such as Did2 [12, 13]. Besides Did2, other ESCRT components have also been identified in *U. maydis* (Fig. 1a) (*see* below) [13]. Infection of this plant pathogenic fungus relies on the switch from yeast-like to hyphal growth [14]. Infectious hyphae expand at the tip and insert septa at the basal pole. Nuclei are positioned in the center (Fig. 1b). Important for efficient hyphal growth is long-distance transport along microtubules [15–17]. The main transport vehicles are Rab5a-positive early endosomes, which are transported bidirectionally along an antiparallel array of microtubule bundles, by the concerted action of molecular motors. Plus- and minus-end-directed motion is mediated by the Kinesin-3-type motor Kin3 and by cytoplasmic dynein Dyn1/2, respectively (Fig. 1c) [18, 19]. Defects in long-distance movement result in the formation of bipolar hyphae that are delayed in the formation of basal septa (Fig. 1b).

Rab5a-positive endosomes carry a plethora of molecular cargo, such as mRNAs, ribosomes, septin complexes, and even organelles such as peroxisomes [17, 20–23]. The RNA-binding protein Rrm4 is the main factor for endosomal transport of mRNAs and associated ribosomes. The protein is linked to endosomes through the FYVE domain protein Upa1. Rrm4 binds thousands of mRNAs and appears to coordinate their transport and translation (Fig. 1c; Olgeiser et al. [24]). Prominent target mRNAs of Rrm4 encode septins, and it has been hypothesized that local translation of septin mRNAs on moving endosomes results in the accumulation of the translation products on the cytoplasmic surface of the vesicles [22]. At this location, septins assemble in heteromeric complexes that are transported toward the growing tip. Upon delivery, they are assembled into high order structures that ultimately form septin filaments [25].

Detailed analysis of the ESCRT regulator Did2 revealed a close link between two endosome functions: protein sorting during endocytosis and long-distance transport of mRNAs, proteins, and

A

	U. maydis		S. cerevisiae	H. sapiens	D. melanogast	A. thaliana
ESCRT-0	Vps27	UMAG_03862	Vps27	HRS	Hrs	-
	Hse1	UMAG_04619	Hse1	STAM1/2	Stam	-
ESCRT-I	Vps23	UMAG_11474	Stp22 (Vps23)	TSG101	TSG101	VPS23A/ELC
	Vps28	UMAG_10687	Vps28	hVPS28	Vps28	VPS28-1 VPS28-2
	Vps37	UMAG_01333	Srn2 (Vps37)	VPS37A, B, C, D	Vps37	VPS37-1 VPS37-2
	-		Mvb12	hMVB12A, B	Mvb12	-
ESCRT-II	Vps36	UMAG_11941	Vps36	EAP45	Vps36	VPS36
	Vps22	UMAG_11539	Snf8 (Vps22)	EAP30	Lsn	VPS22
	Vps25		Vps25	EAP20	Vps25	VPS25
ESCRT-III and accessory components	Vps20	UMAG_04736	Vps20	CHMP6	Vps20	VPS20-1 VPS20-2
	Snf7	UMAG_05062	Snf7	CHMP4A,B,C	Shrb	SNF7-1 SNF7-2
	Vps24	UMAG_10799	Vps24	CHMP3	Vps24	VPS24-1 VPS24-2
	Did4	UMAG_04865	Did4 (Vps2)	CHMP2A,B	Vps2	VPS2-1 VPS2-2 VPS2-3
	Chm7$	UMAG_02077	Chm7$	CHMP7	Chmp7	CHMP7
	Ist1	UMAG_03333	Ist1	IST1	Ist1	ISTL1#
	Did2	UMAG_05607	Did2/Vps46	CHMP1A, B	Chmp1	CHMP1A CHMP1B
	Vps60	UMAG_05282	Vps60	CHMP5	Vps60	VPS60-1 VPS60-2
Vps4 and accessory components	Vps4	UMAG_01669	Vps4	VPS4A,B	Vps4	SKD1
	Vta1	UMAG_01645	Vta1	VTA1	Vta1	LIP5
Bro1/ALIX	Bro1	UMAG_00505	Bro1	PDCD6IP	ALiX	BRO1/ALIX

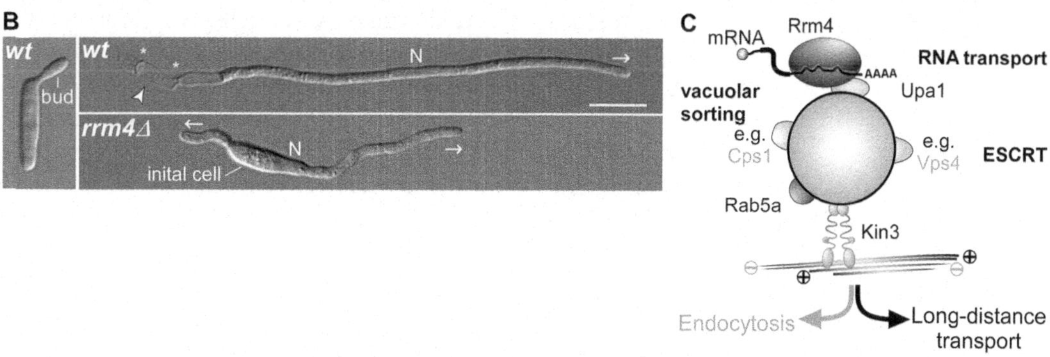

Fig. 1 Overview of ESCRT components and endosomal transport in *U. maydis*. (**a**) ESCRT components encoded in the *U. maydis* genome and homologues found in *S. cerevisiae*, *Homo sapiens*, *D. melanogaster*, and *Arabidospis thaliana*. The associated *U. maydis* gene IDs (UMAG_) are listed ($ similarity to Snf7, # large number of IST1-like proteins). (**b**) DIC images depicting a yeast cell of the wild-type strain (AB33) and hyphae of the wild-type and rrm4D strains, 6 h post induction of hyphal growth. White arrows denote direction of tip growth in hyphae. The arrowhead marks an empty section and basal septae are depicted by asterisks (laboratory strain AB33 serves as wild-type control; scale bar = 10 μm). Loss of Rrm4 disrupts endosomal mRNA transport, resulting in bipolar growth and defective septation. (**c**) Model of a transport endosome in *U. maydis* with identified endosomal cargo. Endosomes are required for endocytosis of vacuolar proteases as well as long-distance transport, for example, the mRNA-binding protein Rrm4 and associated mRNAs

organelles [13]. Loss of Did2 resulted in bipolar hyphae, indicative of defects in long-distance transport (Fig. 1b). Strikingly, markers for stationary late endosomes and vacuoles, such as Rab7 and the vacuolar protease Pcr1, were aberrantly present on shuttling endosomes. Apparently, endosomal maturation was disturbed, resulting in an altered identity of transport endosomes in the absence of Did2. As a result, components that were specific for early endosomes, such as the molecular motor Kin3 or the FYVE protein Upa1, were affected in their endosomal association. Reduced association of Kin3 resulted in disturbed processive endosomal movement and lower amounts of endosomal Upa1 impair Rrm4-mediated mRNA transport. This explains at least in part the observed defects in hyphal growth [13].

Importantly, these insights discovered in model fungi can be transferred to other highly polarized cells, such as neurons, where altered intracellular membrane trafficking is linked to neuronal disease [26, 27]. Notably, axon growth depends on Rab5 [28], and ESCRT-II might control this growth by transport of neuronal mRNAs and regulation of their local translation [29].

In this chapter, we describe the in vivo analysis of fungal ESCRT components combining genetic and cell biological approaches. We outline an efficient strategy to generate ESCRT gene deletion mutants in *U. maydis*, followed by phenotypic analysis of resulting mutants. We also present the experimental design to complement mutant phenotypes with heterologous components by choosing the complementation of *did2Δ* mutants with Chmp1 from *Drosophila melanogaster* as an example. We describe a detailed cell biological study of endocytosis in yeast cells as well as in vivo shuttling of early endosomes in hyphae, including a detailed photobleaching approach to visualize aberrant Rab7 shuttling in *did2Δ* mutants. Combining these approaches results in a straightforward analysis of ESCRT function during the endocytic pathway as well as endosomal transport not only in yeast cells but also in hyphae.

2 Materials

2.1 Culturing of U. maydis [25, 30]

1. Trace elements: 0.06% (w/v) H_3BO_3, 0.14% (w/v) $MnCl_2 \times 4\ H_2O$, 0.4% (w/v) $ZnCl_2$, 0.4% (w/v) $Na_2MoO_4 \times 2\ H_2O$, 0.1% (w/v) $FeCl_3 \times 6\ H_2O$, 0.04% (w/v) $CuSO_4 \times 5\ H_2O$, add ddH_2O, filter sterile.
2. Salt solution: 16% (w/v) KH_2PO_4, 4% (w/v) Na_2SO_4, 8% (w/v) KCl, 1.32% (w/v) $CaCl_2$, 8% (v/v) trace elements, 1% (w/v) $MgSO_4$ (water free), add ddH_2O, filter sterile.
3. Vitamin solution: 0.1% (w/v) thiamine hydrochloride, 0.05% (w/v) riboflavin, 0.05% (w/v) pyridoxine, 0.2% (w/v) calcium pantothenate, 0.05% (w/v) p-aminobenzoic acid, 0.2% (w/v)

nicotinic acid, 0.2% (w/v) choline chloride, 1% (w/v) *myo*-inositol, add ddH_2O, filter sterile.

4. Calcofluor white stock solution (CFW): 50 mM in DMSO. filter sterile.
5. Complete medium (CM): 0.25% (w/v) casamino acids, 0.1% (w/v) yeast extract, 1% (v/v) vitamin solution, 6.25% (v/v) salt solution, 0.05% (w/v) herring sperm DNA, 0.15% (w/v) NH_4NO_3, add ddH_2O, adjust pH with 5 M NaOH to 7.0 and autoclave. For agar plates, add 2 (w/v) bacto agar prior to autoclaving.
6. CM supplemented with glucose (CM-glc): add 1% (w/v) glucose (glc) in autoclaved CM-medium.
7. CM-glc plate media: add 2% (w/v) bacto agar in CM-medium prior to autoclaving. Add 1% (w/v) glucose after autoclaving.
8. CM-glc-Calcofluor White plate media (CM-glc-CFW): add 50 μM Calcofluor White to CM-glc plate medium. Nitrate minimal medium (NM): 0.3% (w/v) KNO_3, 6.25% (v/v) salt solution, add ddH_2O, adjust pH to 7.0 with 5 M KOH and autoclave.
9. NM supplemented with glucose (NM-glc) or arabinose (NM-ara): add 1% (w/v) glucose (glc) or 1% (w/v) arabinose (ara) to autoclaved NM-medium.
10. NM-glc-charcoal plate media: add 2% (w/v) bacto agar and 1% (w/v) charcoal to NM-medium prior to autoclaving. Add 1% (w/v) glucose to NM-charcoal plate medium after autoclaving.
11. Heparin solution: 15 mg/ml heparin in ddH_2O, filter sterile.
12. Sodium citrate sorbitol solution (SCS): Solution 1: 20 mM trisodium citrate, 1 M sorbitol; Solution 2: 20 mM citric acid, 1 M sorbitol; add Solution 2 to Solution 1 until pH 5.8 is reached, autoclave.
13. Sorbitol, Tris–HCl, $CaCl_2$ (STC): 1 M sorbitol, 10 mM Tris–HCl pH 7.5, 100 mM $CaCl_2$, filter sterile.
14. STC/PEG: 40% (v/v) PEG in STC-buffer.
15. RegLight: 1% (w/v) yeast extract, 0.4% (w/v) bacto peptone, 0.4% (w/v) sucrose, 18.22% (w/v) sorbitol, 1.5% (w/v) agar, add ddH_2O and autoclave.
16. Protoplasting solution: 12.5 mg/ml *Trichoderma* lysing enzymes in SCS, filter sterilize through a 22 μm filter; solution needs to be fresh for optimal enzymatic activity.
17. 10 × PBS: 79 mM $Na_2HPO_4 \times 2\ H_2O$, 145 mM KH_2PO_4, 5 mM $MgCl_2 \times 6\ H_2O$, 27 mM KCl, 1.37 M NaCl, adjust pH to 7.2, autoclave. Dilute to 1 × PBS using ddH_2O.
18. Glass reaction tubes, 250 ml and 100 ml baffled flasks, Petri dishes, rotation wheel, and shaker at 28 °C.

2.2 Live Cell Imaging of U. maydis [25]

1. Wide-field fluorescence microscopes. We used two microscopes for imaging. Microscope 1: Zeiss Axio Observer.Z1 in combination with a Photometrics CoolSNAP HQ2 CCD and a Hamamatsu ORCA-Flash 4.0 camera. Microscope 2: Zeiss Imager M.1 in combination with a Spot-Pursuit CCD camera.
2. Fluorescence setup. Microscope 1: laser illumination to excite GFP (488 nm/100 mW) or mCherry (561 nm/150 mW). Microscope 2: HPX halogen lamp.
3. Objectives. Microscope 1: 40× NA 1.3, 63× NA 1.4 or 100× NA 1.3 objectives. Microscope 2: 40× NA 1.3, 6× NA 1.25, 100× NA, 1.3.
4. Filter sets. GFP: ET 470/40×, ET 252/50 m, T495_PXR. mCherry: ET 560/40×, ET 630/75 m, T585lp. DAPI: AT350/50BP, ET400LP, ET460/50BP.
5. Dual-color acquisition system. In combination with microscope 1, we use a VS-LMS4 Laser Merge-System (Visitron Systems).
6. FRAP System. In combination with microscope 1, we use a VisiFRAP 2D-System that includes a 405 nm/80 mW diode laser (Visitron Systems).
7. Microscope imaging and analysis software. We use MetaMorph (Molecular Devices, version 7.7) to control the microscopes, process, and analyze the acquired images.
8. Microscope slides of 76 × 26 mm.
9. High-precision cover slips of 18 × 18 mm, $D = 170 \pm 5$ μm.
10. 2% (w/v) melted agarose in water.
11. CMAC (7-amino-4-chloromethylcoumarin) stock solution: 10 mM in DMSO.
12. FM4-64 stock solution: 8 mM in DMSO.

3 Methods

3.1 Generating Deletion Mutants of Genes Encoding ESCRT Components

Generating deletion mutants of ESCRT components and analyzing aberrant phenotypes is a straightforward approach for functional analysis [13, 31]. A powerful extension to this genetic approach is the complementation of ESCRT mutants with conserved heterologous components. This can be done by expressing the full-length protein or by studying fusion proteins to study individual domains.

Deletion mutants in *U. maydis* can be generated by replacing the endogenous locus with the deletion construct using homologous recombination. For efficient insertion, the homologous flanking regions should be about 1 kb in size and frame a resistance cassette for selection (Fig. 2a). Linearized deletion constructs are

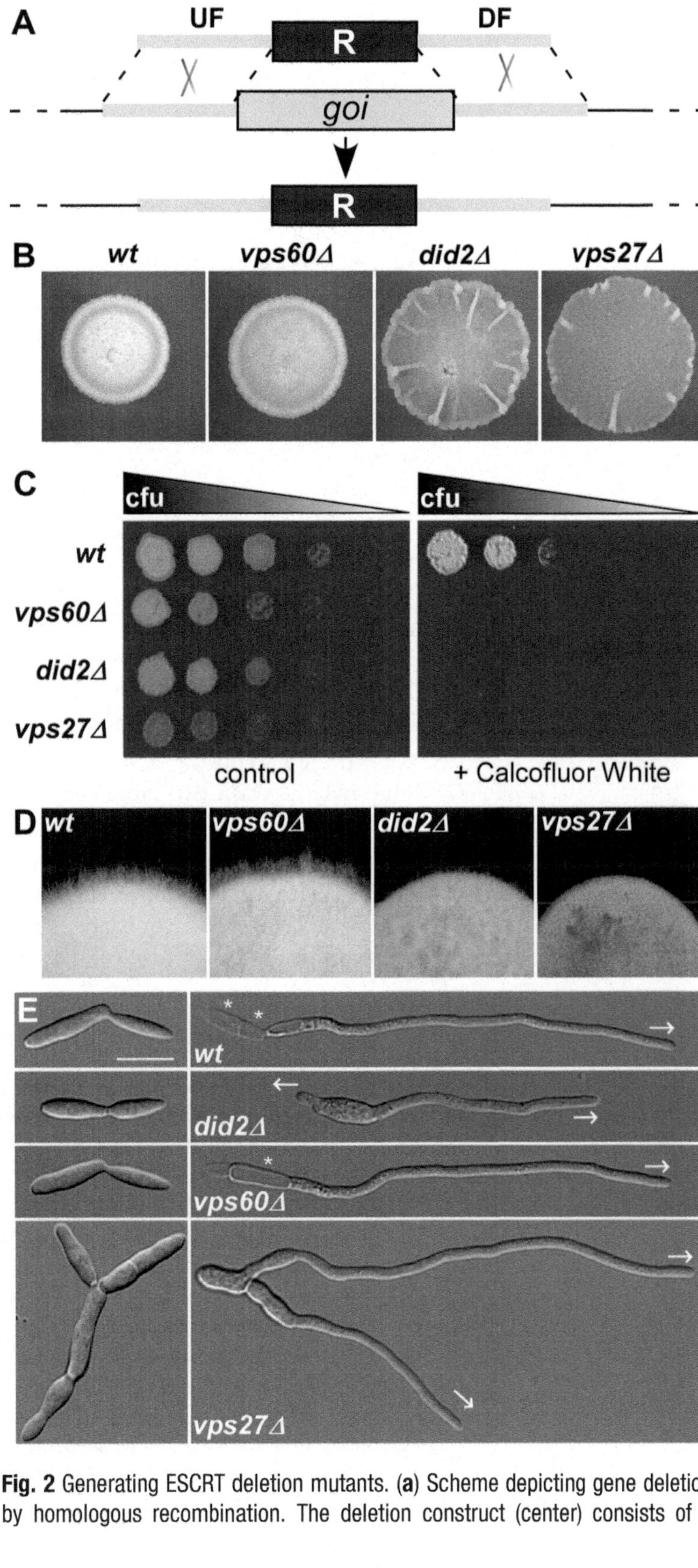

Fig. 2 Generating ESCRT deletion mutants. (**a**) Scheme depicting gene deletion by homologous recombination. The deletion construct (center) consists of a

transformed into protoplasts, and transformants are selected for the appropriate resistance marker. Strain generation in *U. maydis* has also been described in detail elsewhere [32].

1. Generate a deletion construct using standard molecular cloning methods [33, 34] (*see* **Note 1**).
2. Choose an appropriate strain background for gene deletion and start protoplast preparation: inoculate a 3 ml CM-glc preculture in a reagent tube and incubate on a rotating wheel for 18–24 h at 28 °C.
3. Inoculate a 50 ml CM-glc main culture in a 300 ml baffled flask using a 1/2000 dilution of the preculture. Grow the culture overnight at 28 °C for 15–16 h, shaking at 200 rpm, until OD_{600} reaches about 0.8–1.0. Sterile work is a fundamental requirement.
4. Pellet the cells for 5 min with 1500 g at room temperature (RT) and wash the cell pellet in 25 ml SCS.
5. Resuspend the pellet in 2 ml of protoplasting solution and incubate for 5–20 min (RT). Check protoplasting progress under the microscope and stop the process when about 30%–40% of cells are either round or pinheads by adding 10 ml of cold SCS. For all following steps, cells need to be kept on ice and treated gently.
6. Centrifuge at 1000 × *g*, for 5 min, at 4 °C, and discard the supernatant. Wash two more times with 10 ml of cold SCS.
7. Wash the cell pellet once with 10 ml of cold STC and centrifuge at 1000 × *g*, 5 min, 4 °C. Resuspend the cell pellet in 1 ml of cold STC and make 100 μl aliquots in prechilled 2 ml tubes. Aliquots can be stored for later use at −80 °C. Do not snap freeze the cells.

Fig. 2 (continued) resistance marker (R) flanked by the homologous upstream and downstream flanks (UF, DF). Homologous recombination replaces the gene of interest (*goi*) in the original locus with the marker to generate the respective mutant (bottom). (**b**) Example of the colony morphology of ESCRT deletion mutants growing in the sporidial form on solid medium. The plate was incubated for 2 days at 28 °C. The *did2Δ* and *vps27Δ* mutants form crinkled colonies, whereas *vps60Δ* mutants resemble wild type. (**c**) Example of sensitivity of ESCRT mutants toward cell wall stress. Serial dilutions of sporidial cells grown on CM-glc plates with and without 50 μM Calcofluor White (CFW). Cells were serially diluted five-fold, starting from an OD_{600} of 0.5 (cfu, colony-forming units). (**d**) Hyphal growth on NM-glc charcoal plates. Cells of AB33 (*wt*) and the respective mutant strains were spotted onto the plate and grown for 2 days at 28 °C. Fuzziness of colonies is due to aerial hyphae formation. (**e**) Micrographs of yeast cells (left) and hyphae 6 h post induction of filamentous growth (right; symbols as in Fig. 1; scale bar, 10 μm). Laboratory strain AB33 (top) serves as control [42]

8. Prepare two-layered selection regeneration plates for transformation (15 ml of RegLight agar for each layer). The bottom layer contains double the final antibiotic concentration, while no antibiotic is added to the top layer (*see* **Note 2**).
9. Use either fresh protoplasts or thaw protoplasts on ice, and add 1 μl of heparin (15 mg/ml) and 1–5 μg of the linearized deletion construct. Mix carefully and incubate on ice for 10 min.
10. Add 500 μl of STC-PEG, mix well gently and incubate for 15 min on ice.
11. Distribute cells carefully onto transformation plates. To prevent overcrowding, it is recommended to use two plates per transformation.
12. Incubate plates at 28 °C for 5–10 days. Incubation time depends on the gene targeted for deletion and the antibiotic used for selection.
13. Select positive transformants by singling out colonies twice on CM-glc plates containing the appropriate antibiotic (*see* **Notes 3** and **4**).
14. Verify gene deletion by PCR and/or Southern blot. It is recommended to verify strains using both methods. For this, genomic DNA needs to be extracted from each transformant including the progenitor strain [32]).

3.2 Phenotypic Analysis of ESCRT Deletion Mutants

Identify and analyze ESCRT mutants macro- and/or microscopically for one or more of the following phenotypes. The phenotypes are not necessarily shared by all ESCRT mutants.

3.2.1 Screen for Colony Morphology

1. Inoculate and grow precultures. *See* **step 2** in Subheading 3.1.
2. Spot 2 μl onto a CM-glc plate, and incubate the plate for 2 days at 28 °C. Some ESCRT mutants can easily be identified by the formation of crinkled colonies, which are characteristic for strains impaired in cell division (Fig. 2b) (*see* **Note 5**).

3.2.2 Screen for Cell Wall Stress Using Calcofluor White (CFW)

1. Inoculate and grow precultures. *See* **step 2** in Subheading 3.1.
2. Inoculate a 20 ml CM-glc main culture in a 100 ml baffled flask using a 1/2000 dilution of the preculture and grow exponentially for 14–16 h at 28 °C, shaking at 200 rpm, until OD_{600} reaches 0.5.
3. Prepare CM-glc-CFW plates. As control, use CM-glc agar plates.
4. Resuspend 1 ml culture in 1 ml of double-distilled H_2O.
5. Prepare serial dilutions of the cells diluting five-fold four times in sterile ddH_2O.

6. Spot 2 μl of each dilution step onto CM-glc and CM-glc-CFW plates.
7. Incubate plates for 1–3 days at 28 °C and assess plates for growth (Fig. 2c).

3.2.3 Analysis of Hyphae Formation on Plates (See Note 6)

1. Inoculate and grow precultures. *See* **step 2** in Subheading 3.1.
2. Spot 5 μl of the culture onto a fresh NM-glc-charcoal plate, and incubate the plate for 2–3 days at 28 °C.
3. Assess colonies for filament formation by eye and under a microscope. Strains showing filamentous growth are white and fuzzy. Strains impaired in filamentation appear less fuzzy or dull (Fig. 2d).

3.2.4 Microscopic Analysis of Hyphal Growth Polarity

1. Inoculate and grow cultures. *See* **steps 1** and **2** in Subheading 3.2.2.
2. Shift cells to NM-glc minimal medium by washing once with 20 ml NM-glc, centrifuge at 3300 × *g* for 3 min, and resuspend in 20 ml of NM-glc in a fresh 100 ml baffled flask. Continue shaking at 28 °C at 200 rpm for up to 9 h.
3. Prepare an agarose pad for microscopy by pipetting heat-melted 2% agarose onto a slide. Add a second slide perpendicularly on top to create a thin, even pad. Remove the top slide from the solidified agarose carefully after 5–10 min.
4. Pipette 1 μl cell culture on top of the pad center and distribute cells by swaying the slide. Let the drop dry for 1–2 min before adding a cover slip on top.
5. Visualize the cells (here we used differential interference microscopy (DIC)), using a 40× NA 1.3 objective. Acquire a series of images and quantify the number of hyphae growing unipolar or bipolar, with or without septae (*see* **Note 7**).

3.2.5 Microscopic Analysis of Vacuolar Protein Sorting (See Note 8)

1. Insert the coding sequences of the endogenous marker proteins Prc1 and Cps1 into expression vectors for targeted ectopic expression from the ip^S-locus using standard cloning methods. To this end, each vacuolar marker is fused with the red-fluorescent protein mCherry at its C-terminus. Expression is controlled by the constitutive P_{tef}-promoter, and the resulting construct is targeted to the ip^S-locus for ectopic expression. It is important to linearize the plasmid within the ip^R cassette, by choosing an appropriate restriction enzyme with a single recognition site in this region of the plasmid (Fig. 3a; *see* **Notes 9** and **10**) [35, 36].
2. Generate strains. *See* **steps 2–14** in Subheading 3.1.
3. Inoculate and grow cultures. *See* **steps 1** and **2** in Subheading 3.2.2.

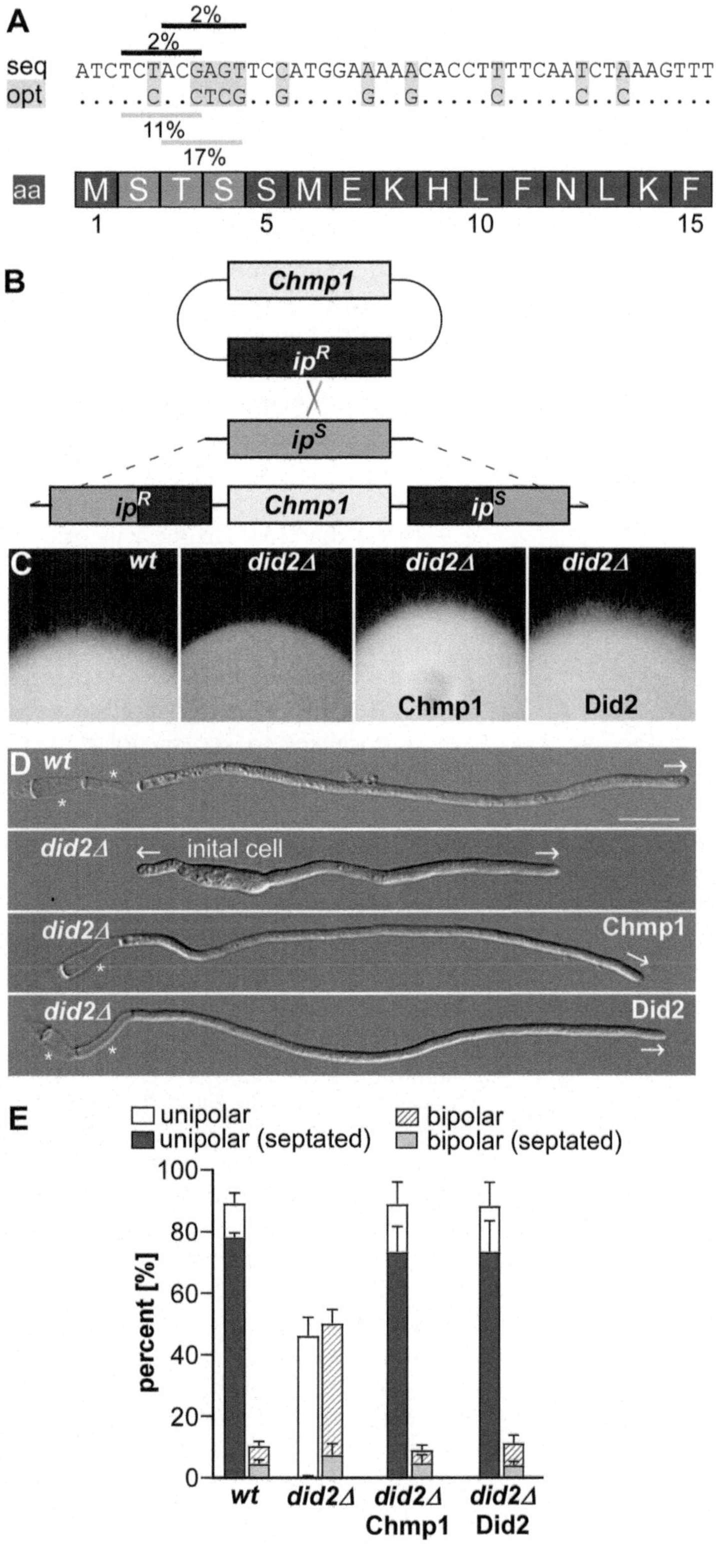

Fig. 3 Heterologous complementation of a *did2Δ* mutant. (**a**) Example of dicodon optimization for *D. melanogaster* Chmp1. Shown are the codons encoding the

4. Stain 1 ml cell suspension in a 2 ml tube with the dye CMAC (7-amino-4-chloromethylcoumarin, final concentration 10 μM, dissolved in DMSO). Protect the cells from light and incubate at 28 °C for 20 min on a rotating wheel. Wash cells once with 1 ml PBS and resuspend in 1 ml PBS.
5. Prepare microscope slides and samples. *See* **steps 3** and **4** under Subheading 3.2.4.
6. Analyze Prc1-mCherry and Cps1-mCherry localization as well as CMAC-stained vacuoles with the fluorescence microscope using the appropriate illumination (mCherry: Ex 587 nm/Em 610 nm; CMAC: Ex 353 nm/Em 466 nm). Here, we used an HXP halogen lamp in conjunction with the filter sets for Gfp and DAPI. Acquire fluorescent images for each channel sequentially, by exposing mCherry images for 200 ms and CMAC images for 35 ms. Note that vacuoles are static structures; therefore, fixing cells prior to image acquisition is not necessary.
7. Generate composite images for further analysis by overlaying corresponding mCherry and CMAC images using appropriate software (here we used MetaMorph, version 7.7). Consistent with results obtained in *S. cerevisiae* [37, 38] Cps1-mCherry mislocalizes to the vacuolar membrane in ESCRT mutants (Fig. 4b).

3.3 Heterologous Complementation of ESCRT Mutants

We recommend to use codon-optimized constructs for functional complementation and target those to the ip^S-locus for ectopic expression (*see* **Note 10**) [35].

Fig. 3 (continued) first 15 amino acids of Chmp1. The original nucleotide sequence is shown on top, with the optimized sequence and the encoded amino acids below. Nucleotide exchanges are highlighted by gray boxes. Shown are the frequencies (top: original; bottom: optimized) for the dicodons corresponding to amino acids 2–4. (**b**) Scheme for targeted expression of Chmp1 from the ip^S-locus. The integrative expression plasmid-carrying promoter, a codon-optimized Chmp1 version, transcriptional terminator and ip^R-allele is inserted into the ip^S-locus by homologous recombination. Upon insertion, the entire plasmid is integrated and flanked by a copy of the ip^S and ip^R gene each. (**c**) Spot assay for hyphal growth on NM-glc charcoal plates. The *did2Δ* mutant was complemented by either expressing codon-optimized Chmp1 or endogenous Did2 from the ip^S-locus. AB33 serves as wild type-like control (wt). (**d**) Micrographs of hyphae (6 h.p.i. of hyphal growth; scale bar, 10 μM). AB33 is shown as wild type control (wt; symbols as in Fig. 1b). (**e**) Quantification of uni- and bipolar growth (6 h.p.i. of hyphal growth) (*see* Fig. 3c; mean with s.e.m., $n = 3$)

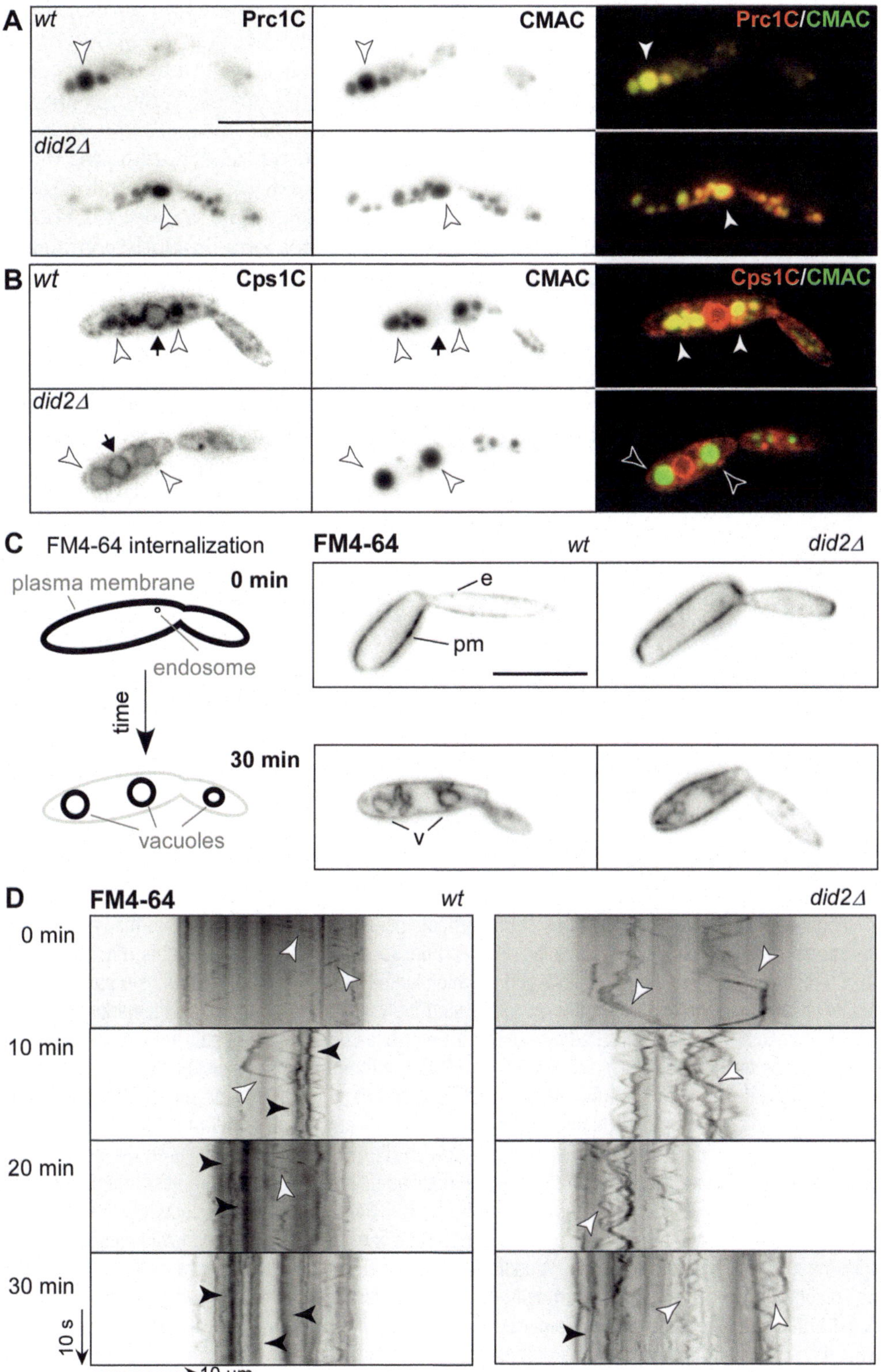

Fig. 4 In vivo analysis of vacuolar sorting and membrane trafficking. (**a**) Fluorescence micrographs of Prc1-mCherry-expressing cells stained with the vacuole marker CMAC. Numerous vacuoles of different sizes are

1. Avoid premature transcriptional termination and ensure efficient expression of heterologous genes in *U. maydis* by optimizing the nucleotide sequence according to a context-dependent codon usage (U_maydis_codon_usage_1.xls; http://www.mikrobiologie.hhu.de/ustilago-community.html) [39]. For every amino acid pair, determine the dicodon with the highest frequency. Decide on the final codon for each amino acid by comparing overlapping dicodons and choose the best option with the highest scores in both combinations (Fig. 3a). Order the resulting optimized sequence from any preferred vendor for custom gene synthesis.
2. Insert the codon-optimized gene into an expression vector targeting the ip^S-locus by using standard molecular cloning procedures (*see* **Note 11**). Generate strains as described in Subheading 3.1. *See* **steps 1** and **2** in Subheading 3.2.5.
3. Analyze functional complementation by screening for the phenotypes mentioned in Subheadings 3.2.1–3.2.5 (Fig. 3c–d).

3.4 Analysis of Endocytic Membrane Trafficking in ESCRT Mutants

Endocytic membrane trafficking can be tracked using the lipophilic dye FM4-64 [40].

1. Inoculate and grow cells. *See* **steps 1** and **2** in Subheading 3.2.2 (*see* **Note 12**).
2. Transfer 1 ml cell suspension to a 2 ml tube and incubate cells for 10 min on ice (*see* **Note 13**).
3. Add FM4-64 (final concentration 4 μM, dissolved in DMSO) to cells and incubate for an additional 10 min on ice, protected from light.

Fig. 4 (continued) stained by the dye. Loss of Did2 affects neither the sorting of Prc1C to the vacuolar lumen nor the number and morphology of vacuoles. The composite image shows both signals (Prc1: red, CMAC: green). Co-localizing signals appear yellow and are highlighted by white arrowheads (scale bar, 10 μm). (**b**) Fluorescence micrographs of Cps1-mCherry-expressing cells stained with the vacuole marker CMAC. The nuclear ER stained by Cps1C is indicated by a black arrow, while co-localizing signals are highlighted by white arrowheads. Loss of Did2 causes missorting of Cps1C to the membrane of vacuoles. The composite image shows both signals (Cps1: red; CMAC: green), and co-localizing signals appear yellow. Mislocalization of Cps1C to the vacuolar membrane is highlighted by white-rimmed arrowheads. (**c**) Uptake of FM4-64 over time in yeast cells. A scheme depicting time-dependent internalization of the lipophilic dye is shown on the left. The dye first stains the plasma membrane and is then internalized by endocytosis and traffics through endosomes to the vacuolar membrane. Fluorescence micrographs of FM4-64-stained cells are shown for each strain and time point on the right. FM4-64 internalization in *did2Δ*-cells is reduced, compared to wild type-like (wt) cells (pm, plasma membrane; e, endosome; v, vacuole; scale bar, 10 μm). (**d**) Pulse-chase experiment of FM4-64 internalization. Shown are kymographs of cells derived from recorded FM4-64 movies. Kymographs enable the visualization of one-dimensional movement of signals over time. Moving endosomes are marked by white arrowheads, while static vacuoles are marked by black arrowheads

4. Spin down cells at 2400 × *g* for 3 min at 4 °C, and resuspend cells in 1 ml of ice-cold CM-glc medium. Keep cells on ice until ready to start FM4-64 uptake.
5. Shift cells to a thermoblock set to 28 °C and shaking at 1100 rpm to start FM4-64 uptake.
6. Prepare microscope slide and sample. *See* **steps 3** and **4** in Subheading 3.2.4.
7. Visualize FM4-64 uptake over time under the fluorescence microscope using the appropriate illumination (FM4-64: Ex 515 nm/Em 640 nm) and a mCherry filter set. Here, we used a 63× objective, set the 561 nm laser intensity to 30% (100 mW laser), and exposed each image for 200 ms (Fig. 3c, d).
8. Set the camera to stream mode and record a movie of 100 frames with 200 ms exposure time for each frame to visualize uptake of FM4-64 into moving endosomes.
9. Convert the movie into a kymograph to plot movement of signals over time in a single graph by using appropriate software (here we use MetaMorph, version 7.7) (Fig. 3d). In ESCRT mutants, sorting to the vacuole may be delayed; therefore, the accumulation within static vacuoles is reduced, and prolonged dynamic movement is visible (Fig. 3d, *see* **Note 14**).

3.5 Analysis of the Intracellular Localization of ESCRT Components

The intracellular localization of ESCRT components can be analyzed by tagging the proteins with fluorescent proteins, such as the green fluorescent protein (Gfp, enhanced version, Clontech). Corresponding genetic constructs can be targeted to the endogenous loci and should contain homologous flanking regions of 1 kb in size, the *gfp* gene, a transcriptional terminator, and an appropriate resistance cassette [33]. Alternatively, constructs of the genes fused to *gfp* can be assembled in a plasmid for ectopic insertion and targeted to the ip^S-locus [35].

1. Generate a fusion construct for either endogenous or ectopic insertion by using standard cloning methods. For C-terminal Gfp fusions, insert the gene encoding *gfp* in-frame, directly downstream of the coding sequence, thereby deleting the stop codon (*see* **Note 11**).
2. Generate the strain. *See* Subheading 3.1 for endogenous insertion. For ectopic insertion, *see* **steps 1** and **2** in Subheading 3.2.5.
3. Prepare slides and samples for microscopic analysis, *see* **steps 3** and **4** in Subheading 3.2.4. For the intracellular localization in hyphae, *see* **steps 1–4** in Subheading 3.2.4.
4. Analyze the intracellular localization under the fluorescence microscope using the appropriate illumination (eGfp: Em 488 nm/Ex 509 nm) and Gfp filter set. Here, we used a 63×

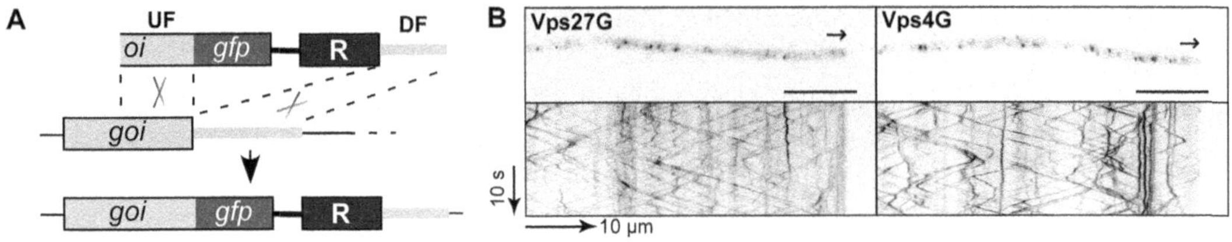

Fig. 5 In vivo imaging of the intracellular localization of ESCRT components. (**a**) Schematic representation of the strategy to generate C-terminal Gfp-fusions using homologous recombination in the genomic loci of the respective *goi*. The upstream flank (UF) needs to be placed such that it aligns with the 3′ end of the *goi* open reading frame excluding the stop codon. The downstream flank (DF) is placed downstream of the *goi*. (**b**) Fluorescence micrographs and the respective kymographs of hyphae expressing Vps27-Gfp (Vps27G) or Vps4-Gfp (Vps4G) under the native promoter. Only a part of each hypha is shown. The positions of the hyphal tips are marked by black arrows (scale bar, 10 µm)

NA 1.4 objective and set the 488 nm laser intensity to 30% (100 mW laser). For dynamic analysis of endosomal movement, set the camera to stream mode and record a movie of 150 frames with 150 ms exposure time for each frame.

5. Generate kymographs from the recorded movies. *See* **step 9** in Subheading 3.4. Analyze processivity, velocity, and intracellular localization (Figs. 4b, c and 5a, b).

3.6 Analysis of ESCRT Function During Endosomal Transport

Molecular markers like Rab5a can be used to track endosomal transport in fungal cells (Fig. 6). Constructs for marker gene fusions with fluorescent proteins are typically inserted ectopically in the ip^S-locus.

1. Generate strains expressing fluorescence reporter fusions of endosomal marker proteins using standard molecular cloning procedures. A common marker for early endosomes is Rab5a, to which Gfp is fused N-terminally. It is also possible to use other reporters such as the Phox domain of the endosomal SNARE Yup1 (amino acids 1–148) fused C-terminally to Gfp, as biosensors for endosomal phospholipids.
2. Delete *did2* in the strains expressing endosomal markers to investigate ESCRT function during endosomal transport. *See* **steps 1–14** in Subheading 3.1.
3. Prepare slides and samples for microscopic analysis under the fluorescence microscope, as described for hyphae in Subheading 3.2, **steps 13–16**.
4. Choose the appropriate illumination for eGfp and use the same settings as those described in Subheading 3.4, **step 4**. Record movies of 150 frames with an exposure time of 150 ms.
5. Generate kymographs from the recorded movies using appropriate software. Analyze the kymographs for number of shuttling signals, processivity, directionality, and intracellular localization.

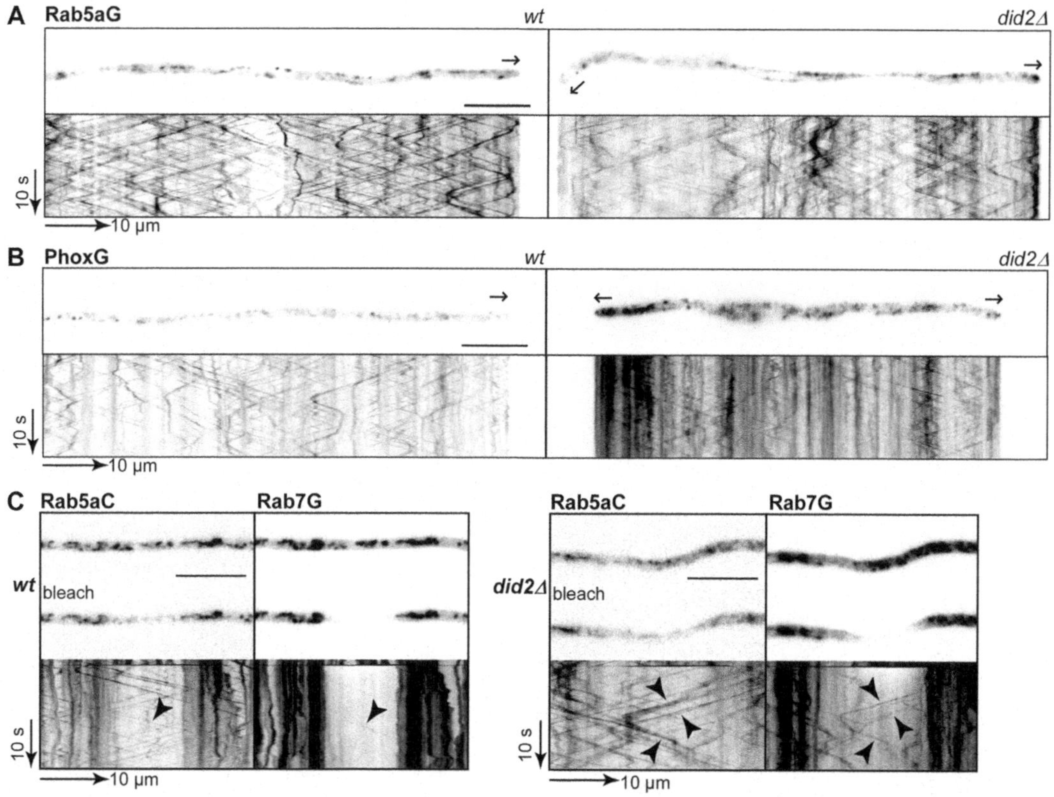

Fig. 6 In vivo imaging of endosomal transport and maturation. (**a**) Fluorescence micrographs and corresponding kymographs of strains expressing Gfp-Rab5a (Rab5aG). The strain carrying *did2* serves as wild type control (wt). Positions of the hyphal tips are marked by black arrows (scale bar, 10 μm). (**b**) Fluorescence micrographs and corresponding kymographs of strains expressing a C-terminal Gfp-fusion to the Phox-domain of the SNARE Yup1 (amino acid position 1–148) as a biosensor for endosomal lipids (PhoxG; scale bar, 10 μm). (**c**) Dual-color acquisition of mCherry-Rab5a (Rab5aC) and Gfp-Rab7 (Rab7G) coupled with bleaching. Shown here are fluorescence micrographs with the corresponding kymographs of parts of the hyphae. An area of 10 μm was bleached in each hypha, and the movement of Rab5aC and Rab7G across the bleached area was analyzed by kymographs. Co-localizing and moving signals are highlighted by filled arrowheads (scale bar: 10 μm). Figure used under the Creative Commons Attribution (CC BY) license from [13]; doi: https:/doi.org/10.1371/journal.pgen.1006734.g006

3.7 Analysis of the Effect of ESCRT Mutants on Endosomal Maturation

In *U. maydis*, Rab7 can be used as a late endosomal marker to visualize intracellular localization of late endosomal compartments. Analyzing strains co-expressing tagged Rab5a enables the simultaneous analysis of early and late endosomal components.

1. Clone an N-terminal Gfp-fusion of endogenous Rab7 using standard cloning methods. Design a construct for ectopic insertion into the *pep4*-locus (*see* **Note 15**) [41]. Constructs should contain 1 kb flanking regions, the fusion gene with a suitable promoter and terminator, and an appropriate resistance cassette.

2. Insert the construct into a strain that expresses an N-terminal mCherry-fusion of Rab5a for simultaneous detection of early and late endosomal compartments *see* **steps 1–14** in Subheading 3.1.
3. Prepare slides and samples for fluorescence microscopy as described in Subheading 3.2, **steps 18** and **19**.
4. Acquire a single Gfp-image and define the area to be bleached. Here, we bleached a region of 10 μm between the nucleus and hyphal tip.
5. Co-detect Rab7-Gfp and Rab5a-mCherry using dual-view acquisition in conjunction with a 100× NA 1.3 objective [30]. Set the laser intensities to the following values: 405 nm laser (30%; 50 mW laser), 488 nm laser (20%; 100 mW laser), 561 nm laser (50%; 150 mW laser). Set camera to stream mode and record a movie of both signals prior, during, and after the bleaching process. The movie should comprise 155 frames with an exposure of 150 ms/frame, with bleaching starting after the fifth frame.
6. Generate kymographs from recorded movies and quantify the number of moving signals across the bleached area and the extent of co-localization between Rab7 and Rab5a for passing signals.

4 Notes

1. As selection markers, resistance cassettes allowing growth on the fungicides carboxin, hygromycin B, nourseothricin, and Geneticin (G418) are available. The following antibiotics are used at the final concentrations: carboxin 2 μg/ml, hygromycin B 200 μg/ml, phleomycin 5 μg/ml, nourseothricin 150 μg/ml, and G418 500 μg/ml.
2. This increases regeneration time for protoplasts before subjecting cells to selection pressure by the antibiotic.
3. Since core ESCRT components are important for fungal growth, recombination efficiencies may be very low, and therefore, numerous transformants must be checked for homologous recombination events.
4. In certain cases, the chance for identification of ESCRT mutants can be enhanced by screening for phenotypes, for example, colony morphology (*see* Subheading 3.2).
5. Use freshly prepared CM-glc plates for best results.
6. In laboratory strain, AB33 expression of the heterodimeric b-transcription factor required for filamentous growth can be induced in NM minimal medium. On NM-glc minimal medium plates, AB33 hyphae form white, fuzzy colonies [42].

7. For some ESCRT mutants, an increased rate of bipolar growth is only visible at an early time point post induction (e.g., 4 h post induction, h.p.i.). At later time points, one pole takes over and unipolar hyphae are formed (e.g., 9 h.p.i.) [13].
8. Vacuolar protease sorting and membrane internalization can be best analyzed in yeast cells, as vacuoles tend to be larger in size and fewer in number in yeasts compared to hyphae.
9. Vacuoles have an acidic luminal pH. It is therefore recommended to tag proteins localizing to the vacuolar lumen with pH-insensitive fluorescent probes such as mCherry rather than with Gfp.
10. The ip^R-allele contains a single-nucleotide exchange and confers resistance against carboxin, making the sensitive ip^S gene locus usable for homologous recombination for targeted ectopic expression [35].
11. Tagging ESCRT proteins at the N- or C-terminus may interfere with their functionality; for complementation: check untagged proteins first, if possible, before generating fusion proteins.
12. For direct comparison, perform the experiments with a mixed culture consisting of an ESCRT-deletion strain and a wild-type-like strain expressing, for example, cytoplasmic Gfp, for distinguishing the two cell types by fluorescence microscopy.
13. FM4-64 is taken up rapidly and can be found in endosomes already 1–2 min after adding the dye. To stall membrane internalization, cells need to be kept on ice.
14. This can also be seen in hyphae.
15. Deletion of *pep4* has no discernible phenotype on sporidial or hyphal growth; therefore, it was chosen as an additional locus for targeted ectopic insertion [41].

Acknowledgments

We thank lab members for critical comments on the manuscript. Our research was in part financed by grants from the German Science Foundation through DFG-FOR2333, DFG-EXC1024 CEPLAS, and DFG-CRC1208.

References

1. Christ L, Raiborg C, Wenzel EM, Campsteijn C, Stenmark H (2017) Cellular functions and molecular mechanisms of the ESCRT membrane-scission machinery. Trends Biochem Sci 42:42–56. https://doi.org/10.1016/j.tibs.2016.08.016
2. Campsteijn C, Vietri M, Stenmark H (2016) Novel ESCRT functions in cell biology:

spiraling out of control? Curr Opin Cell Biol 41:1–8. https://doi.org/10.1016/j.ceb.2016.03.008

3. Hanson PI, Cashikar A (2012) Multivesicular body morphogenesis. Annu Rev Cell Dev Biol 28:337–362. https://doi.org/10.1146/annurev-cellbio-092910-154152
4. McCullough J, Clippinger AK, Talledge N, Skowyra ML, Saunders MG, Naismith TV, Colf LA, Afonine P, Arthur C, Sundquist WI, Hanson PI, Frost A (2015) Structure and membrane remodeling activity of ESCRT-III helical polymers. Science 350:1548–1551. https://doi.org/10.1126/science.aad8305
5. Babst M, Katzmann DJ, Snyder WB, Wendland B, Emr SD (2002) Endosome-associated complex, ESCRT-II, recruits transport machinery for protein sorting at the multivesicular body. Dev Cell 3:283–289
6. Kranz A, Kinner A, Kolling R (2001) A family of small coiled-coil-forming proteins functioning at the late endosome in yeast. Mol Biol Cell 12:711–723
7. Azmi IF, Davies BA, Xiao J, Babst M, Xu Z, Katzmann DJ (2008) ESCRT-III family members stimulate Vps4 ATPase activity directly or via Vta1. Dev Cell 14:50–61. https://doi.org/10.1016/j.devcel.2007.10.021
8. Bankaitis VA, Johnson LM, Emr SD (1986) Isolation of yeast mutants defective in protein targeting to the vacuole. Proc Natl Acad Sci U S A 83:9075–9079
9. Raymond CK, Howald-Stevenson I, Vater CA, Stevens TH (1992) Morphological classification of the yeast vacuolar protein sorting mutants: evidence for a prevacuolar compartment in class E vps mutants. Mol Biol Cell 3:1389–1402
10. Hecht KA, O'Donnell AF, Brodsky JL (2014) The proteolytic landscape of the yeast vacuole. Cell Logist 4:e28023. https://doi.org/10.4161/cl.28023
11. Gabriely G, Kama R, Gerst JE (2007) Involvement of specific COPI subunits in protein sorting from the late endosome to the vacuole in yeast. Mol Cell Biol 27:526–540. https://doi.org/10.1128/MCB.00577-06
12. Penalva MA (2010) Endocytosis in filamentous fungi: cinderella gets her reward. Curr Opin Microbiol 13:684–692. https://doi.org/10.1016/j.mib.2010.09.005
13. Haag C, Pohlmann T, Feldbrugge M (2017) The ESCRT regulator Did2 maintains the balance between long-distance endosomal transport and endocytic trafficking. PLoS Genet 13:e1006734. https://doi.org/10.1371/journal.pgen.1006734
14. Lanver D, Tollot M, Schweizer G, Lo Presti L, Reissmann S, Ma LS, Schuster M, Tanaka S, Liang L, Ludwig N, Kahmann R (2017) *Ustilago maydis* effectors and their impact on virulence. Nat Rev Microbiol 15:409–421. https://doi.org/10.1038/nrmicro.2017.33
15. Steinberg G, Penalva MA, Riquelme M, Wosten HA, Harris SD (2017) Cell biology of hyphal growth. Microbiol Spectr 5. https://doi.org/10.1128/microbiolspec.FUNK-0034-2016
16. Vollmeister E, Schipper K, Feldbrügge M (2012) Microtubule-dependent mRNA transport in the model microorganism *Ustilago maydis*. RNA Biol 9:1–8
17. Haag C, Steuten B, Feldbrügge M (2015) Membrane-coupled mRNA trafficking in fungi. Annu Rev Microbiol 69:265–281. https://doi.org/10.1146/annurev-micro-091014-104242
18. Schuster M, Kilaru S, Fink G, Collemare J, Roger Y, Steinberg G (2011) Kinesin-3 and dynein cooperate in long-range retrograde endosome motility along a non-uniform microtubule array. Mol Biol Cell 22:3645–3657. https://doi.org/10.1091/mbc.E11-03-0217
19. Baumann S, Pohlmann T, Jungbluth M, Brachmann A, Feldbrügge M (2012) Kinesin-3 and dynein mediate microtubule-dependent co-transport of mRNPs and endosomes. J Cell Sci 125:2740–2752. https://doi.org/10.1242/jcs.101212
20. Steinberg G (2012) The transport machinery for motility of fungal endosomes. Fungal Genet Biol 49:675–676. https://doi.org/10.1016/j.fgb.2012.01.009
21. König J, Baumann S, Koepke J, Pohlmann T, Zarnack K, Feldbrügge M (2009) The fungal RNA-binding protein Rrm4 mediates long-distance transport of *ubi1* and *rho3* mRNAs. EMBO J 28:1855–1866. https://doi.org/10.1038/emboj.2009.145
22. Baumann S, König J, Koepke J, Feldbrügge M (2014) Endosomal transport of septin mRNA and protein indicates local translation on endosomes and is required for correct septin filamentation. EMBO Rep 15:94–102. https://doi.org/10.1002/embr.201338037
23. Guimaraes SC, Schuster M, Bielska E, Dagdas G, Kilaru S, Meadows BR, Schrader M, Steinberg G (2015) Peroxisomes, lipid droplets, and endoplasmic reticulum "hitchhike" on motile early endosomes. J Cell Biol 211:945–954. https://doi.org/10.1083/jcb.201505086

24. Olgeiser L, Haag C, Boerner S, Ule J, Busch A, Koepke J, König J, Feldbrügge M, Zarnack K (2019) The key protein of endosomal mRNP transport Rrm4 binds translational landmark sites of cargo mRNAs. EMBO reports 20(1), e46588. https://doi.org/10.15252/embr.201846588
25. Zander S, Baumann S, Weidtkamp-Peters S, Feldbrügge M (2016) Endosomal assembly and transport of heteromeric septin complexes promote septin cytoskeleton formation. J Cell Sci 129:2778–2792. https://doi.org/10.1242/jcs.182824
26. Hunn BH, Cragg SJ, Bolam JP, Spillantini MG, Wade-Martins R (2015) Impaired intracellular trafficking defines early Parkinson's disease. Trends Neurosci 38:178–188. https://doi.org/10.1016/j.tins.2014.12.009
27. Lee JA, Gao FB (2012) Neuronal functions of ESCRTs. Exp Neurobiol 21:9–15. https://doi.org/10.5607/en.2012.21.1.9
28. Falk J, Konopacki FA, Zivraj KH, Holt CE (2014) Rab5 and Rab4 regulate axon elongation in the Xenopus visual system. J Neurosci 34:373–391. https://doi.org/10.1523/JNEUROSCI.0876-13.2014
29. Konopacki FA, Wong HH, Dwivedy A, Bellon A, Blower MD, Holt CE (2016) ESCRT-II controls retinal axon growth by regulating DCC receptor levels and local protein synthesis. Open Biol 6:150218. https://doi.org/10.1098/rsob.150218
30. Baumann S, Takeshita N, Grün N, Fischer R, Feldbrügge M (2015) Live cell imaging of endosomal trafficking in fungi. In: Tang BL (ed) Methods in molecular biology: membrane trafficking, 2nd edn. Springer, New York, pp 347–363
31. Hervas-Aguilar A, Rodriguez-Galan O, Galindo A, Abenza JF, Arst HN Jr, Penalva MA (2010) Characterization of *Aspergillus nidulans* DidB Did2, a non-essential component of the multivesicular body pathway. Fungal Genet Biol 47:636–646. https://doi.org/10.1016/j.fgb.2010.03.010
32. Bösch K, Frantzeskakis L, Vranes M, Kamper J, Schipper K, Göhre V (2016) Genetic manipulation of the plant pathogen *Ustilago maydis* to study fungal biology and plant microbe interactions. J Vis Exp. https://doi.org/10.3791/54522
33. Brachmann A, König J, Julius C, Feldbrügge M (2004) A reverse genetic approach for generating gene replacement mutants in *Ustilago maydis*. Mol Genet Genomics 272:216–226
34. Terfrüchte M, Joehnk B, Fajardo-Somera R, Braus G, Riquelme M, Schipper K, Feldbrügge M (2014) Establishing a versatile Golden Gate cloning system for genetic engineering in fungi. Fungal Genet Biol 62:1–10. https://doi.org/10.1016/j.fgb.2013.10.012
35. Loubradou G, Brachmann A, Feldbrügge M, Kahmann R (2001) A homologue of the transcriptional repressor Ssn6p antagonizes cAMP signalling in *Ustilago maydis*. Mol Microbiol 40:719–730
36. Stock J, Sarkari P, Kreibich S, Brefort T, Feldbrügge M, Schipper K (2012) Applying unconventional secretion of the endochitinase Cts1 to export heterologous proteins in *Ustilago maydis*. J Biotechnol 161:80–91. https://doi.org/10.1016/j.jbiotec.2012.03.004
37. Nickerson DP, West M, Odorizzi G (2006) Did2 coordinates Vps4-mediated dissociation of ESCRT-III from endosomes. J Cell Biol 175:715–720. https://doi.org/10.1083/jcb.200606113
38. Henne WM, Stenmark H, Emr SD (2013) Molecular mechanisms of the membrane sculpting ESCRT pathway. Cold Spring Harb Perspect Biol 5:a016766. https://doi.org/10.1101/cshperspect.a016766
39. Zarnack K, Maurer S, Kaffarnik F, Ladendorf O, Brachmann A, Kämper J, Feldbrügge M (2006) Tetracycline-regulated gene expression in the pathogen *Ustilago maydis*. Fungal Genet Biol 43:727–738
40. Vida TA, Emr SD (1995) A new vital stain for visualizing vacuolar membrane dynamics and endocytosis in yeast. J Cell Biol 128:779–792
41. Sarkari P, Reindl M, Stock J, Müller O, Kahmann R, Feldbrügge M, Schipper K (2014) Improved expression of single-chain antibodies in *Ustilago maydis*. J Biotechnol 191:165–175. https://doi.org/10.1016/j.jbiotec.2014.06.028
42. Brachmann A, Weinzierl G, Kämper J, Kahmann R (2001) Identification of genes in the bW/bE regulatory cascade in *Ustilago maydis*. Mol Microbiol 42:1047–1063

Chapter 20

Genetic Suppressor Screen Using an Inducible *FREE1-RNAi* Line to Detect ESCRT Genetic Interactors in *Arabidopsis thaliana*

Qiong Zhao, Ying Zhu, Wenhan Cao, Jinbo Shen, Yong Cui, Shuxian Huang, and Liwen Jiang

Abstract

FREE1 (FYVE domain protein required for endosomal sorting 1), a newly identified component of endosomal sorting complex required for transport I (ESCRT I), plays multiple roles in regulating protein trafficking and organelle biogenesis in *Arabidopsis*. Similar to other ESCRT components, FREE1 is essential for plant growth and development because *free1* mutant is seedling lethal. To identify key components that genetically interact with FREE1, we performed forward genetic suppressor screening using a dexamethasone (DEX)-inducible *FREE1-RNAi* line. Here we describe the detailed protocol of identifying novel FREE1 regulators using an inducible *FREE1-RNAi* line for the genetic suppressor screen. This protocol represents a whole procedure for identifying ESCRT genetic interactors in *Arabidopsis thaliana*.

Key words ESCRT, FREE1, Inducible, RNAi, EMS mutagenesis, Suppressor, Forward genetic screening, Mapping

1 Introduction

The endomembrane system is highly conserved in eukaryotes to regulate protein membrane sorting and homeostasis. Many regulators for the conventional trafficking pathway in plants have been identified through reverse genetic studies by searching for the evolutionary counterpart from yeast [1, 2]. Through a study of the FYVE domain–containing proteins in *Arabidopsis*, FREE1 (FYVE domain protein required for endosomal sorting 1) has been identified as a key endosomal sorting complex required for transport I (ESCRT I) component regulating multivesicular body (MVB) biogenesis and vacuole protein trafficking [3–5]. Loss-of-function *free1* T-DNA insertional mutants and dexamethasone (DEX-inducible *FREE1-RNAi* mutants are seedling lethal and

Emmanuel Culetto and Renaud Legouis (eds.), *The ESCRT Complexes: Methods and Protocols*, Methods in Molecular Biology, vol. 1998, https://doi.org/10.1007/978-1-4939-9492-2_20,

display various defects in MVB biogenesis, vacuolar transport, vacuole biogenesis, and autophagic degradation [3, 6]. FREE1 directly interacts with VPS23A, which is a key ESCRT I component, to regulate MVB biogenesis [3]. Also, FREE1 forms a complex with SH3P2 (an autophagosome regulator) and ATG6/VPS30 (a key PI3K component essential for PI3P production) to regulate autophagic degradation [6, 7]. However, it remains elusive how FREE1 performs its multiple functions to regulate endomembrane trafficking pathways. We hypothesize that FREE1 may genetically regulate downstream players to exert its role. In wild-type conditions, FREE1 may inhibit downstream negative regulator X and/or activate downstream positive regulator Y (Fig. 1a). In FREE1-depleted cells, either amplified effects of X or reduced effects of Y would result in defects in FREE1-mediated downstream cellular events, thus causing lethal seedling (Fig. 1b). To identify these potential downstream genetic regulators, we employed the *suppressor of free1* (*sof*) screening approach by using the DEX-inducible *FREE1-RNAi* transgenic plant as the genetic background [8]. We expect that in some *sof* mutants, SOF may function as negative regulator X. In *sof* loss-of-function mutant cells, loss of SOF eliminates the amplified effect of X in free1, so that seedlings grow normally with little FREE1 (Fig. 1c). In other *sof* mutants, SOF may negatively regulate positive regulator Y. In these *sof* loss-of-function mutant cells, loss of SOF releases the inhibition effect on Y, so that seedlings grow normally with little FREE1 (Fig. 1d). Although occurring with a much lower frequency, we also expect that in some *sof* mutants, the *sof* mutation may cause gain of function for SOF. SOF may function as positive regulator Y. In *sof* gain-of-function cells, overactivation of SOF enhances the reduced effect of Y in *free1*, so that seedlings grow normally with reduced protein levels of FREE1 (Fig. 1e).

In the fungi *Aspergillus nidulans*, new players on the ESCRT pathways have been successfully isolated via suppressor screening [9]. However, in plants, it is difficult to perform suppressor screening using lethal mutants. By using DEX-inducible *FREE1-RNAi* plants as genetic background in which to perform mutagenesis, combined with a further FREE1 protein-level detection, we have successfully isolated FREE1-related *sof* mutants [8]. In this chapter, we present a comprehensive workflow for suppressor of *free1* (*sof*) mutants' screening. This protocol is an expansion of that described by Zhao et al. [8].

The whole protocol includes four parts.

1.1 Establishment of the Mutant Pool

The mutant pool is established using the *Arabidopsis thaliana* transgenic line containing a single-insertion dexamethasone-inducible *FREE1-RNAi*. Seeds are treated with ethylmethanesulfonate

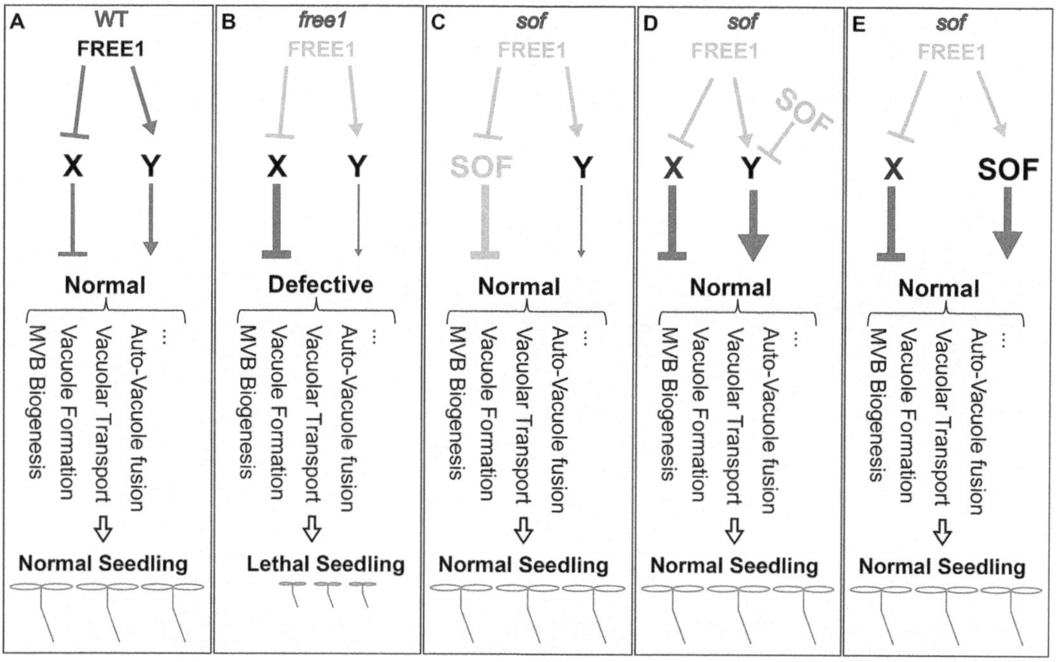

Fig. 1 Working model of SOF as potential FREE1 genetic regulators. FREE1 may genetically regulate downstream player (negative regulator X and/or positive regulator Y) to exert its role in regulating MVB biogenesis, vacuole formation, and autophagosome vacuole fusion. (**a**) In wild-type cells, FREE1 may inhibit downstream negative regulator X and/or activate downstream positive regulator Y. (**b**) In *free1*-depleted cells, either amplified effect of X or reduced effect of Y results in defects in FREE1-mediated downstream cellular events, thus causing lethal seedlings. (**c**) In some *sof* mutants, SOF may function as negative regulator X. In *sof* loss-of-function mutant cells, loss of *SOF* eliminates the amplified effect of X, so that seedlings grow normally with little FREE1. (**d**) In some *sof* mutants, SOF may negatively regulate positive regulator Y. In *sof* loss-of-function mutant cells, loss of SOF releases the inhibition effect on Y, so that seedlings grow normally with little FREE1. (**e**) In some *sof* mutants, *sof* mutation may cause gain of function for SOF. SOF may function as positive regulator Y. In *sof* gain-of-function cells, overactivation of SOF enhances the reduced effect of Y in *free1*, so that seedlings grow normally with little FREE1

(EMS) which induces either G-to-A transition or C-to-T transition to nuclear genomic DNA (Fig. 2). Then, mutagenized seeds are divided into subfamilies, giving M1 plants.

1.2 The sof *Mutant Isolation at M2 Stage*

M2 seeds from each subfamily are collected and treated with dexamethasone to induce FREE1 ds RNA interfering, which leads to nearly an elimination of FREE1 proteins and a seedling lethal phenotype. Survived *sof* (suppressors of *free1*) seedlings are screened using M2 seeds 7 days after germination and then transferred to the soil for setting individual M3 seeds (Fig. 2).

1.3 Identification of FREE1-Related sof *Mutants at M3*

Based on FREE1 protein-level detection using M3 seeds, putative *sof* mutants are further classified into FREE1-related (reduced/barely detection of FREE1 protein) and RNAi-related (no change

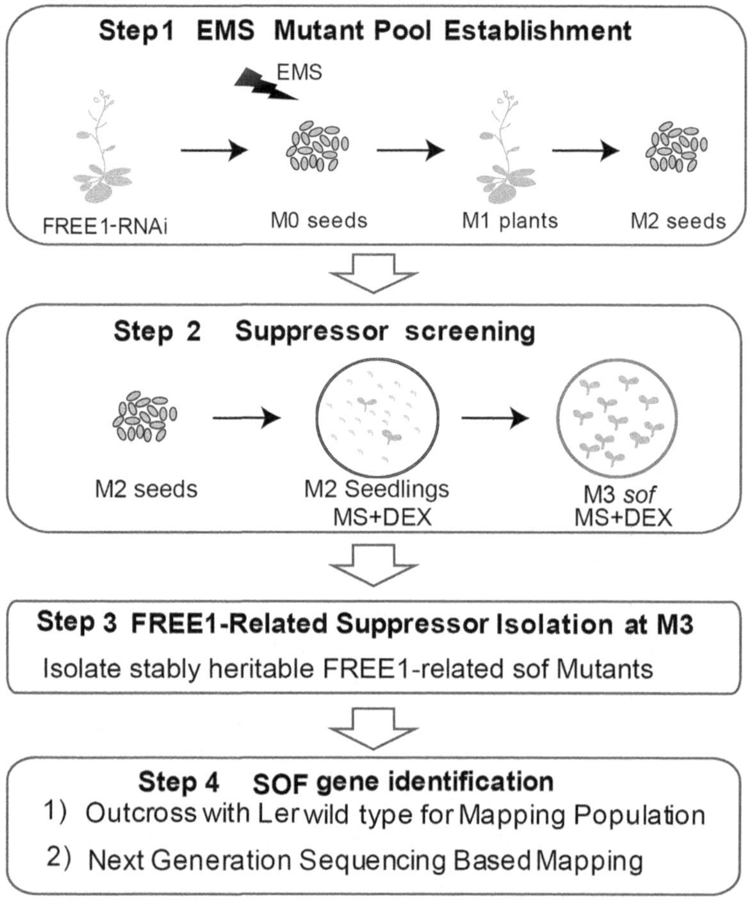

Fig. 2 Overview workflow for screening *sof* mutants. (Step 1) The mutant pool was established using a single-insertion Dex-inducible *FREE1-RNAi* line with EMS treatment. (Step 2) Survived suppressors of free1 (*sof*) seedlings are screened using M2 seeds. (Step 3) Based on FREE1 protein levels, putative M3 seeds are further classified into FREE1-related (reduced/barely detection of FREE1 protein) and RNAi-related (no change of FREE1 protein level) groups. (Step 4) FREE1-related *sof* mutants are outcrossed with *Ler* to establish the mapping populations. Then next generation sequencing (NGS)–based mapping is used for mutated SOF gene identification

of FREE1 protein level) groups. For isolation of FREE1-related *sof* mutants at M3, first we confirm the seedling survival phenotype under DEX induction; only mutants that show heritable survival phenotypes will be processed for subsequent FREE1 protein-level detection. Then we detect FREE1 protein levels using FREE1 antibodies; only mutants showing reduced/barely detection of FREE1 proteins are further studied as FREE1-related *sof* mutants. For these FREE1-related *sof* mutants, we detect the vacuole morphology using FM-dye staining; FREE1-related *sof* mutants are expected to have normal vacuole formation. Then, we observe the

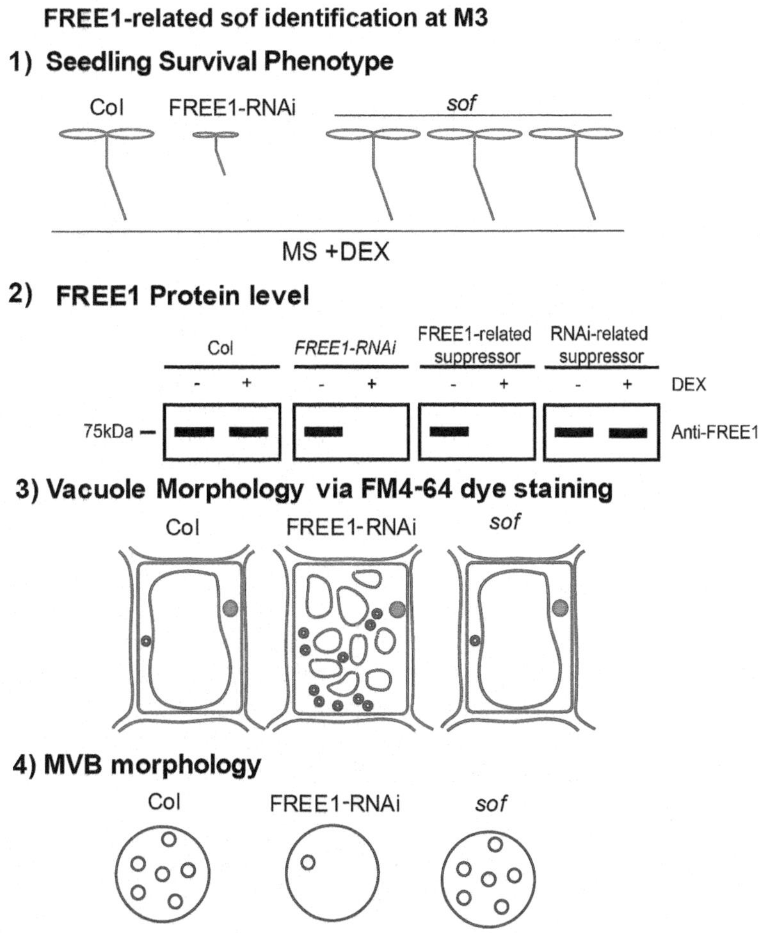

Fig. 3 Identification of FREE1-related *sof* mutants at M3. (1) Confirm seedling survival phenotype under DEX induction. (2) Detect FREE1 protein level using FREE1 antibody, only mutants showing reduced/barely detection of FREE1 protein are further studied as FREE1-related *sof* mutants. (3) Detect vacuole morphology using FM-dye staining, FREE1-related *sof* mutants are expected to have normal vacuole formation. (4) Observe the MVB morphology by using high-pressure freeze and freeze substitution electron microscopy analysis, FREE1-related *sof* mutants are expected to have normal MVB morphology

MVB morphology by using high-pressure freeze and freeze substitution electron microscopy analysis; FREE1-related *sof* mutants are expected to have normal MVB morphology (Fig. 3).

1.4 SOF *Gene Identification*

For mutated SOF gene identification, FREE1-related *sof* mutants are outcrossed with *Landsberg erecta* (*Ler*), and the F1 hybrid seeds self-pollinate to generate F2 seeds. F2 seeds are sown on Murashige and Skoogs (MS) dexamethasone and hygromycin B agar plates. The survived mutant seedlings (3 of 16 F2) are pooled together and used as mapping population (Fig. 4). Then, next-generation

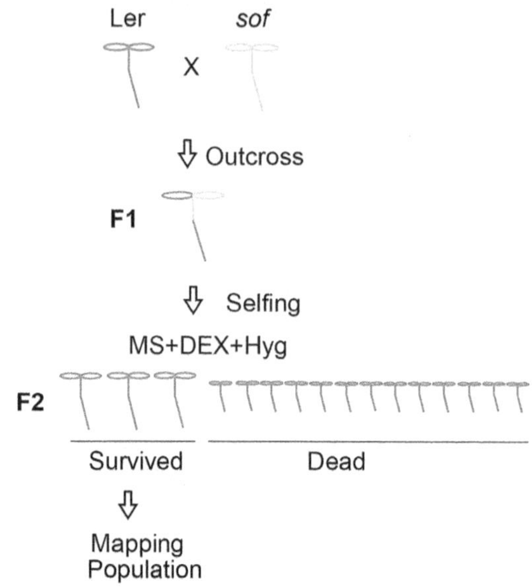

Fig. 4 Schematic illustration of mapping population establishment. FREE1-related *sof* mutants (*Col-0* ecotype) are outcrossed with *Ler* ecotype *A. thaliana*, and the F1 hybrid seeds self-pollinate to generate F2 seeds. F2 seeds are sown on MS dexamethasone and hygromycin B agar plates. Survived mutant seedlings (3 of 16 F2) are pooled together as mapping population

sequencing (NGS)–based mapping is used for mutated SOF gene identification (Fig. 5). The *sof* mutation site (m site) was selected during the mapping sample pooling process. They inherit the parental genomes (Columbia [*Col*] and *Ler*). Resequencing reads are then aligned to *Col* genome using the BWA program [10]. Consensus is called using the SAMtools program based on SNP genetic marker in between *Col* and *Ler* [11, 12]. *Col* allele frequency estimations (AFEs) are calculated and plotted using SHOREmap outcross function [13]. Detailed EMS-induced mutation information is found using SHOREmap annotate function [13, 14].

2 Materials

2.1 Plant Materials

1. *Arabidopsis thaliana* Columbia-0 (*Col*-0) ecotype.
2. *FREE1-RNAi* mutant is generated by transforming pTA7002 vector of which hairpin FREE1 RNAi fragments are driven under the DEX-inducible promoter into the Columbia-0-ecotype plants, and hygromycin-resistant single-insertion *FREE1-RNAi* transgenic plants are used for mutagenesis [3, 8].
3. *Arabidopsis thaliana* Landsberg *erecta* (*Ler*) ecotype.

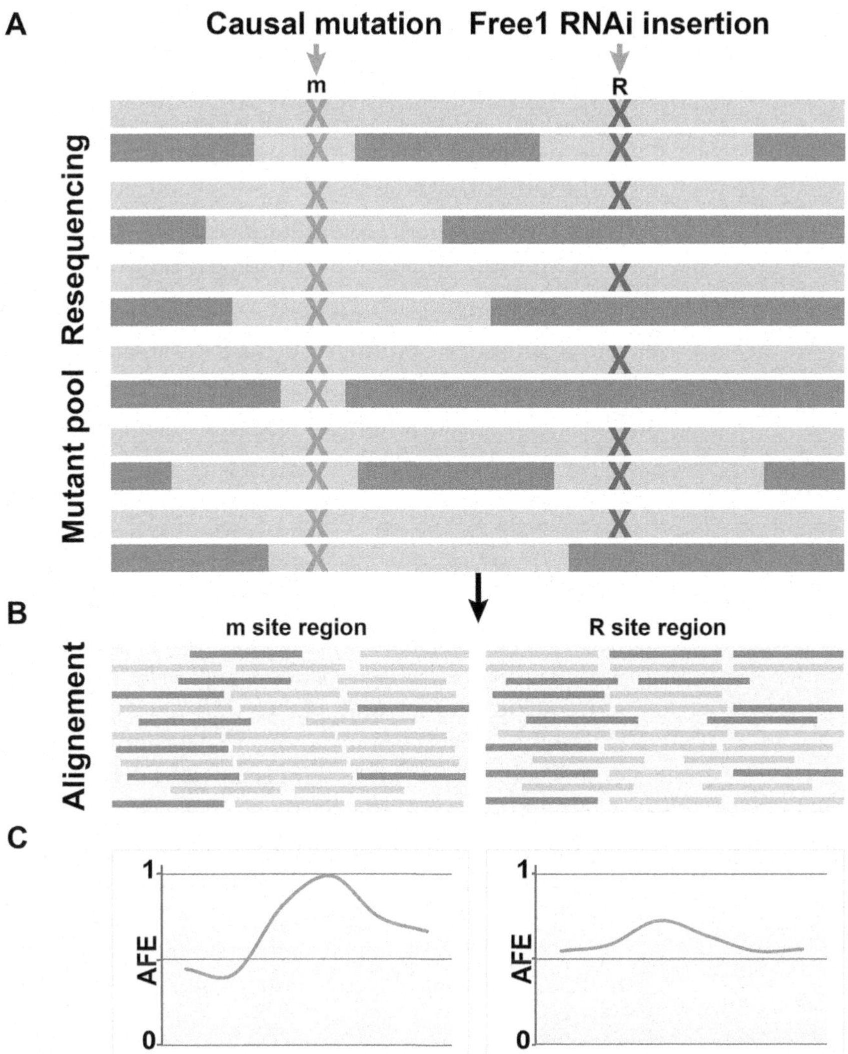

Fig. 5 Schematic illustration of NGS-based mapping. (**a**) Schematic of mapping mutants and whole-genome resequencing. The *sof* mutation site (m site) was selected during the mapping sample pooling process. They are colored after the parental genomes. Reads in dark grey stand for *Ler* genome and reads in light grey stand for *Col* genome. The cross-marks resequencing reads hit along with the genome. (**b**) Simulative reads aligned to m-site region and R-site (T-DNA bearing *FREE1-RNAi* insertion site) region. (**c**) Allele frequency estimations (AFEs) of both m-site region and R-site region. One *Col* allele frequency peak should correspond to the causal m-site region, while a mild *Col* allele frequency peak may appear at R-site region

2.2 Reagents, Buffers, and Molecular Tools

1. Distilled water (dH_2O).
2. Ethyl methanesulfonate (EMS) solution: 0.3% (v/v) in dH_2O. In the hood, add 120 μL of EMS to 40 mL of dH_2O.-Prepare this solution just before using it (EMS, F.W. 124.16; *see* **Note 1**).

3. 0.5 M NaOH: dissolve 20 g NaOH in 500 mL water, let it cool down to room temperature, and fill it up to 1 L.
4. 0.1% liquid agar: 0.1% (w/v) agar in dH_2O. Add 1 g of Bacto agar to 1 L of distilled water (dH_2O), and autoclave the liquid agar solution at 121 °C for 15 min. Cool down to room temperature before use.
5. Seeds' surface sterilization solution: 70% (v/v) ethanol and 0.05% (v/v) Triton-X100 in sterilized dH_2O.
6. 100% ethanol.
7. Dexamethasone solution: 10 mM dexamethasone (DEX) in ethanol.
8. Hygromycin B solution: 50 mg/mL hygromycin B in dH_2O (Bioworld 40810004-2).
9. MS agar medium: 4.33 g/L Murashige and Skoogs (MS) medium, 30 g/L sucrose, adjust to pH 5.8 with KOH, and add 0.8% (w/v) Bacto agar before autoclave. Autoclave the medium at 121 °C for 15 min. Cool the autoclaved medium to about 55 °C before adding heat-labile supplements.
10. MS-dexamethasone agar plates: Cool down the autoclaved MS agar medium to around 55 °C, add the 10 mM DEX stock solution to a final concentration at 10 μM, mix thoroughly, and pour about 20 mL into each plastic petri dish.
11. MS dexamethasone and hygromycin B agar plates: Cool down the autoclaved MS agar medium to around 55 °C, add the 10 mM DEX stock solution to a final concentration at 10 μM, and then add the 50 mg/mL hygromycin B stock solution to a final concentration at 50 μg/mL. Finally, mix the above medium thoroughly and pour ~20 mL into each plastic petri dish.
12. MS liquid medium: 4.33 g/L Murashige and Skoogs (MS) medium, 30 g/L sucrose, and adjust to pH 5.8 with KOH. Autoclave the medium at 121 °C for 15 min. Cool the autoclaved medium to room temperature before use.
13. DNeasy® Plant Mini Kit (Qiagen, Cat No. 69104, USA).
14. Lysis buffer: 25 mmol/L Tris–HCl pH 7.5, 1 mmol/L EDTA, 150 mmol/L NaCl, 1 × Complete Protease Inhibitor Cocktail (Roche, USA), and 1% Triton-X100 in dH_2O.
15. FM4-64 solution: 12 mmol/L in DMSO are prepared and stored at −20 °C. Dilute the FM4-64 with MS liquid medium to the working solution of 12 μmol/L just before use.
16. 5× loading buffer: 5% (v/v) β-mercaptoethanol, 0.02% (w/v) bromophenol blue, 30% (v/v) glycerol, 10% (w/v) sodium dodecyl sulfate (SDS), and 250 mmol/L Tris–Cl, pH 6.8.
17. 100% acetone.

18. Uranyl acetate solutions: 0.1% (w/v) uranyl acetate in acetone.
19. Lowicryl HM20 resin mixture: Monomer E 17.02g, Cross-linker D 2.98 g, and Initiator 0.10 g. The resins should be protected from direct light.
20. Antibodies: rabbit anti-VSR [15]; rabbit anti-cFBPase (Agrisera, AS04 043); rabbit anti-FREE1 [3]; and gold-coupled secondary antibody.

2.3 Equipment and Supplies

1. Bench-top centrifuge (capable of 20,000 × *g* at 4 °C).
2. Water bath or heating block capable of heating at 65 °C and 95 °C.
3. Mortars and pestles.
4. Liquid nitrogen.
5. Vacuum centrifuge.
6. Plastic petri dishes (10 cm diameter).
7. 1.5 mL Eppendorf tubes.
8. Parafilm.
9. Fine vermiculite soil.
10. Large plastic tray (45 cm × 30 cm × 6 cm).
11. 7 oz transparent disposable cups.
12. Leica SP8 laser scanning confocal system.
13. The Leica EM PACT2 high-pressure freezer.
14. Leica EM AFS machine: freeze substitution and low-temperature embedding system for light and electron microscopy AFS2.
15. HPF aluminum planchettes.
16. Stereo microscope.

2.4 Software

1. BWA software (http://bio-bwa.sourceforge.net/).
2. SAMtools software (http://samtools.sourceforge.net/).
3. Picard Tools software (https://broadinstitute.github.io/picard/).
4. SHOREmap software (http://bioinfo.mpipz.mpg.de/shoremap/).

3 Methods

3.1 Sterilization of Arabidopsis Seeds

1. Aliquot about 100 seeds into 1.5 mL autoclaved Eppendorf tube.
2. Place the above seeds in a sterile flow hood, add 1 mL of seeds' surface sterilization solution, and shake well for 10–12 min.

3. Let the tube stand still so that the seeds settle naturally. Then pipette out the liquid carefully.
4. Add 1 mL of 100% ethanol, and shake well for 2 min.
5. Let the tube stand still so that the seeds settle naturally. Then, pipette out the liquid carefully.
6. Dry the seeds using the vacuum centrifuge at room temperature via evaporation of ethanol.

3.2 Plant Growth on Plates

1. Surface sterilize the seeds (following steps in Subheading 3.1).
2. Spray the sterilized seeds on desired MS agar plates in a sterile flow hood.
3. Seal the plates with the parafilm.
4. Put the plates at 4 °C for 2–4 days for stratification.
5. Place the plates horizontally onto the growth shelf with top light (about 150 μE/m^2/s) at 22 °C under a long-day (16 h light/8 h dark) photoperiod.

3.3 Plant Growth in Soil

1. Make 2–3 holes at the bottom of 7 oz transparent disposable cups.
2. Put 20 transparent disposable cups into one large plastic tray.
3. Fill autoclaved wet soil into transparent disposable cups.
4. Plant the selected *Arabidopsis* seedlings into the soil in transparent disposable cups (2–4 individual seedlings each cup).
5. Place the plastic tray-holding seedlings onto the growth shelf with top light (about 150 μE/m^2/s) at 22 °C under a long-day (16 h light/8 h dark) photoperiod.
6. Water the plants by directly adding water into the plastic tray.

3.4 EMS Mutagenesis

1. Soak seeds with good-quality dH_2O at 4 °C. Use 500 μL seeds, which is around 10,000 seeds, and soak them into 40 mL of dH_2O for 60 h (*see* **Note 2**).
2. Shake at room temperature at 140 rpm for 1 h.
3. Wash the seeds twice with 50 mL of dH_2O each time.
4. Treat the seeds with the EMS (0.3%, v/v) solution. Seal well and shake for 8 h at room temperature. Clearly label so that the lab mates know that it is seeds treated with EMS (*see* **Note 3**).
5. Pour off the EMS into a container with 7.5-folds volume of 0.5 M NaOH (*see* **Note 4**).
6. Wash the seeds ten times with 50 mL of dH_2O each time.
7. Resuspend seeds in 50 mL of 0.1% agar and mix every 2 mL with dry fine vermiculite soil.

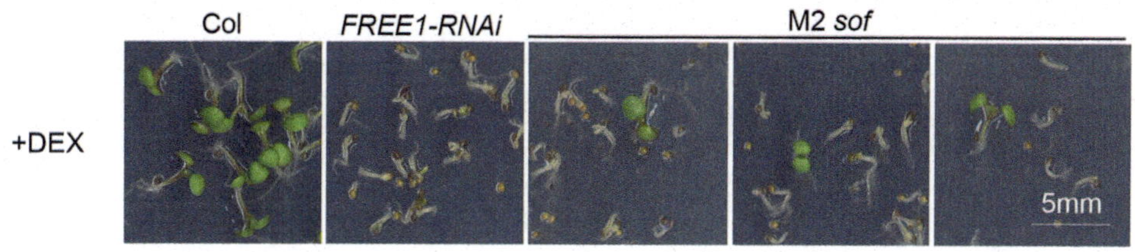

Fig. 6 Example of *sof* mutants' screening at M2. Representative survived *sof* seedlings are indicated with arrows. Survived *sof* seedlings are picked 5 days after germination and then transferred to the soil for setting individual M3 seeds. M2 seeds are plated on MS-dexamethasone agar plates. Scale bar: 5 mm

8. Prepare soil in the large plastic tray (following **steps 1–3** in Subheading 3.3).
9. Spray every 2 mL per large plastic tray.
10. For M2 seeds collection, M_1 plants are divided into about 200 subfamilies (*see* **Note 5**).

3.5 Screening of sof Suppressor at M2 Stage

1. Surface sterilize the M2 seeds (following steps in Subheading 3.1).
2. Grow the sterilized M2 seeds on MS dexamethasone agar plates (following steps in Subheading 3.2).
3. Screen for survived seedlings based on seedling growth phenotypes using 5-day-old seedlings (Fig. 6).
4. Plant selected M2 seedlings in the soil for individual M3 seeds collection following Subheading 3.3 (*see* **Note 6**).

3.6 FREE1-Related sof Mutant Identification at M3 Stage

3.6.1 Seedling Survival Phenotypes

1. Surface sterilize ~100 individual M3 seeds (following steps in Subheading 3.1).
2. Grow half the sterilized M3 seeds on MS agar-medium plates and the other half on MS dexamethasone agar plates (*see* **Note** 7).
3. Check the seedling phenotype with the 5-day-old seedlings by comparing the seedling grown on MS agar-medium plates and on MS dexamethasone agar plates (Fig. 2 *see* **Note 8**).

3.6.2 FREE1 Protein-Level Detection

1. Collect 0.1 g survived 5-day-old *sof* seedlings (*see* Subheading 3.6.1) from MS dexamethasone agar plates into one 1.5 mL Eppendorf tube, label with "+DEX", and place on ice.
2. Collect 0.1 g survived 5-day-old *sof* seedlings (*see* Subheading 3.6.1) from MS agar plates into one Eppendorf tube, label with "−DEX", and place on ice.
3. Grind whole seedling tissues of above-individual *sof* mutants in liquid nitrogen.
4. Add 100 μL of lysis buffer into each tube and mix well.
5. Centrifuge for 15 min at 10,000 × *g* at 4 °C.

6. Transfer 80 μL of the supernatant to a new well labeled 1.5 mL Eppendorf tube, add 20 μL of 5× loading buffer, and boil at 95 °C for 5 min.
7. Load 20 μL of above-boiled protein sample using 10% SDS-PAGE gel and perform Western blot following previous described protocol [16]
8. Hybrid the protein blot with rabbit FREE1 antibodies. Use rabbit cFBPase antibodies as loading control. Only mutants showing reduction or barely detection of FREE1 are considered as FREE1-related *sof* mutants.

3.6.3 Vacuole Morphology Phenotype

1. Wash five seedlings of 5-day-old FREE1-related *sof* mutants grown on MS dexamethasone agar plates in 3 mL MS liquid medium for 15 min.
2. Stain the seedlings for 3 h with FM4-64 working solution by replacing the MS liquid medium with FM4-64 working solution.
3. Wash the seedlings twice with fresh MS liquid medium before confocal imaging.
4. Acquire confocal fluorescence images using the Leica SP8 laser-scanning confocal system with a 63× water lens. Set the emission bandwidth of the PMT to the 580–650-nm range and use a laser of 552 nm for excitation (*see* **Note 9**).

3.6.4 MVB Morphology Phenotype

1. Cut root tips of 5-day-old wild types and *sof* mutants grown on MS dexamethasone agar plates.
2. Place the root tip into HPF aluminum planchettes filled with liquid MS solution. Immediately freeze the planchette sandwich carrying *Arabidopsis* root tips in the high-pressure freezer at 2.1 kbar.
3. Perform freeze substitution in dry acetone containing 0.1% uranyl acetate at −85 °C for 48 h in the AFS freeze-substitution system. Then set the AFS program to warm up from −85 °C to −35 °C at 5 °C h^{-1}.
4. After three 30-min rinses in pure ethanol, infiltrate samples with Lowicryl HM20/ethanol mixture stepwise with HM20 concentration at 33%(v/v) for 1 h, at 66%(v/v) for 1 h, and at 100% for 1 h at −35 °C.
5. Exchange to 100% Lowicryl HM20 resin for 16 h at −35 °C.
6. Wash samples two times with 100% Lowicryl HM20 resin. Separate root tip samples and transfer the samples to flat-bottomed embedding substrates at −35 °C.

7. Polymerize the HM20 resin–embedded sample under UV illumination by replacing the AFS cap with UV illumination at −35 °C.
8. Set the AFS2 chamber to warm up to room temperature once the samples become hard.
9. Perform ultra-sectioning and immunogold labeling with anti-VSR antibodies at 40 μg/mL and gold-coupled secondary antibodies at 1:50 dilution following previous well-described protocols [17].
10. Observe the MVB structures under transmission electron microscopy to see whether MVB phenotype is restored in FREE1-related *sof* mutants (*see* **Note 10**).

3.7 Establishment of Mapping Population

1. Grow both FREE1-related *sof* mutants and *Ler* wild-type plants in good condition at growth room for outcross.
2. Choose an inflorescence and remove all the flowers that are either too young or have already finished self-pollination.
3. Expose the stigma (from either *Ler* or *sof*) as the recipient by carefully removing the sepals, petals, and anthers under stereo microscopes.
4. Brush pollen from fully mature flowers (from either *Ler* or *sof*) to the exposed stigma.
5. Repeat hand pollination several times for a successful outcross.
6. Mark the crossed flowers and obtain F1 hybrid seeds.
7. Grow F1 hybrid seeds to self-pollinate for F2 seeds (*see* **Note 11**).
8. Sow F2 seeds on MS dexamethasone and hygromycin B agar plates.
9. Pool survived mutant seedlings (total seedling number >250) together as mapping population for DNA extraction (*see* **Note 12**).

3.8 Genomic DNA Extraction from Pooled Mapping Population

Genomic DNA was prepared using DNeasy Plant Mini Kit following the manufacturer's instruction with mild modifications.

1. Preheat a water bath or heating block to 65 °C.
2. Harvest the abovementioned survived seedlings (Subheading 3.5, **step 9**) and sop up excess water.
3. Grind the seedlings in liquid nitrogen and then weigh the ground tissue powder, about 800 mg, for each sample. Do not allow the sample to thaw.
4. Add 3.2 mL of buffer AP1 and 32 μL of RNase A. Vortex and incubate for 10 min at 65 °C. Invert the tube 2–3 times during incubation. Centrifuge the lysate for 5 min at 20,000 × *g* (14,000 rpm) (*see* **Note 13**).

5. Transfer the supernatant into a new tube without disturbing the pellet. Add 1040 mL of buffer P3. Mix and incubate for 5 min on ice.
6. Add 1.5 volumes of buffer AW1 and mix by pipetting.
7. Transfer 650 μL of the mixture into a DNeasy Mini spin column placed in a 2 mL collection tube (eight columns will be needed for each sample). Centrifuge for 1 min at ≥6000 × *g* (≥8000 rpm). Discard the flow-through. Repeat this step with the remaining sample.
8. Place the spin column into a new 2 mL collection tube. Add 500 μL of buffer AW2, and centrifuge for 1 min at ≥6000 × *g*. Discard the flow- through. Add another 500 μL of buffer AW2. Centrifuge for 2 min at 20,000 × *g* (8 mL of AW2 will be needed in total) (*see* **Note 14**).
9. Transfer the spin column to a new 1.5 mL microcentrifuge tube.
10. Add 40 μL of 65 °C ddH$_2$O for elution. Incubate for 5 min at room temperature (15–25 °C). Centrifuge for 1 min at ≥6000 × *g*. Repeat once.

3.9 Identification of Mutated SOF Gene

1. Subject extracted DNA (>3 μg) to Illumina library (~300 bp) construction and whole-genome resequencing. Details of how Illumina paired-end sequencing works can be found at the Illumina technology webpage (*see* **Note 15**).
2. Align the qualified sequencing reads against the reference genome sequence (TAIR10) using BWA [10] without mismatches via default parameters.
3. Remove duplicated reads using Picard Tools (see detailed commands in Picard manual) and call consensus using SAMtools program (*see* detailed commands in SAMtools manual) [11].
4. Plot the next generation sequence–based mapping result based on 461,070 SNP markers using SHOREmap outcross function [12, 13]. A detailed procedure on the usage of SHOREmap v3.x has been described [14]. Relatively reliable loci are filtered as below: consensus quality >20 (error rate, 1%) and total depth >5. Only EMS-induced C/G to T/A SNP mutations are further considered as candidate SOF mutations (*see* **Note 16**).
5. Confirm the mutated gene with Sanger sequencing by PCR amplification of the mutated region.

4 Notes

1. Ethyl methanesulfonate can be purchased from Sigma-Aldrich, product number M0880.
2. Before EMS treatment, the *Arabidopsis FREE-RNAi* seeds should be derived from well-cultivated plants. The seeds

should be of high germination rates and should be homozygous.

3. For a saturated screening, one generally needs 500,000 M2 seeds derived from 125,000 M1 seeds.
4. EMS is a powerful mutagen and that is why we use it here. Wear double gloves, lab coat, and safety glasses and always handle it behind the fume hood. Solutions are decontaminated either by disposal onto the solid sodium thiosulfate or 2.5 × vol of 0.5 M NaOH. Let them stand overnight and then dispose down the drain in the hood with 30 min of rinsing afterwards. You may also treat the used EMS solution while complying with the safety and security procedures of your institution.
5. Phenotypes including half-pale leaves should be observed when M1 plants are at 2–4 weeks. Ideally, for M2 seeds collection, the more subfamilies the better.
6. Some mutants die at this stage because some essential genes may get mutated.
7. No less than 30 seeds should be analyzed at this stage.
8. Compared to *FREE1-RNAi* plants, survived seedlings show better growth with green cotyledons and elongated main roots. Only mutants showing heritable and uniform survived seedling phenotypes are further selected as putative *sof* candidates.
9. Both roots and cotyledons should be checked for vacuole morphology phenotype.
10. For identification of MVB, it is easy to identify from the characteristic multivesicular morphology. Classical normal mature MVBs have a round shape with diameters that range from 200 nm to 300 nm, and they are filled with multiple intraluminal vesicles with diameters around 30 nm. Better do immune-labeling using VSR antibody to further confirm the identity of MVB [15].
11. Better grow 8–12 individual F1 seeds in the soil and collect the selfing seeds from individual plants separately. Keep the F1 plant in a good growth condition for high-quality F2 seeds.
12. For higher mapping resolution, better collect a pool of F2 mutants of more than 250 individuals.
13. Do not mix buffer AP1 and RNase A before use.
14. Remove the spin column from the collection tube carefully so that the column does not come into contact with the flow-through.
15. To gain enough material for DNA, better cultivate the isolated F2 mutants from MS + DEX + Hyg onto pure MS plates for another week. The sequencing depth should be more than 15-fold. Detailed sequencing methods can be found at

http://www.illumina.com/technology/next-generation-sequencing/paired-end-sequencing_assay.html

16. For reference genome, the *Col*-0 (TAIR10) sequence was used. Both *sof* and *FREE1-RNAi* are in *Col* background.

Acknowledgments

This work was supported by grants from the Research Grants Council of Hong Kong (CUHK466313 and 14130716 and CUHK2/CRF/11G, C4011-14R, C4012-16E, and AoE/M-05/12), the National Natural Science Foundation of China (31270226 and 31470294), and the Shenzhen Peacock Project (KQTD201101) (to L.J.). The authors declare no conflict of interest.

References

1. Cui Y, Shen J, Gao C, Zhuang X, Wang J, Jiang L (2016) Biogenesis of plant prevacuolar multivesicular bodies. Mol Plant 9(6):774–786. https://doi.org/10.1016/j.molp.2016.01.011
2. Gao C, Zhuang X, Shen J, Jiang L (2017) Plant ESCRT complexes: moving beyond endosomal sorting. Trends Plant Sci 22(11):986–998. https://doi.org/10.1016/j.tplants.2017.08.003
3. Gao C, Luo M, Zhao Q, Yang R, Cui Y, Zeng Y, Xia J, Jiang L (2014) A unique plant ESCRT component, FREE1, regulates multivesicular body protein sorting and plant growth. Curr Biol 24(21):2556–2563. https://doi.org/10.1016/j.cub.2014.09.014
4. Kolb C, Nagel MK, Kalinowska K, Hagmann J, Ichikawa M, Anzenberger F, Alkofer A, Sato MH, Braun P, Isono E (2015) FYVE1 is essential for vacuole biogenesis and intracellular trafficking in Arabidopsis. Plant Physiol 167 (4):1361–1373. https://doi.org/10.1104/pp.114.253377
5. Belda-Palazon B, Rodriguez L, Fernandez MA, Castillo MC, Anderson EA, Gao C, Gonzalez-Guzman M, Peirats-Llobet M, Zhao Q, De Winne N, Gevaert K, De Jaeger G, Jiang L, Leon J, Mullen RT, Rodriguez PL (2016) FYVE1/FREE1 interacts with the PYL4 ABA receptor and mediates its delivery to the vacuolar degradation pathway. Plant Cell. https://doi.org/10.1105/tpc.16.00178
6. Gao C, Zhuang X, Cui Y, Fu X, He Y, Zhao Q, Zeng Y, Shen J, Luo M, Jiang L (2015) Dual roles of an Arabidopsis ESCRT component FREE1 in regulating vacuolar protein transport and autophagic degradation. Proc Natl Acad Sci U S A 112(6):1886–1891. https://doi.org/10.1073/pnas.1421271112
7. Zhuang X, Wang H, Lam SK, Gao C, Wang X, Cai Y, Jiang L (2013) A BAR-domain protein SH3P2, which binds to phosphatidylinositol 3-phosphate and ATG8, regulates autophagosome formation in Arabidopsis. Plant Cell 25 (11):4596–4615. https://doi.org/10.1105/tpc.113.118307
8. Zhao Q, Gao C, Lee P, Liu L, Li S, Hu T, Shen J, Pan S, Ye H, Chen Y, Cao W, Cui Y, Zeng P, Yu S, Gao Y, Chen L, Mo B, Liu X, Xiao S, Zhao Y, Zhong S, Chen X, Jiang L (2015) Fast-suppressor screening for new components in protein trafficking, organelle biogenesis and silencing pathway in *Arabidopsis thaliana* using DEX-inducible *FREE1-RNAi* plants. J Genet Genomics 42(6):319–330. https://doi.org/10.1016/j.jgg.2015.03.012
9. Calcagno-Pizarelli AM, Hervas-Aguilar A, Galindo A, Abenza JF, Penalva MA, Arst HN Jr (2011) Rescue of Aspergillus nidulans severely debilitating null mutations in ESCRT-0, I, II and III genes by inactivation of a salt-tolerance pathway allows examination of ESCRT gene roles in pH signalling. J Cell Sci 124(Pt 23):4064–4076. https://doi.org/10.1242/jcs.088344
10. Li H, Durbin R (2009) Fast and accurate short read alignment with burrows-wheeler transform. Bioinformatics 25(14):1754–1760. https://doi.org/10.1093/bioinformatics/btp324

11. Li H, Handsaker B, Wysoker A, Fennell T, Ruan J, Homer N, Marth G, Abecasis G, Durbin R, Genome Project Data Processing S (2009) The sequence alignment/map format and SAMtools. Bioinformatics 25 (16):2078–2079. https://doi.org/10.1093/bioinformatics/btp352
12. Galvao VC, Nordstrom KJ, Lanz C, Sulz P, Mathieu J, Pose D, Schmid M, Weigel D, Schneeberger K (2012) Synteny-based mapping-by-sequencing enabled by targeted enrichment. Plant J 71(3):517–526. https://doi.org/10.1111/j.1365-313X.2012.04993.x
13. Schneeberger K, Ossowski S, Lanz C, Juul T, Petersen AH, Nielsen KL, Jorgensen JE, Weigel D, Andersen SU (2009) SHOREmap: simultaneous mapping and mutation identification by deep sequencing. Nat Methods 6 (8):550–551. https://doi.org/10.1038/nmeth0809-550
14. Sun H, Schneeberger K (2015) SHOREmap v3.0: fast and accurate identification of causal mutations from forward genetic screens. Methods Mol Biol 1284:381–395. https://doi.org/10.1007/978-1-4939-2444-8_19
15. Tse YC, Mo B, Hillmer S, Zhao M, Lo SW, Robinson DG, Jiang L (2004) Identification of multivesicular bodies as prevacuolar compartments in *Nicotiana tabacum* BY-2 cells. Plant Cell 16(3):672–693. https://doi.org/10.1105/tpc.019703
16. Chen L, Li F, Xiao S (2017) Analysis of plant autophagy. Methods Mol Biol 1662:267–280. https://doi.org/10.1007/978-1-4939-7262-3_24
17. Karahara I, Kang BH (2014) High-pressure freezing and low-temperature processing of plant tissue samples for electron microscopy. Methods Mol Biol 1080:147–157. https://doi.org/10.1007/978-1-62703-643-6_12

Chapter 21

Screening of Interactions with the ESCRT Machinery by a *Gaussia princeps* Split Luciferase-Based Complementation Assay

Rina Barouch-Bentov, Yves Jacob, and Shirit Einav

Abstract

The endosomal sorting complex required for transport (ESCRT) machinery comprises five complexes that act sequentially to recruit and cluster transmembrane cargo proteins (ESCRT-0), drive membrane curving (ESCRT-I and II), catalyze fission of cargo-containing vesicles (ESCRT-III and VPS/VTA1), and disassemble and recycle the ESCRT-III complex (VPS/VTA1). Since interactions between ESCRT components and cellular or microbial proteins are typically weak, transient, and involve membrane proteins, they are often difficult to study by standard technologies. Here, we describe the utility of high-throughput protein-fragment complementation assays based on the reconstitution of a split luciferase reporter to screen for interactions between any protein and a library of ESCRT proteins in mammalian cells and provide a detailed protocol for these assays.

Key words ESCRT, Proteomics, Protein-protein interactions, Protein-fragment complementation assays

1 Introduction

The ESCRT is a mobile machinery that is recruited to various cellular membranes to mediate cargo recognition, sorting, and membrane sculpting in various cellular processes, such as multivesicular bodies (MVBs) biogenesis, cell abscission, exosome secretion, and autophagy [1]. In addition, the ESCRT machinery plays a critical role in the life cycle of multiple viruses [2] and other intracellular pathogens, such as mycobacteria tuberculosis [3].

Most standard proteomic technologies are somewhat limited in their ability to map interactions between cargo proteins and the ESCRT machinery. Among the common limitations is poor sensitivity, particularly in detecting weak and/or transient interactions, such as typical cargo-ESCRT interactions, with dissociation constants [K_ds] in the micromolar range [4, 5]. This is associated with limited overlap between sets of interactions independently

Emmanuel Culetto and Renaud Legouis (eds.), *The ESCRT Complexes: Methods and Protocols*, Methods in Molecular Biology, vol. 1998, https://doi.org/10.1007/978-1-4939-9492-2_21,

measured [6, 7]. The reported specificity is low as well [6, 7]. Moreover, the ability to detect interactions involving membrane proteins by most of the standard methods is limited [8]. This is relevant for both cellular transmembrane cargo proteins and RNA viruses, such as hepatitis C virus (HCV), whose proteins are largely transmembrane. Novel proteomic technologies are therefore required to map cargo-ESCRT interactions with high fidelity.

Protein-fragment complementation assay (PCA) is an advanced proteomic technology that overcomes these challenges. This assay is based on the fusion of bait (A) and prey (B) proteins to two complementary fragments of a reporter protein, such as the enzymes dihydrofolate reductase, β-lactamase, tobacco etch virus (TEV) protease, luciferase, or fluorescent proteins [9–11]. If proteins A and B bind each other, the reporter fragments are brought into close proximity, reassemble, and fold into the native structure, leading to reconstitution of the reporter activity. Protein-protein interactions (PPIs) are detected by measuring the reconstituted activity of the reporter via the corresponding signal, for example, luminescence or fluorescence. PCAs have been used to study PPIs in various models including cells, bacteria, plants, and animals [12, 13].

We have recently adapted a PCA format that is based on reconstitution of the humanized *Gaussia princeps* luciferase (Fig. 1a) and used it to screen for interactions between the ESCRT machinery and the proteome of hepatitis C virus (HCV), an RNA virus from the Flaviviridae family [14]. This assay strategy was developed in the Michnick laboratory [12, 15] and further optimized for high-throughput application at the Jacob laboratory [16]. Isolated from the marine copepod *Gaussia princeps*, this luciferase reporter is a monomeric protein of 185 aa (19.9 kDa) that catalyzes the oxidation of coelenterate luciferin (coelenterazine) substrate in a reaction that emits blue light with a peak at 480 nm [17]. Coelenterazine effectively permeates cell membranes and diffuses into all intracellular compartments [15]. This assay can therefore be conducted in any biologically relevant cell type and allows detection of PPIs, including those involving membrane proteins, in their appropriate subcellular compartment (unlike most yeast two-hybrid systems where the interaction occurs in the nucleus) [14, 16, 18]. Moreover, the humanized form of *Gaussia princeps* luciferase (hGLuc) generates a bioluminescent signal that is 100-fold higher than the humanized forms of the *Photinus pyralis* (firefly) and *Renilla* luciferases [15]. In addition, since hGLuc is among the smallest known coelenterazine-utilizing luciferases, its fusion minimizes alterations in protein folding relative to larger reporters. The optimal design of a short (20 aa), flexible polypeptide linker between the protein of interest and the reporter fragment (Fig. 1b) further facilitates proper protein folding and enables binding independently of the size of the interacting proteins

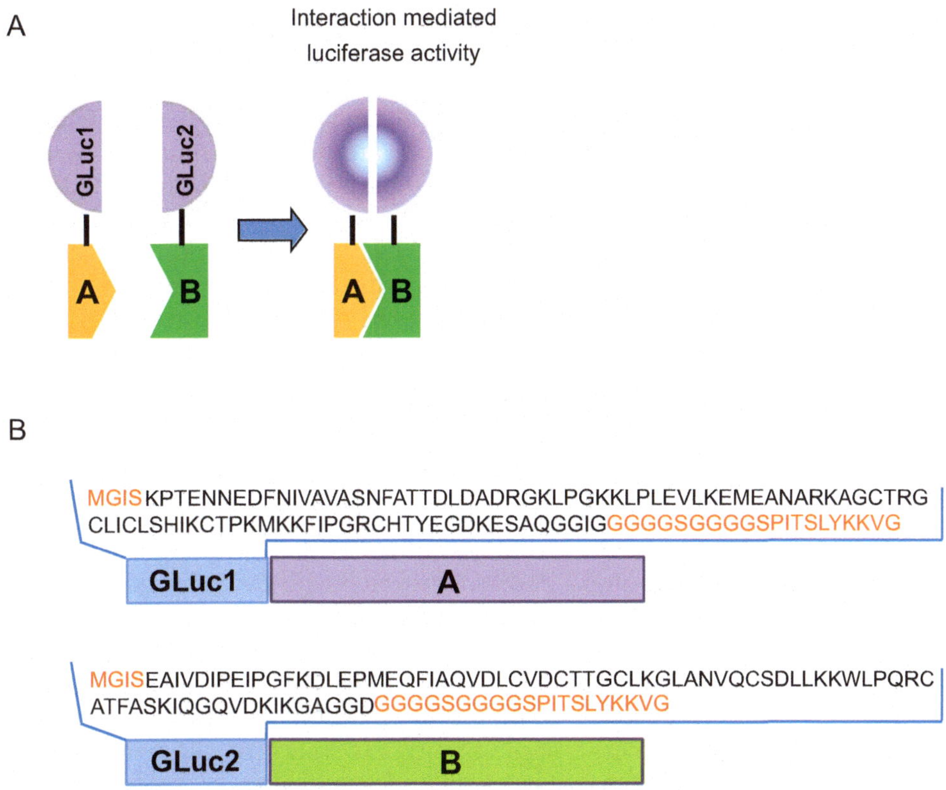

Fig. 1 Schematics of the PCA format (**a**) and the constructs used (**b**). The letters A and B represent protein pairs tested for interaction fused to the GLuc1 and GLuc2 luciferase fragments. The GLuc1 and GLuc2 *Gaussia* luciferase fragments used in the pSPICA-N1 and pSPICA-N2 plasmids, respectively, are displayed in the lower panel. The amino acid sequence of the reporter tag (black) and flexible linker (orange) are indicated. Panel B was adapted from [16]

[16]. Together, these properties enable measurements of PPIs with a high sensitivity and very high reproducibility. Since the activation of the reporter does not occur spontaneously but is rather dependent on the binding of the prey and bait proteins, either directly or via a bridging protein(s), the specificity of this PCA format is similarly high.

Another important quality of the reconstituted hGLuc reporter is its reversibility. This property eliminates trapping of protein complexes, thereby allowing studies of PPI dynamics as well as kinetics and equilibrium of protein complex assembly and disassembly in living cells [12]. This reversibility also facilitates monitoring the effects of pharmacological inhibitors and signals, such as nutritional, environmental, or hormonal [12], which induce or suppress PPIs [18]. Additionally, the strength of the signal positively correlates with the affinity of the binding and represents apparent affinity. This important feature can be utilized to precisely map the interaction surfaces and characterize the mechanism of binding and its regulation, such as via introduction of point mutations within the interacting proteins or depletion of cellular regulatory proteins [14, 18].

Lastly, the design of Gateway-compatible (pSPICA-N1 and pSPICA-N2) plasmids that introduce N-terminal hGLuc fusion [16] enables large-scale cloning via recombination of individual open reading frames (ORFs) derived from ORF libraries, such as the human ORFeome library [19]. Such high-throughput *Gaussia princeps* protein-fragment complementation assays (HT-GPCAs) are performed in 96-well plates and provide results within 24 h post-transfection [16]. The throughput of these assays, combined with relatively easy and quick performance, makes them attractive for screening interactions involving targeted protein collections, such as that of the ESCRT machinery.

The results of HT-GPCAs are expressed as a normalized luminescence ratio (NLR), which is the ratio between the signal in cells transfected with the pair of the studied interactors divided by the average signal in cells transfected with the individual proteins and the reciprocal empty GLuc plasmid. Importantly, the accuracy and sensitivity of HT-GPCA screens can be benchmarked by analysis of a random reference set (RRS) and a positive reference set (PRS) composed of multiple non-interacting and interacting human proteins, respectively. These two reference sets have demonstrated a clear separation based on NLR values [14, 16]. We have shown that this PCA format provides a high fidelity means to measure weak and transient interactions, such as between cargo and ESCRT components or clathrin adaptor proteins [14, 18].

We have recently assembled a library of 24 ESCRT proteins by recombining the relevant ORFs from the human ORFeome library into the pSPICA-N1 plasmid for fusion with one fragment of the hGLuc reporter [14] (Fig. 2). Eight HCV proteins were recombined into the pSPICA-N2 plasmid, encoding the second reporter fragment [14]. We used HT-GPCAs to screen for interactions between the ESCRT library and the HCV proteome. This screen revealed nine novel interactions [14]. Hepatocyte growth factor-regulated tyrosine kinase substrate (HRS), an ESCRT-0 complex component, bound four of the eight studied HCV proteins with the highest apparent affinity, thereby emerging as an important entry point for HCV into the ESCRT pathway. The interactions of HRS with the transmembrane HCV nonstructural (NS) 2 and membrane-bound NS5A proteins were validated in membrane fractions derived from HCV-infected cells by co-immunoprecipitation assays [14], thereby supporting the high specificity of the PCA platform for the detection of interactions, including those involving transmembrane proteins. The same PCA platform was then used to decipher the mechanism by which HCV proteins bind HRS. Specifically, point mutations within the ubiquitin interacting motif (UIM) of HRS disrupted binding to HCV proteins,

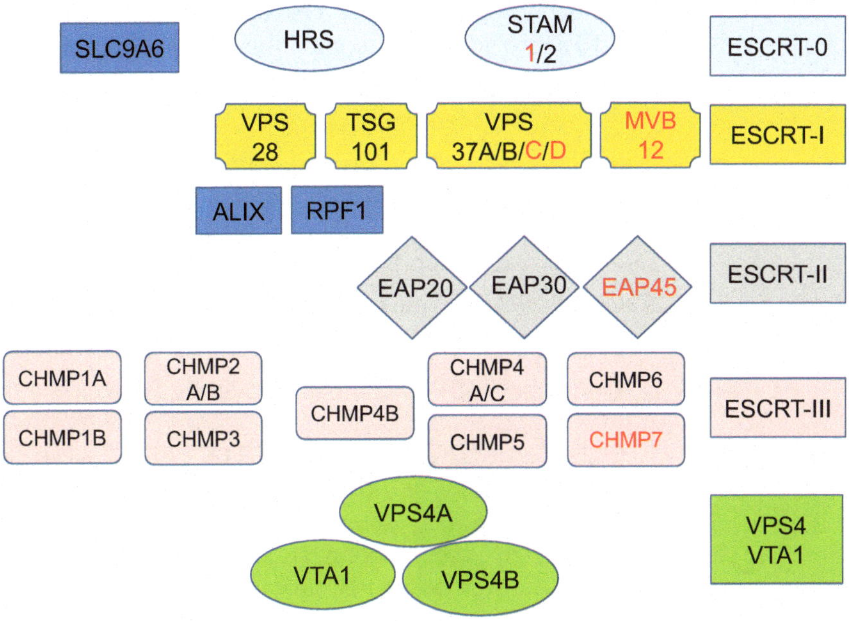

Fig. 2 Schematics of the ESCRT complex network used in the PCA screen. ESCRT components labeled in red were unavailable in the ORFeome library (version V5.1). Blue rectangles represent accessory ESCRT proteins

suggesting a role for ubiquitination in mediating HRS binding [14]. These results led to the discovery that ubiquitination of NS2 mediates binding of HRS and that the NS2-HRS interaction is essential for the budding of HCV into intracellular compartments [14].

Taken together, our studies demonstrate the advantages of the PCA platform for mapping PPIs and its utility for screening of interactions with the ESCRT machinery, deciphering binding mechanisms, and elucidating novel roles of PPIs in the life cycle of viruses. While we have thus far utilized this platform to screen for ESCRT interactions with viral proteins, this platform can be used to screen for interactions with other microbial or cellular proteins [20–22].

2 Materials

2.1 Proteomic Library Assembly

1. Open reading frames (ORFs) of ESCRT proteins: Gateway-compatible pDonor plasmids harboring ORFs of 24 distinct ESCRT components are available in the human ORFeome library [19] (Fig. 2) (*see* **Note 1**).
2. ORFs of studied protein(s): ORFs of cellular or microbial protein(s) to be screened for interactions with ESCRT proteins.

3. Primers: Forward and reverse primers for cloning the studied ORFs into a Gateway-compatible pDonor vector (*see* **Note 2**).
4. pDonor vector: pDonor207 (Invitrogen).
5. Gateway®-compatible destination vectors: pSPICA-N1 and pSPICA-N2 vectors encode a protein of interest fused in its N terminus to a flexible 20 amino acid polypeptide hinge and one of the two fragments of the humanized *Gaussia princeps* luciferase reporter, GLuc1 and GLuc2, respectively [16, 20].
6. Gateway™ BP Clonase™ Enzyme Mix (Life Technologies).
7. Gateway™ LR Clonase™ Enzyme Mix (Life Technologies).
8. One Shot® TOP10 competent cells (Invitrogen).

2.2 Transfections

1. Cells: Grow HEK-293 T cells (ATCC) in Dulbecco's modified Eagle's medium (DMEM) supplemented with 10% fetal calf serum, 2 mM L-glutamine, 100 units/ml penicillin, and 100 μg/ml streptomycin at 37 °C with 5% CO_2 and 95% humidity. Split cells three times a week (*see* **Notes 3** and **4**).
2. Transfection reagent: Linear polyethylenimine HCl (PEI MAX 40000) (Polysciences Inc). Prepare a 1 mg/ml stock solution according to the manufacturer's instructions. Briefly, weigh 500 mg of PEI powder, mix in 450 ml of water, stir until powder is completely dissolved, and adjust the pH to 7 by adding 10 M NaOH. Add water for a final concentration of 1 mg/ml and filter via a 0.22 μm membrane (*see* **Note 5**).
3. Plates: Use 96-well PCR plates for mixing the PEI with plasmid DNA and a PCR plate-sealing film to cover these plates. Use white-bottom 96-well plates with lids for seeding cells.
4. Medium for DNA mixtures: DMEM with no supplements.
5. M-Per protein extraction reagent (Thermo Fisher Scientific).
6. 4–12% Bis-Tris gels.
7. PVDF membranes.
8. Blocking buffer (5% milk, 5% bovine serum albumin (BSA), or 1% casein (*see* **Note 6**)).
9. Anti-GLuc antibody (New England BioLabs).
10. Anti–β-actin antibody.
11. Horseradish peroxidase (HRP)-conjugated secondary antibody.

2.3 Luciferase Assays

1. Phosphate buffer saline (PBS, 137 mM NaCl, 10 mM phosphate, and 2.7 mM KCl, pH of 7.4) supplemented with 1 mm of $CaCl_2$ and 1 mm $MgCl_2$ (*see* **Note** 7).
2. Lysis buffer: *Renilla* lysis buffer (Promega).

3. Luciferase reagent: *Renilla* luciferase assay reagent (Promega).
4. Luminescence plate reader: Tecan multimode plate readers (Infinite® M1000 PRO) or others.

3 Methods

3.1 Assembly of ESCRT and Studied ORF Libraries

1. To clone the ESCRT library plasmids (pSPICA-N1(GLuc1)-ESCRT), recombine the ORFs encoding ESCRT proteins from the pDonor plasmid individually into the pSPICA-N1 plasmid using the Gateway™ LR Clonase™ Enzyme Mix (*see* **Note 8**). Transform 10 μl of the reaction into 50 μl One Shot® TOP10 competent cells and amplify the plasmids by mini preparation.
2. To clone the studied protein(s) plasmids (pSPICA-N2(GLuc2-studied ORF)), amplify the ORF(s) of interest via PCR to introduce flanking attB recombination sites (*see* **Note 2** for primers). Recombine the PCR product(s) individually into the pDonor207 vector using the Gateway™ BP Clonase™ Enzyme Mix (*see* **Note 9)**. Transform and amplify the plasmid(s) as described above. Recombine the studied ORF(s) into the pSPICA-N2 vector using the Gateway™ LR Clonase™ Enzyme Mix (*see* **Note 8**). Transform and amplify the plasmid(s) as described above (*see* **Note 10**).
3. To confirm recombination of the right ORF and at the right orientation, perform standard sequencing of the expression plasmids (*see* **Note 11**).
4. To determine protein expression, transfect the individual plasmids encoding the ESCRT and studied proteins into HEK-293T cells (as described below) and incubate for 24 h at 37 °C. Lyse the cells in M-Per protein extraction reagent. Run cell lysates on 4–12% Bis-Tris gels, transfer onto PVDF membranes, block the membrane with a blocking buffer, and blot with anti-GLuc and anti-β-actin antibodies (*see* **Note 12**). Detect signal with horseradish peroxidase (HRP)-conjugated secondary antibodies (1:3000 dilution, 1 h incubation at room temperature or overnight incubation at 4 °C). Quantify band intensity with ImageJ software (NIH) (*see* **Note 13**).

3.2 Transfection (Fig. 3)

1. To determine optimal PEI to DNA ratio, transfect a plasmid encoding a fluorescent tag (e.g., GFP) at various PEI to DNA ratios ranging from 2.5:1 (2.5 μg PEI: 1 μg DNA) to 3.5:1 (3.5 μg PEI: 1 μg DNA) (*see* **Note 14**). Assess protein expression level by a fluorescent microscope. This step has to be performed only once upon preparation of a new PEI stock.

2. Seed 35,000 HEK-293T cells per well in 96-well plates in triplicates, add growth medium to a total volume of 100 μl, and incubate at 37 °C for 24 h (*see* **Note 15**).
3. Prepare DNA mixtures, each sufficient for transfection into three wells, by preparing mixtures of the following pair-wise plasmid(s): (1) individual GLuc2-studied ORF(s) and individual GLuc1-ESCRT ORFs; (2) individual GLuc2-studied ORF (s) and control empty GLuc1; and (3) GLuc1-ESCRT and control empty GLuc2. To do so, for each well dilute 100 ng of each plasmid in 6 μl of DMEM (with no supplements) per well (*see* **Note 16**). Dispense 6 μl of the individual GLuc1 plasmids and 6 μl of the individual respective GLuc2 plasmids per well in PCR 96-well plates (*see* **Note 17**).
4. Prepare the PEI solution by mixing the PEI with DMEM to a final volume of 8 μl/well (*see* **Note 18**).
5. Add 8 μl of the PEI solution to each well in the 96-well plates containing the DNA mixtures with a multichannel pipette. Pipette up and down several times and incubate at room temperature for 20 min (*see* **Note 19**).
6. Pipette the DNA/PEI mixtures (volume: 20 μl per well) onto the cells (*see* **Note 20**) and incubate at 37 °C for 24 h.

3.3 Luciferase Assays

1. Aspirate the media from all wells.
2. Wash wells once with 100 μl of PBS (with calcium and magnesium) and aspirate.
3. Add 40 μl of the Renilla lysis buffer and place the plate on a shaker for 30 min at room temperature.
4. Measure *Renilla* luciferase activity with a Luminometer reader by injecting 50 μl of the Renilla luciferase assay reagent into the cell lysates and counting luminescence over 10 s (*see* **Note 21**).

3.4 Data Analysis

Data from two or three independent PCA experiments (each in triplicates) should be combined (*see* **Note 22**).

1. Calculate normalized luminescence ratio (NLR) for each studied interaction pair. Divide the luminescence signal in cells co-transfected with GLuc1-ESCRT and GLuc2-studied ORF by the average signal in wells co-transfected with GLuc1-ESCRT and an empty GLuc2 plasmid and those co-transfected with GLuc2-studied ORF and an empty GLuc1 plasmid (Fig. 3).
2. To benchmark the accuracy and sensitivity of the screen, consider analyzing a random reference set (RRS) of interactions (*see* **Note 23**). Calculate the means and standard deviations (SDs) of log-transformed NLRs of the studied set and RRS of interactions. Calculate z scores for each interaction by

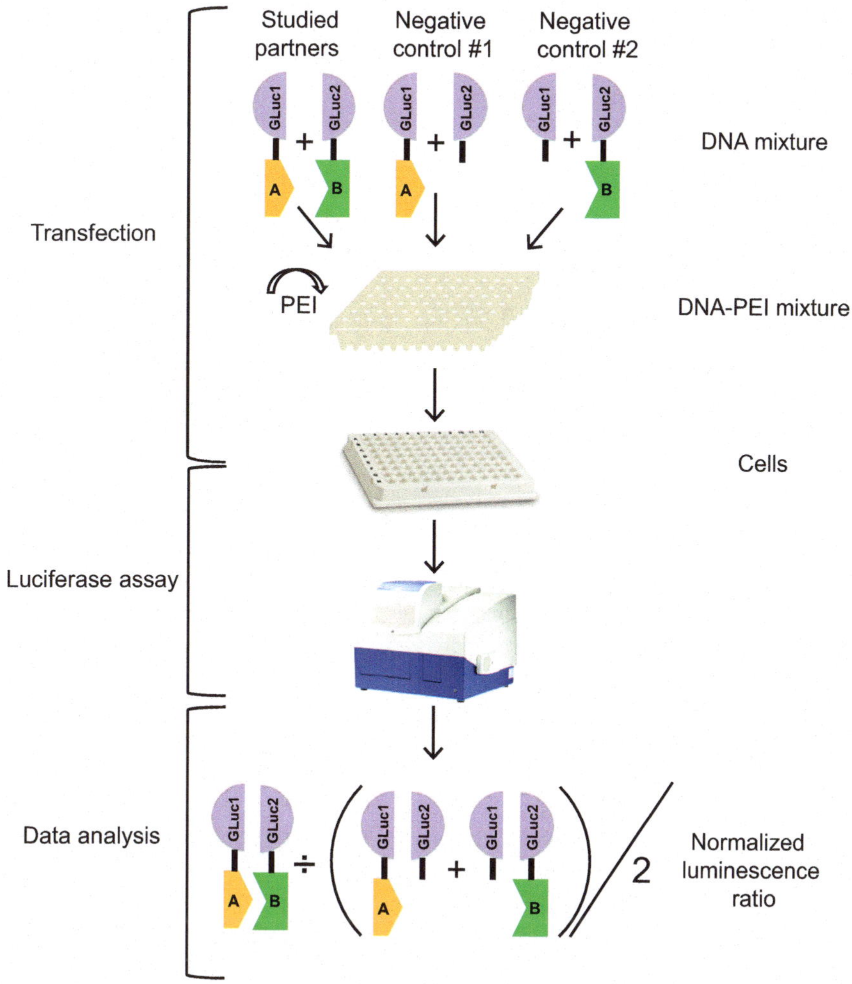

Fig. 3 Schematics of the PCAs' protocol

subtracting the mean value of the RRS from the interaction signal and dividing the resulting value by the SD of the RRS. Assess the separation between the mean *z* score values of the studied set and the RRS by a histogram distribution curve and select an SD cutoff value as the threshold to define positive interactions (Fig. 4) (*see* **Note 24**). As an alternative to the RRS, include 5–10 human proteins not known to interact with the studied protein in the screen and use this NLR data to calculate *z* scores, as described above.

3. Present the data by a histogram distribution curve (Fig. 4), heat map, and/or interaction network [14].

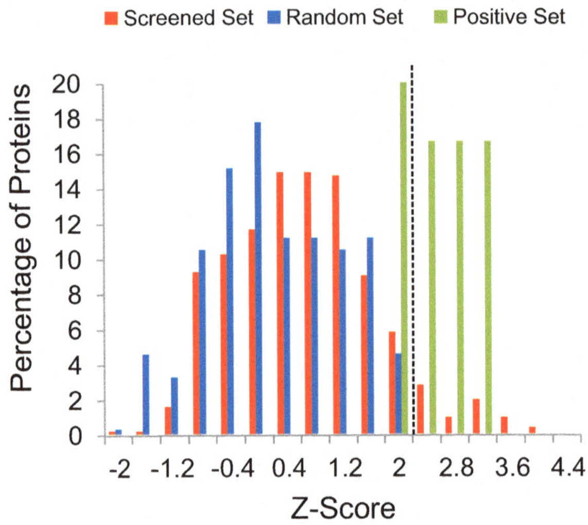

Fig. 4 Interactions between HCV proteins and ESCRT components measured by PCAs (adapted from [14]). Histogram of the mean *z* score values of the screened set, random reference set (RRS), and positive reference set (PRS) of interactions obtained from three independent experiments. The dotted line defines the cutoff used for positive interactions

4 Notes

1. We selected ESCRT ORFs from the human ORFeome library (version V5.1) (http://horfdb.dfci.harvard.edu/hv5/). Additional ESCRT components may be available in newer releases of this library. Other sources for Gateway-compatible ORFs include GeneCopoeia and Addgene.
2. Sequences of forward and reverse primers for cloning the studied ORF(s) into a Gateway-compatible pDonor vector should contain 18–25 gene-specific nucleotides (X) in frame with the attB1 or attB2 sequence, respectively, as follows: (1) attB1-forward primer: 5′GGGGACAACTTTGTACAAAAAAGTT GGCATGXXXX3′ and (2) attB1-reverse primer: 5′GGGAC AACTTTGTACAAGAAAGTTGGGTATTAXXXX3′.
3. Avoid passaging HEK-293 T cells for longer than 3 weeks since the efficiency of transfection is reduced over time. Passage the cells at least once upon thawing a new cell vial. Split the cells 24 h prior to seeding them for PCAs.
4. The use of a more biologically relevant mammalian cell line should be considered (e.g., Huh7.5 (human hepatoma cells) for the study of ESCRT interactions with HCV proteins). Nevertheless, protein expression levels may vary in different cell lines.

5. Aliquot the PEI stock in freezing tubes and store for longer-term use at −80 °C. A "working stock" could be stored at −20 °C. The reagent can be frozen and thawed multiple times.
6. The manufacture instructions for anti-GLuc antibody recommend using 50% milk and 50% BSA (5%) as a blocking buffer. Nevertheless, in our hands this antibody works equally well with various blocking buffers, including 5% milk, 5% BSA, or 1% casein.
7. Calcium and magnesium are used here to reduce cell detachment during the washing step.
8. In a PCR tube, mix 4 μl of buffer 5×, destination vector (200 ng), 1 μl of the pDONOR–ORF plasmid (300 ng), and 12 μl H_2O. Vortex the LR-Clonase Enzyme twice for 2 s and add 2 μl to the reaction mixture. Swirl the mix with a tip and incubate at 25 °C (in a PCR machine) for 2 h. Use 3 μl of this reaction to transform 50 μl chemically competent cells, such as One Shot® TOP10. Following transformation, add 250 μl SOC medium, shake bacteria for 1 h at 37 °C, and spin down at maximum speed for 1 min. Discard the supernatant, resuspend the pellet with the spent medium left in the tube, and plate all on 100 μg/ml ampicillin agar plates.
9. In a PCR tube, mix 7 μl PCR product with 1 μl of the pDonor plasmid (150 ng). Vortex the BP-Clonase II Enzyme, mix twice for 2 s, and add 2 μl to the reaction mixture. Swirl the mix with a tip and incubate at 25 °C (in a PCR machine) for 1–2 h. Use 5–10 μl of the reaction to transform 50 μl chemically competent cells, such as One Shot® TOP10. Following transformation, add 250 μl SOC medium, shake bacteria for 1 h at 37 °C, and spin down at maximum speed for 1 min. Discard the supernatant, resuspend the pellet with the spent medium left in the tube, and plate all on 10 μg/ml gentamycin plates.
10. Use gentamycin and chloramphenicol for selection when amplifying the pDonor207 plasmid and ampicillin and chloramphenicol when amplifying the pSPICA plasmids.
11. It is important to confirm that the ORF is recombined into the expression vector at the right orientation, as in ~5% of the cases, an opposite orientation is detected.
12. The recommended dilution of the anti-GLuc antibodies for Western blotting is 1:3000.
13. A method for using ImageJ to quantitate bands' intensity of Western blots is described here: blots http://www.lukemiller.org/ImageJ_gel_analysis.pdf
14. In our hands, the optimal PEI to DNA ratio has been 3:1.

15. To account for differences between cell lines, we propose assessing cellular viability 24 h post-transfection by incubating the cells for 2–4 h at 37 °C in the presence of 10% alamarBlue reagent (TREK Diagnostic Systems). Detect fluorescence using a reader, such as FLEXstation II 384 (Molecular Devices, Inc.). Once the assay is optimized, there is no need to repeat this procedure.
16. Ideally, prepare a stock solution based on the total number of wells, for example, 1 μg of DNA in 60 μl DMEM for 10 wells plus ~20% extra for volume loss during pipetting.
17. The DNA mixtures can be prepared in advance and stored for several days at −20 °C.
18. Prepare a stock solution based on the number of wells, for example, for ten wells with a PEI to DNA ratio of 3:1, with 0.2 μg of DNA/well (=2 × 100 ng), use 6 μg of PEI (=0.6 μg × 10) in a final volume of 80 μl DMEM (=6 μl of PEI in 74 μl of DMEM). Prepare a stock that is ~20% larger than needed to account for volume loss during pipetting.
19. The PEI must be added to the mixtures right before the transfection and cannot be stored at −20 °C along with the DNA mixtures.
20. There is no need to aspirate the medium from the well. Gently add 20 μl of the PEI/DNA mixture into the 100 μl complete medium already present in the well. Do not mix once the PEI/DNA mixture is added to the cells.
21. Luminescence must be read immediately after cell lysis. The signal dramatically drops if cell lysates are incubated for longer or stored at −20 °C or −80 °C.
22. While the relative light units (RLUs) may vary between experiments, the NLRs are typically very reproducible, with Pearson correlation coefficients of ~0.99 [20].
23. Examples of RRSs composed of 50–100 non-interacting human protein pairs were previously described [14, 16].
24. We typically observe a clear separation between the studied set and the RRS (with P values lower than 10^{-6} by a student t test). We selected a standard deviation cutoff value of >2.2 (corresponding to an NLR of >25) as the threshold to define positive interactions (Fig. 4).

References

1. Henne William M, Buchkovich Nicholas J, Emr Scott D (2011) The ESCRT pathway. Dev Cell 21(1):77–91. https://doi.org/10.1016/j.devcel.2011.05.015
2. Votteler J, Sundquist Wesley I (2013) Virus budding and the ESCRT pathway. Cell Host Microbe 14(3):232–241. https://doi.org/10.1016/j.chom.2013.08.012

3. Mehra A, Zahra A, Thompson V, Sirisaengtaksin N, Wells A, Porto M, Köster S, Penberthy K, Kubota Y, Dricot A, Rogan D, Vidal M, Hill DE, Bean AJ, Philips JA (2013) *Mycobacterium tuberculosis* type VII secreted effector EsxH targets host ESCRT to impair trafficking. PLoS Pathog 9(10):e1003734. https://doi.org/10.1371/journal.ppat.1003734
4. Hirano S, Kawasaki M, Ura H, Kato R, Raiborg C, Stenmark H, Wakatsuki S (2006) Double-sided ubiquitin binding of Hrs-UIM in endosomal protein sorting. Nat Struct Mol Biol 13(3):272–277
5. Raiborg C, Bache KG, Gillooly DJ, Madshus IH, Stang E, Stenmark H (2002) Hrs sorts ubiquitinated proteins into clathrin-coated microdomains of early endosomes. Nat Cell Biol 4(5):394–398. http://www.nature.com/ncb/journal/v4/n5/suppinfo/ncb791_S1.html
6. Parrish J, Gulyas K, Finley R (2006) Yeast two-hybrid contributions to interactome mapping. Curr Opin Biotechnol 17 (4):387–393
7. Cusick ME, Klitgord N, Vidal M, Hill DE (2005) Interactome: gateway into systems biology. Hum Mol Genet 14:R171–R181. https://doi.org/10.1093/hmg/ddi335
8. Garbis S, Lubec G, Fountoulakis M (2005) Limitations of current proteomics technologies. J Chromatogr A 1077(1):1–18
9. Remy I, Michnick SW (2004) A cDNA library functional screening strategy based on fluorescent protein complementation assays to identify novel components of signaling pathways. Methods 32(4):381–388
10. Capdevila-Nortes X, López-Hernández T, Ciruela F, Estévez R (2012) A modification of the split-tobacco etch virus method for monitoring interactions between membrane proteins in mammalian cells. Anal Biochem 423 (1):109–118. https://doi.org/10.1016/j.ab.2012.01.022
11. Remy I, Michnick SW (1999) Clonal selection and in vivo quantitation of protein interactions with protein-fragment complementation assays. Proc Natl Acad Sci U S A 96 (10):5394–5399
12. Remy I, Michnick SW (2007) Application of protein-fragment complementation assays in cell biology. BioTechniques 42(2):137–145
13. Morell M, Ventura S, Avilés FX (2009) Protein complementation assays: approaches for the in vivo analysis of protein interactions. FEBS Lett 583(11):1684–1691. https://doi.org/10.1016/j.febslet.2009.03.002
14. Barouch-Bentov R, Neveu G, Xiao F, Beer M, Bekerman E, Schor S, Campbell J, Boonyaratanakornkit J, Lindenbach B, Lu A, Jacob Y, Einav S (2016) Hepatitis C virus proteins interact with the endosomal sorting complex required for transport (ESCRT) machinery via ubiquitination to facilitate viral envelopment. MBio 7(6). https://doi.org/10.1128/mBio.01456-16
15. Remy I, Michnick S (2006) A highly sensitive protein-protein interaction assay based on Gaussia luciferase. Nat Methods 3 (12):977–979
16. Cassonnet P, Rolloy C, Neveu G, Vidalain PO, Chnatier T, Pellet J, Jones L, Muller M, Demeret C, Gaud G, Vuillier F, Lotteau V, Tangy F, Favre M, Jacob Y (2011) Benchmarking a luciferase complementation assay for detecting protein complexes. Nat Methods 8 (12):990–992
17. Tannous BA, Kim D-E, Fernandez JL, Weissleder R, Breakefield XO (2005) Codon-optimized gaussia luciferase cDNA for mammalian gene expression in culture and in vivo. Mol Ther 11(3):435–443. https://doi.org/10.1016/j.ymthe.2004.10.016
18. Neveu G, Barouch-Bentov R, Ziv-Av A, Gerber D, Jacob Y, Einav S (2012) Identification and targeting of an interaction between a tyrosine motif within hepatitis C virus core protein and AP2M1 essential for viral assembly. PLoS Pathog 8(8):e1002845
19. Rual JF, Hirozane Kishikawa T, Hao T, Bertin N, Li S, Dricot A, Li N, Rosenberg J, Lamesch P, Vidalain PO, Clingingsmith T, Hartley J, Esposito D, Cheo D, Moore T, Simmons B, Sequerra R, Bosak S, Doucette-Stamm L, Le Peuch C, Vandenhaute J, Cusick M, Albala J, Hill D, Vidal M (2004) Human ORFeome version 1.1: a platform for reverse proteomics. Genome Res 14 (10B):2128–2135
20. Neveu G, Cassonnet P, Vidalain P-O, Rolloy C, Mendoza J, Jones L, Tangy F, Muller M, Demeret C, Tafforeau L, Lotteau V, Rabourdin Combe C, TravÃƒÂ© G, Al D, Hill D, Vidal M, Favre M, Jacob Y (2012) Comparative analysis of virus-host interactomes with a mammalian high-throughput protein complementation assay based on *Gaussia princeps* luciferase. Methods 58(4):349–359
21. Hill SJ, Rolland T, Adelmant G, Xia X, Owen MS, Dricot A, Zack TI, Sahni N, Jacob Y, Hao T, McKinney KM, Clark AP, Reyon D, Tsai SQ, Joung JK, Beroukhim R, Marto JA, Vidal M, Gaudet S, Hill DE, Livingston DM (2014) Systematic screening reveals a role for BRCA1 in the response to transcription-

associated DNA damage. Genes Dev 28 (17):1957–1975. https://doi.org/10.1101/gad.241620.114

22. Sahni N, Yi S, Taipale M, Fuxman Bass JI, Coulombe-Huntington J, Yang F, Peng J, Weile J, Karras GI, Wang Y, Kovács IA, Kamburov A, Krykbaeva I, Lam MH, Tucker G, Khurana V, Sharma A, Liu Y-Y, Yachie N, Zhong Q, Shen Y, Palagi A, San-Miguel A, Fan C, Balcha D, Dricot A, Jordan DM, Walsh JM, Shah AA, Yang X, Stoyanova A, Leighton A, Calderwood MA, Jacob Y, Cusick ME, Salehi-Ashtiani K, Whitesell LJ, Sunyaev S, Berger B, Barabási A-L, Charloteaux B, Hill DE, Hao T, Roth FP, Xia Y, Walhout AJM, Lindquist S, Vidal M (2015) Widespread macromolecular interaction perturbations in human genetic disorders. Cell 161(3):647–660. https://doi.org/10.1016/j.cell.2015.04.013

Chapter 22

RNA Interference-Mediated Inhibition of ESCRT in Mammalian Cells

Katherine Bowers

Abstract

Specific depletion of proteins from cultured cells using RNA interference (RNAi) has been a useful technique in assessing protein function for many years. RNAi allows the degradation of specific, targeted mRNA, allowing the effects of protein depletion on cellular processes to be examined. Here, I present a protocol for the depletion of proteins from cultured HeLa cells and list specific reagents and considerations for targeting the endosomal sorting complexes required for transport (ESCRT).

Key words ESCRT, Endosomes, siRNA, RNAi, Protein depletion

1 Introduction

The depletion of ESCRT proteins using RNA interference (RNAi) has been a widely used technique over the past 17 years to investigate the function of ESCRT complexes in mammalian cells. Using RNAi, we have studied the role of ESCRT0, I, II, and III, as well as associated proteins, in the trafficking of the epidermal growth factor receptor (EGFR), on overall endosomal morphology and in the fate of virally ubiquitinated major histocompatibility complex (MHC) class I [1, 2]. We have also used the same techniques to show that the sodium/proton exchanger NHE8 plays a role in endosomal protein trafficking at multivesicular bodies [3].

RNAi can be used to silence genes with the transfection of short, interfering RNA duplexes (siRNA). The siRNAs are 21 base-pair duplexes that are a perfect match to the target sequence [4, 5]. The siRNAs allow an RNA-induced silencing complex (RISC) to bind to the target mRNA, leading a component of RISC, argonaute 2 (ago-2), to unwind the siRNA [6]. The sense strand of the siRNA is degraded and activated RISC, with the antisense strand of the siRNA targeted to the complementary mRNA [7]. The mRNA of the gene of interest is degraded, leading

Emmanuel Culetto and Renaud Legouis (eds.), *The ESCRT Complexes: Methods and Protocols*, Methods in Molecular Biology, vol. 1998, https://doi.org/10.1007/978-1-4939-9492-2_22, © Springer Science+Business Media, LLC, part of Springer Nature 2019

to a depletion of protein levels [7–10]. RNAi therefore provides a simple way of depleting a protein of interest and studying the resultant phenotype. However, there are many factors that must be considered in setting up any RNAi experiment. Firstly, how effective will the siRNA depletion be? This can obviously be checked by real-time (RT) quantitative PCR (to assess mRNA levels) and western blot (to study protein levels). To maximize the chance of effective silencing, a mix of four different siRNA duplexes is often used. Secondly, how specific will the process be, using the chosen siRNA duplexes or pool of duplexes? This second consideration is perhaps the most difficult to address. siRNA and micro RNAs (small, endogenous, regulatory RNAs) share the same RNAi effector, ago-2, and thus small RNAs loaded onto RISC may perform the functions of siRNA and micro RNA. Micro RNA suppresses partially complementary mRNA sequences, and thus transfection of siRNA may result in many off-target effects via silencing unintended and partially complementary target sequences [11]. Steps can be taken to minimize and control for off-target effects. In addition, new techniques are being developed to minimize or eliminate off-target effects, as reviewed in [11].

Recent advances in gene-editing techniques (e.g., the clustered regulatory interspaced short palindromic repeats [CRISPR] and CRISPR-associated protein 9 [Cas9], CRISPR/Cas9 system) have led to the development of other ways to knock out or mutate ESCRT genes in mammalian tissue culture cells. CRISPR/Cas9 systems are being refined and improved and are already a very powerful tool for the analysis of gene function in mammalian systems. Indeed, CHMP2B-, TSG101-, VPS36-, and ALIX-edited cell lines have been generated via CRIPSR/Cas9 [12–16]. A recent review of RNAi compared to CRISPR/Cas9 can be found here [17]. When choosing a method for ESCRT analysis, your own facilities, budget, time, and ultimate goals will need to be considered carefully. RNAi has the advantage of being quick and simple, with many reagents commercially available. Gene editing by CRISPR/Cas9 is likely to be more complicated in standard laboratory cell lines (e.g., HeLa) that have chromosomal duplications, depletions, or rearrangements, making RNAi a more straightforward choice. However, off-target effects are a major concern with RNAi, and controls to validate any results must be performed. Here, I provide a detailed protocol for siRNA depletion of ESCRTs, as used in our studies [1, 2]. This procedure is based on that used by Motley et al. [18]. It is our experience that siRNA depletion of ESCRT proteins is most efficient using a "2 hit" protocol. In summary, the cells are plated on day 0, with transfections on days 1 and 3 and harvesting of the cells for further experiments on day 5 (but *see* **Note 1**). This protocol is for the transfection of HeLa M cells [19] and should give enough transfected cells to perform initial western blots and immunofluorescence experiments. The protocol can be scaled up or down as necessary for different experiments.

Before starting an siRNA protocol, it is extremely important to think about the controls required. Initial experiments should include controls to confirm that any effect of the transfection is due to protein depletion and not a nonspecific effect of the transfection procedure itself. This can be assessed by performing a mock transfection (i.e., performing the transfection procedure exactly as stated but with RNase-free water in place of the siRNA oligonucleotide duplexes) or by purchasing specialized negative control oligonucleotides, designed to determine the baseline cellular response that occurs when transfecting nonspecific siRNA oligonucleotides. You may also consider a positive control for the siRNA-transfection procedure and for this I recommend either using one of the oligonucleotides or pools of oligonucleotides listed here (Table 1) or purchasing a control set of oligonucleotides (e.g., targeting cyclophilin B or GAPD). Once initial experiments confirm a depletion of the protein of interest, validation of any associated phenotype by controlling for off-target effects is essential.

If possible, it is best to start with a pool of siRNA oligonucleotide duplexes, as this maximizes the chances that the protein of interest will be depleted [20]. In addition, the concentration of each individual duplex within the pool is lower (usually one-fourth concentration) than would normally be used for individual duplexes. This lowers the risks of off-target effects, as these increase with the concentration of each duplex [21]. A table of pools and individual oligonucleotide duplexes are provided for ESCRTs and related proteins (Table 1).

2 Materials

2.1 Cell Growth Medium for Routine Cell Culture and Selection of Stable Cell Lines (See Note 2)

1. Routine culture medium for HeLa M cells: Roswell Park Memorial Institute (RPMI 1640) medium containing 2 g/L sodium bicarbonate, supplemented with 2 mM L-glutamine (*see* **Note 3**), 100 units/mL penicillin, 0.1 mg/mL streptomycin (*see* **Note 4**), and 10% (v/v) fetal bovine serum (FBS) (*see* **Note 5**).

   Store at 4 °C.

   Alternative routine culture medium for HeLa M cells: Dulbecco's modified eagle's medium (DMEM) with 4.5 g/L glucose, 0.11 g/L sodium pyruvate, and 3.7 g/L sodium bicarbonate, supplemented with 2 mM L-glutamine (*see* **Note 3**), 100 units/mL penicillin, 0.1 mg/mL streptomycin (*see* **Note 4**), and 10% (v/v) fetal bovine serum (FBS) (*see* **Note 5**).

   Store at 4 °C.

Table 1
siRNA oligonucleotide sequences targeting ESCRTs and other relevant genes (human sequences) that we have used

Gene	Type of oligonucleotide	Target sequence	5′ modification	3′ modification	Dharmacon catalogue number	Accession number	Reference
ESCRT 0							
HRS (HGS)	ON-TARGET plus SMART pool		Phosphate	UU	L-016835-00	NM_004712	[2]
ESCRT I							
TSG101 (VPS23)	Custom	CCTCCAGTCTTCTC TCGTC	None	dTdT	N/A	NM_006292	[1, 5, 25]
	siGENOME SMART pool		Phosphate	UU	M-003549-01		[2]
ESCRT II							
VPS22 (EAP30, SNF8)	Custom SMART pool	Sequences unknown	Phosphate	UU	[a]	NM_001317192	[1]
VPS25 (EAP20)	Custom SMART pool	Pool of sequences 1–4 listed below	Phosphate	UU	[a]	NM_032353	[1]
	Custom SMART pool #1	CAGAACAACTCCG TCTTTA	Phosphate	UU			[1]
	Custom SMART pool #2	GCACAAGGCCGAGA TCATC	Phosphate	UU			[1]
	Custom SMART pool #3	GGGAAACTCATCTA TCAGT	Phosphate	UU			[1]
	Custom SMART pool #4	GTCGATCCAGATTG TATTA	Phosphate	UU			[1]
VPS36 (EAP45)	Custom SMART pool	Sequences unknown	Phosphate	UU	[a]	NM_001282168	[1]
ESCRT III							

VPS20 (CHMP6)	siGENOME SMART pool	Pool of sequences 1–4 listed below	Phosphate	UU	[a]	NM_024591	[2]
	siGENOME SMART pool #1		Phosphate	UU	D-005060-01		[2]
	siGENOME SMART pool #2	GGACAAGGCCATCC TGCAA	Phosphate	UU	[a]		[2]
	siGENOME SMART pool #3		Phosphate	UU	D-005060-03		[2]
	siGENOME SMART pool #4		Phosphate	UU	D-005060-04		[2]
VPS24 (CHMP3)	siGENOME SMART pool	Pool of four oligonucleotides	Phosphate	UU	M-004696-01	NM_001005753	[2]
VPS2A (CHMP2A)	siGENOME SMART pool	Pool of four oligonucleotides	Phosphate	UU	M-020247-00	NM_014453	[2]
VPS2B (CHMP2B)	siGENOME SMART pool	Pool of four oligonucleotides	Phosphate	UU	M-004700-00	NM_001244644	[2]
ESCRTIII-associated proteins							
Alix (AIP1, PDCD6IP)	Custom	GCCGCTGGTGAAG TTCATC	Phosphate	dTdT	N/A	NM_001162429	[1, 26, 27]
De-ubiquitinating enzymes							
AMSH (STAMBP)	Custom	TTACAAATCTGCTG TCATT	Phosphate	dTdT	N/A	NM_006463	[28]
UBPY (USP8)	siGENOME SMART pool	Pool of four oligonucleotides	Phosphate	UU	[a]	NM_001128610	[1]

All siRNA duplexes were purchased from Dharmacon (Horizon Discovery). The order numbers are given for the oligonucleotides available. For those that have been discontinued or were custom orders ([a]), the target sequence is given.

2. Dulbecco's phosphate buffered saline (DPBS): 8 g sodium chloride, 0.2 g potassium phosphate, monobasic, 1.15 g sodium phosphate, dibasic, and 0.2 g potassium chloride. Made up to 1 L with distilled water and sterilized.
   Store at 4 °C.
3. Hank's balanced salt solution with phenol red: 1 g/L D-glucose, 400 mg/L KCl, 60 mg/L KH_2PO_4, 8 g/L NaCl, 350 mg/L $NaHCO_3$, 48 mg/L Na_2HPO_4 (anhydrous), and 10 mg/L phenol red (sodium salt).
4. Trypsin solution: 0.5 g/L porcine trypsin and 0.2 g/L tetrasodium EDTA in Hank's balanced salt solution with phenol red.
   Store at 4 °C.
5. Geneticin antibiotic (G418 sulfate): 50 mg/mL stock in distilled water, filter- sterilized (*see* **Note 6**). This is used for the maintenance of stable cell lines generated using vectors with neomycin resistance genes (e.g., pIRESneo2).
   Store at 4 °C.

2.2 siRNA Transfection Reagents and Oligonucleotides

1. Oligofectamine 2000 reagent and OptiMEM I reduced serum medium with L-glutamine and phenol red (*see* **Note** 7).
   Store at 4 °C.
2. 20% FBS medium: RPMI or DMEM routine culture medium (Subheading 2.1, **item 1**) with L-glutamine but with an increase of the FBS to 20% and without antibiotics (*see* **Note 8**).
3. Prepare solutions A and B for the transfection protocol, as shown in Table 2. Volumes for different sizes of culture dish are given.
4. siRNA is targeted against ESCRT components (Table 1). siRNA is supplied dry and should be made up at 50 μM in RNase-free water. Aliquot and store at −20 °C.

2.3 Solutions for Fixation and Cell Lysis

1. PBS: 137 mM NaCl, 2.7 mM KCl, 10 mM Na_2HPO_4, and 2 mM KH_2PO_4, in distilled water (*see* **Note 10**).
2. Fixative: 3% (w/v) paraformaldehyde in PBS (*see* **Note 11**).
3. Cell lysis solution: 0.5% (v/v) Nonidet P-40, 150 mM NaCl, 2 mM EDTA, and 20 mM Tris, pH 8. This solution is ideally made fresh and chilled on ice before use. Add 1× protease inhibitors (*see* **Note 12**) prior to use.

Table 2
Solutions for siRNA transfection of HeLa M cells

Dish size (area)	Solution A		Solution B		Additions after 20 min	Additions after 4 h
	OptiMEM I (μL)	siRNA, 50 μM stock (μL)[a]	OptiMEM I (μL)	Oligo-fectamine (μL)	OptiMEM I (mL)	20% FBS medium (mL)
One well of 6-well plates (8.87 cm^2)	166	4	20	10	0.8	1
6 cm (21 cm^2)	332	8	40	20	1.6	2
10 cm (58 cm^2)	830	20	100	50	4.0	5

[a]This concentration of siRNA gives a final concentration of an oligonucleotide pool of 100 nM (25 nM of each oligonucleotide duplex), after all additions (*see* **Note 9**)

3 Methods

Work in a tissue culture hood, using sterile equipment and good sterile technique throughout.

3.1 Routine Cell Culture

1. Trypsinize cells from a 10 cm tissue culture dish as follows: Warm all media to 37 °C (*see* **Note 13**). Remove medium from cell culture plate, wash cells gently with 5 mL DPBS, remove DPBS, and add 2 mL trypsin solution. Incubate at 37 °C until the cells round up and begin to float off the dish (approximately 5 min). Add 8 mL routine culture medium and gently pipette up and down to mix and break up clumps.
2. To new 10 cm dishes, add 1–2 mL cell suspension and make up to 10 mL with culture medium (*see* **Note 14**). If the cell line is a stable, G418-resistant line, add G418 as appropriate (*see* **Note 6**).

3.2 Day 0: Plating the Cells Prior to Transfection

1. Trypsinize cells (Subheading 3.1, **step 1**).
2. Count the cells using a hemocytometer following the manufacturer's instructions. Plate 1.5×10^5 cells per well of a 6-well plate, in a total volume of 2 mL routine culture medium (*see* **Note 15**). Set up one well per transfection, and incubate overnight in a standard tissue culture incubator at 37 °C with 5% CO_2. Transfection volumes for other-sized dishes are given in Table 1 for reference.

3.3 Day 1: siRNA Transfection 1

1. For the first transfection, make up enough solution B (Table 2) for all the transfections (*see* **Note 16**), and incubate for 5 min at room temperature.

2. Make up solution A in individual sterile tubes. Use tubes that can take a total volume of at least 1 mL.
3. Add solution B to solution A tubes and mix by gentle pipetting. Incubate at room temperature for 20 min.
4. Toward the end of the 20 min incubation in **step 3**, wash the cells once with 1.5 mL prewarmed DPBS to remove traces of FBS (*see* **Note 17**).
5. Add 0.8 mL OptiMEM I to the mixed solutions A and B from **step 3**, and mix by gentle pipetting.
6. Remove the DPBS from the cells and add the whole mix from **step 5**. Incubate for 4 h at 37 °C in a 5% CO_2 incubator.
7. Add 1 mL 20% FBS medium and rock the plate very gently to mix. Incubate overnight at 37 °C, 5% CO_2.

3.4 Days 2 and 3: Split Cells and Complete siRNA Transfection 2

1. In the morning, replace the transfection medium on the cells with routine culture medium (containing antibiotics, as necessary).
2. In the afternoon of day 2 (4–5 h after **step 1**), trypsinize the cells as described in Subheading 3.1, **step 1**, using 1.5 mL DPBS, 0.5 mL trypsin solution, and 2.5 mL routine culture medium. Transfer all cells into 6 cm culture dishes and add a further 2 mL routine culture medium. Rock the plates very gently to mix and then incubate overnight at 37 °C, 5% CO_2.
3. On day 3, repeat the transfection procedure as in Subheading 3.3, using the volumes in Table 1 for 6 cm dishes.

3.5 Day 4: Plate Out Cells

1. Replace the transfection medium on the cells with routine culture medium (containing antibiotics, as necessary).
2. After 4–5 h, wash the cells once with 2.5 mL prewarmed DPBS, trypsinize as in Subheading 3.1, **step 1**, using 1 mL trypsin solution, and resuspend with 1 mL routine growth medium (to give 2 mL cell suspension). Add 100 μL cell suspension per well to a 4-well (or 24-well) plate (growth area: 1.82 cm^2) containing a sterile glass coverslip for immunostaining (*see* **Note 18**). Add 1 mL routine culture medium per well. The remaining cell suspension can be added to a 6 cm culture dish and the volume made up to 5 mL with routine culture medium.

3.6 Day 5: Test Efficiency of Knockdown (See Note 19)

1. Fix cells on coverslips as follows (*see* **Note 20**): Wash three times with PBS (1 mL per wash), incubate in fixative for 15 min at room temperature, and wash twice further with PBS. Continue with a standard procedure for immunostaining adherent, fixed cells to localize the protein(s) of interest [1, 22].

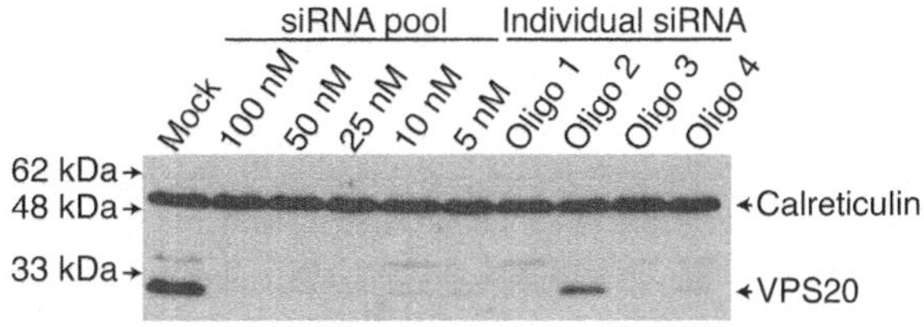

Fig. 1 RNAi depletion of VPS20 (CHMP6) in HeLa M cells. The double-transfection RNAi protocol described was used, with either a pool of siRNAs targeting VPS20 at varying concentrations or each of the individual siRNAs at 25 nM (*see* Table 1). Mock-transfected cells are included for comparison. A total of 33 μg protein lysate was loaded per lane on SDS-PAGE, the gel blotted to nitrocellulose, and the resulting western blot cut between the 33 kDa and 48 kDa markers. The upper half was probed with anti-calreticulin as a loading control and the lower half with anti-VPS20 [2]

2. Wash cells in 6 cm dishes three times in 5 mL cold PBS, on ice. Remove PBS and add 2.5 mL cold PBS. Using a plastic cell scraper, scrape the cells into the PBS and transfer to a 15 mL tube. Further add 2.5 mL PBS and scrape the cells again, adding this to the same tube (5 mL cell suspension). Spin down cells gently at 4 °C for 5 min at a 1000 × *g spin*. Remove PBS and resuspend the cells in cold cell lysis solution. Incubate 15 min on ice. Remove cell debris and nuclei by a further spin at 16,000 × *g* for 20 min at 4 °C, and remove the post-nuclear supernatant (lysate) to a fresh tube. Discard the pellet. Carry out a protein assay on the lysate (*see* **Note 21**), and load equal amounts of protein per lane for SDS-polyacrylamide gel electrophoresis (PAGE) and western blot, using standard procedures [1, 22], (*see* Fig. 1).

3.7 Validation (See Note 22)

1. Repeat the siRNA transfections using individual oligonucleotides (if previously a pool was used) (*see* **Note 23** and Fig. 1).
2. Titrate the siRNA oligonucleotides to the lowest possible concentration that still produces the phenotype, to minimize off-target effects (*see* **Note 24** and Fig. 1).
3. Rescue the phenotype seen using an siRNA-resistant version of the cDNA of interest (*see* **Note 25**). To do this, choose one siRNA oligonucleotide (usually the most efficient for knockdown). Generate a mutant cDNA by site-directed mutagenesis with silent mutations (i.e., mutations that do not change the amino acid sequence in the protein) along the length of the oligonucleotide binding site (*see* **Note 26**). This mutated version of the cDNA can then be transfected into the cells prior to siRNA transfection (we would normally create a stable cell line expressing the mutant protein). Repeat siRNA transfections in the cell line expressing the siRNA-resistant mRNA and determine whether the phenotype seen is corrected or "rescued" by the exogenous protein.

4 Notes

1. For TSG101 (VPS23, part of ESCRTI), transfections were carried out on days 2 and 3 to avoid the extensive cell death seen after the longer transfection protocol. For VPS2A (CHMP2A, part of ESCRTIII), only one transfection was carried out, on day 1, again to avoid extensive cell death.
2. Media for tissue culture, including DMEM, RPMI 1640, trypsin solution, DPBS, G418, and other listed supplements, can be purchased from cell culture suppliers.
3. 200 mM L-glutamine is purchased as a stock in 50 mL quantities. Store at −20 °C until required and then thaw at 37 °C in a water bath (some agitation is usually required to get all the glutamine into solution). 5 mL aliquots (sufficient for a 500 mL bottle of medium) are prepared under sterile conditions in sterile tubes for storage at −20 °C. Thaw at 37 °C before use.
4. A stock of 10,000 units penicillin and 10 mg streptomycin per mL is purchased in 50 mL quantities. Store at −20 °C and thaw at 37 °C in a water bath. As with L-glutamine (*see* **Note 3**), prepare 5 mL aliquots and store at −20 °C until required.
5. FBS is purchased in 500 mL quantities and stored at −20 °C. Thaw (in a 37 °C water bath), and filter the FBS through a 0.2 μm syringe filter (several are usually required for a 500 mL bottle of FBS, as they clog easily). Prepare 50 mL aliquots in sterile tubes. The aliquots are frozen and thawed as required. 50 mL is sufficient for one 500 mL bottle of complete medium with 10% FBS.
6. The concentration of G418 antibiotic is titrated for each batch and each cell line but usually ranges from 0.2 to 0.5 mg/mL in routine culture medium, to select and culture stable cell lines.
7. Oligofectamine 2000 and OptiMEM I reduced serum medium with L-glutamine and phenol red are proprietary formulations, supplied by Thermo Fisher Scientific. I have not tested other, similar reagents.
8. For 100 mL of 20% FBS medium, add 20 mL FBS and 1 mL L-glutamine stock to 79 mL DMEM or RPMI culture medium.
9. Use sterile filter tips on pipettes for the siRNA transfection. The siRNA concentration given is a good starting point, but pools and individual duplexes should be titrated so that the lowest concentration is used to achieve depletion of the protein. Volumes are given for different-sized wells or dishes. The dish/well chosen will depend on how many cells are required for further analysis.

10. Prepare a 10× stock of PBS, sterilize, and dilute to 1× prior to use.
11. Prepare 6% (w/v) paraformaldehyde stock in PBS. This solution should be prepared in a fume hood and requires heating to about 80 °C to dissolve. Once cooled, add 100 mg/L $MgCl_2$ and 100 mg/L $CaCl_2$. Store in aliquots at −20 °C. Thaw and dilute 1:1 with PBS just prior to use.
12. Protease inhibitor tablets (cOmplete™, Mini, EDTA-free protease inhibitor cocktail) are a proprietary formulation from Roche.
13. Prewarming the media speeds up the trypsinization process and avoids possible cold stress to the cells.
14. HeLa M cells should be maintained by splitting 1:5 to 1:10 twice per week. Use a plate that is just confluent (i.e., the cells are just touching each other) and trypsinize and plate onto new plates as described. Maintain some cells for future use and plate some for siRNA transfection, as required.
15. If necessary, the cells can be plated on day 1 (rather than day 0). In this case, add 5×10^5 cells per well of a 6-well plate and leave for at least 3 h to adhere to the plastic before transfection. For other plate or dish sizes, scale the number of cells plated depending on the surface area.
16. Make up a little more solution B than required to ensure you have enough.
17. Traces of FBS will decrease the transfection efficiency.
18. To prepare sterile, circular glass coverslips (13 mm diameter): Wash in 70% ethanol and dry with tissue. Ensure coverslips are completely dry. Place in a glass petri dish in single layers divided by filter paper. Bake at 180 °C in an oven for at least 2 h and cool before use.
19. The efficiency of the siRNA knockdown must be assessed before further experiments can be conducted. This is perhaps best done by a combination of SDS-PAGE/western blotting and immunofluorescent staining, assuming an antibody specific to the protein of interest is available. In this case, a western blot will give an overall idea of the percentage protein depletion, compared to mock or control-transfected cells and with a suitable loading control (e.g., anti calreticulin). Immunofluorescence will allow you to assess the extent of knockdown and in which cells specifically the protein is depleted. For example, a 90% depletion of a particular protein may be the result of 90% depletion levels in all cells or may be the result of 90% of the cells having no detectable protein but 10% of the cells having near wild-type levels. The latter is more common, in my experience. Most of the siRNA oligonucleotides or oligonucleotide

pools in Table 1 should give 90% or greater depletion efficiency, but this should be assessed for each transfection. If an antibody to the protein of interest is not available, then siRNA efficiency may be studied using real-time quantitative polymerase chain reactions to assess the depletion of mRNA. Transfection in 6-well plates moving to 6 cm dishes should give adequate numbers of cells for an initial investigation of depletion of the protein of interest. The cells can also be used for a variety of other assays, including real-time quantitative PCR (RT-PCR, to assess the depletion of mRNA levels), fluorescence-activated cell sorting (FACS) analysis, biochemical assays (including ^{125}I-EGF uptake assays), and electron microscopy [1, 2], although the procedure may need scaling up to give satisfactory cell numbers on day 5. Also *see* **Note 1**.

20. Once the siRNA transfections are complete and the cells are removed from the tissue culture room for fixation and lysate preparation, there is no need to maintain sterile technique.
21. Many protein assay methods are available. We use a bicinchoninic acid-based method. Whichever method is chosen, it is essential to generate a standard curve and to ensure that the sample measures are within the linear range of the assay.
22. Following siRNA-depletion of a protein of interest, it is essential that any phenotype observed is validated to ensure that it is a result of depletion of that protein (directly or indirectly) and not a consequence of nonspecific effects of the oligonucleotide duplexes used. I recommend reading an editorial in *Nature Cell Biology* regarding validation of siRNA [23].
23. The first step in validation is to determine which of the four oligonucleotides used (if using a pool) give rise to efficient depletion and which also give rise to the phenotype of interest. One would expect three or four of the siRNA duplexes used to give rise to efficient mRNA and protein depletion, and it goes some way to validating the procedure if all siRNAs that deplete the protein also show the observed phenotype.
24. It is important to titrate the concentration of siRNA to the lowest possible to achieve effective depletion, to minimize off-target effects [21]. An example of VPS20 (CHMP6) depletion using both a pool and individual duplexes at various concentrations is shown in Fig. 1. As shown, efficient depletion is seen even with 5 nM of the pool (i.e., 1.25 nM of each oligonucleotide duplex). Of the individual oligonucleotide duplexes, number 2 is not particularly effective but numbers 1, 3 and 4 give effective protein knockdown.
25. The rescue experiment is the best validation of an observed effect of protein depletion. However, it may not be possible for a variety of reasons. For example, a transfected cell line (even a

stable cell line) will often express much higher levels of the protein of interest than is expressed endogenously, and thus rescue may be partial or possibly not apparent. Overexpression of the construct of interest may in itself also cause a phenotype (this has been documented for ESCRTs, e.g., expression of N-terminally tagged TSG101, ALIX [AIP1], VPS2A [CHMP2A], VPS32A [CHMP4A], VPS32B [CHMP4B], VPS32C [CHMP4C], CHMP1A, CHMP1B and CHMP5 disrupts HIV-1 production [24]). Ultimately, the controls should validate the effects of protein depletions seen by the best methods available to you. This may, of course, include complementary techniques (such as CRISPR-Cas 9 gene editing) to mutate the protein of interest and validate the data.

26. Any translational effects of silent mutations can be avoided by using an siRNA that targets the 3′UTR.

Acknowledgments

I would like to thank Dr. Paul Pryor and Prof. Paul Luzio for critical reading of the manuscript.

References

1. Bowers K, Piper SC, Edeling MA, Gray SR, Owen DJ, Lehner PJ, Luzio JP (2006) Degradation of endocytosed epidermal growth factor and virally ubiquitinated major histocompatibility complex class I is independent of mammalian ESCRTII. J Biol Chem 281 (8):5094–5105. https://doi.org/10.1074/jbc.M508632200
2. Parkinson MD, Piper SC, Bright NA, Evans JL, Boname JM, Bowers K, Lehner PJ, Luzio JP (2015) A non-canonical ESCRT pathway, including histidine domain phosphotyrosine phosphatase (HD-PTP), is used for down-regulation of virally ubiquitinated MHC class I. Biochem J 471(1):79–88. https://doi.org/10.1042/BJ20150336
3. Lawrence SP, Bright NA, Luzio JP, Bowers K (2010) The sodium/proton exchanger NHE8 regulates late endosomal morphology and function. Mol Biol Cell 21(20):3540–3551. https://doi.org/10.1091/mbc.E09-12-1053
4. Elbashir SM, Harborth J, Lendeckel W, Yalcin A, Weber K, Tuschl T (2001) Duplexes of 21-nucleotide RNAs mediate RNA interference in cultured mammalian cells. Nature 411 (6836):494–498. https://doi.org/10.1038/35078107
5. Garrus JE, von Schwedler UK, Pornillos OW, Morham SG, Zavitz KH, Wang HE, Wettstein DA, Stray KM, Cote M, Rich RL, Myszka DG, Sundquist WI (2001) Tsg101 and the vacuolar protein sorting pathway are essential for HIV-1 budding. Cell 107(1):55–65
6. Rand TA, Ginalski K, Grishin NV, Wang X (2004) Biochemical identification of Argonaute 2 as the sole protein required for RNA-induced silencing complex activity. Proc Natl Acad Sci U S A 101(40):14385–14389. https://doi.org/10.1073/pnas.0405913101
7. Matranga C, Tomari Y, Shin C, Bartel DP, Zamore PD (2005) Passenger-strand cleavage facilitates assembly of siRNA into Ago2-containing RNAi enzyme complexes. Cell 123 (4):607–620. https://doi.org/10.1016/j.cell.2005.08.044
8. Ameres SL, Martinez J, Schroeder R (2007) Molecular basis for target RNA recognition and cleavage by human RISC. Cell 130 (1):101–112. https://doi.org/10.1016/j.cell.2007.04.037
9. Rand TA, Petersen S, Du F, Wang X (2005) Argonaute2 cleaves the anti-guide strand of siRNA during RISC activation. Cell 123 (4):621–629. https://doi.org/10.1016/j.cell.2005.10.020

10. Whitehead KA, Langer R, Anderson DG (2009) Knocking down barriers: advances in siRNA delivery. Nat Rev Drug Discov 8 (2):129–138. https://doi.org/10.1038/nrd2742
11. Seok H, Lee H, Jang ES, Chi SW (2017) Evaluation and control of miRNA-like off-target repression for RNA interference. Cell Mol Life Sci. https://doi.org/10.1007/s00018-017-2656-0
12. Meng B, Ip NC, Prestwood LJ, Abbink TE, Lever AM (2015) Evidence that the endosomal sorting complex required for transport-II (ESCRT-II) is required for efficient human immunodeficiency virus-1 (HIV-1) production. Retrovirology 12:72. https://doi.org/10.1186/s12977-015-0197-x
13. Sanyal S, Ashour J, Maruyama T, Altenburg AF, Cragnolini JJ, Bilate A, Avalos AM, Kundrat L, Garcia-Sastre A, Ploegh HL (2013) Type I interferon imposes a TSG101/ISG15 checkpoint at the Golgi for glycoprotein trafficking during influenza virus infection. Cell Host Microbe 14(5):510–521. https://doi.org/10.1016/j.chom.2013.10.011
14. Zhang Y, Schmid B, Nielsen TT, Nielsen JE, Clausen C, Hyttel P, Holst B, Freude KK (2016) Generation of a human induced pluripotent stem cell line via CRISPR-Cas9 mediated integration of a site-specific heterozygous mutation in CHMP2B. Stem Cell Res 17 (1):148–150. https://doi.org/10.1016/j.scr.2016.06.004
15. Zhang Y, Schmid B, Nikolaisen NK, Rasmussen MA, Aldana BI, Agger M, Calloe K, Stummann TC, Larsen HM, Nielsen TT, Huang J, Xu F, Liu X, Bolund L, Meyer M, Bak LK, Waagepetersen HS, Luo Y, Nielsen JE, Consortium FR, Holst B, Clausen C, Hyttel P, Freude KK (2017) Patient iPSC-derived neurons for disease modeling of frontotemporal dementia with mutation in CHMP2B. Stem Cell Reports 8(3):648–658. https://doi.org/10.1016/j.stemcr.2017.01.012
16. Fedele AO, Isenmann S, Kamei M, Snel MF, Trim PJ, Proud CG, Hopwood JJ (2018) Lysosomal N-acetyltransferase interacts with ALIX and is detected in extracellular vesicles. Biochim Biophys Acta 1865(10):1451–1464. https://doi.org/10.1016/j.bbamcr.2018.07.001
17. Unniyampurath U, Pilankatta R, Krishnan MN (2016) RNA interference in the age of CRISPR: will CRISPR interfere with RNAi? Int J Mol Sci 17(3):291. https://doi.org/10.3390/ijms17030291
18. Motley A, Bright NA, Seaman MN, Robinson MS (2003) Clathrin-mediated endocytosis in AP-2-depleted cells. J Cell Biol 162 (5):909–918. https://doi.org/10.1083/jcb.200305145
19. Kusari J, Sen GC (1986) Regulation of synthesis and turnover of an interferon-inducible mRNA. Mol Cell Biol 6(6):2062–2067
20. Parsons BD, Schindler A, Evans DH, Foley E (2009) A direct phenotypic comparison of siRNA pools and multiple individual duplexes in a functional assay. PLoS One 4(12):e8471. https://doi.org/10.1371/journal.pone.0008471
21. Semizarov D, Frost L, Sarthy A, Kroeger P, Halbert DN, Fesik SW (2003) Specificity of short interfering RNA determined through gene expression signatures. Proc Natl Acad Sci U S A 100(11):6347–6352. https://doi.org/10.1073/pnas.1131959100
22. Harlow E, Lane D (1999) Using antibodies: a laboratory manual. Cold Spring Harbor Laboratory Press, Cold Spring Harbor, NY
23. Whither RNAi? (2003) Nat. Cell Biol 5 (6):489–490. https://doi.org/10.1038/ncb0603-490
24. Martin-Serrano J, Yarovoy A, Perez-Caballero D, Bieniasz PD (2003) Divergent retroviral late-budding domains recruit vacuolar protein sorting factors by using alternative adaptor proteins. Proc Natl Acad Sci U S A 100 (21):12414–12419. https://doi.org/10.1073/pnas.2133846100
25. Hewitt EW, Duncan L, Mufti D, Baker J, Stevenson PG, Lehner PJ (2002) Ubiquitylation of MHC class I by the K3 viral protein signals internalization and TSG101-dependent degradation. EMBO J 21(10):2418–2429. https://doi.org/10.1093/emboj/21.10.2418
26. Cabezas A, Bache KG, Brech A, Stenmark H (2005) Alix regulates cortical actin and the spatial distribution of endosomes. J Cell Sci 118(Pt 12):2625–2635. https://doi.org/10.1242/jcs.02382
27. Matsuo H, Chevallier J, Mayran N, Le Blanc I, Ferguson C, Faure J, Blanc NS, Matile S, Dubochet J, Sadoul R, Parton RG, Vilbois F, Gruenberg J (2004) Role of LBPA and Alix in multivesicular liposome formation and endosome organization. Science 303 (5657):531–534. https://doi.org/10.1126/science.1092425
28. McCullough J, Clague MJ, Urbe S (2004) AMSH is an endosome-associated ubiquitin isopeptidase. J Cell Biol 166(4):487–492. https://doi.org/10.1083/jcb.200401141

INDEX

Emmanuel Culetto and Renaud Legouis (eds.), *The ESCRT Complexes: Methods and Protocols*, Methods in Molecular Biology, vol. 1998, https://doi.org/10.1007/978-1-4939-9492-2, © Springer Science+Business Media, LLC, part of Springer Nature 2019

K

L

M

N

O

P

R

S

Printed by Printforce, the Netherlands